**REMOTE SENSING:
THE QUANTITATIVE APPROACH**

McGRAW-HILL
BOOK COMPANY

New York
St. Louis
San Francisco
Auckland
Bogotá
Hamburg
Johannesburg
Lisbon
London
Madrid
Mexico
Montreal
New Delhi
Panama
Paris
São Paulo
Singapore
Sydney
Tokyo
Toronto

Edited by
PHILIP H. SWAIN
and
SHIRLEY M. DAVIS

Contributing authors
SHIRLEY M. DAVIS   **ROGER M. HOFFER**
DAVID A. LANDGREBE   **JOHN C. LINDENLAUB**
TERRY L. PHILLIPS   **LeROY F. SILVA**
PHILIP H. SWAIN
All from Laboratory for Applications of Remote Sensing
Purdue University,
West Lafayette, Ind., U.S.A.

Remote Sensing:
The Quantitative Approach

This book was set in Times Roman 327

British Library Cataloguing in Publication Data

Remote sensing.
  1. Remote sensing—Data processing
  I. Swain, Philip H II. Davis, Shirley M
  621.36′7   G70.4   77-30577

  ISBN 0-07-062576-X

REMOTE SENSING

  3 4 5 MPMP 8 3 2 1

Printed and bound in the United States of America

CONTENTS

Chapter 3 Fundamentals of Pattern Recognition in Remote Sensing   136
Philip H. Swain

Remote sensing is not a new technology. For many decades man has been ascending above the earth in order to observe it from a distance and thus learn more about its condition. Aerial photography has been used extensively for this purpose and over the years a sophisticated technology has evolved using photographic sensors for remote sensing.

The recent development of satellites carrying earth-observational sensor systems has made available enormous quantities of photographic and other forms of data about the surface of the earth, data which have great potential for helping to solve many human problems: for relieving critical food shortages; for monitoring and controlling environmental pollution; for augmenting shrinking supplies of natural resources; and for planning the orderly growth of cities. In view of these needs, these satellite data are of great human value, provided they can be reduced to useful information both quickly and economically. Modern, high-speed digital computers are well suited to this data-reduction task, and the synthesis of computer technology with the new observation systems has already revolutionized our ability to obtain accurate and current information about the world we live in. The product of this synthesis is the quantitative approach to remote sensing.

To an extent, we have indulged our prejudices in using the phrase "the quantitative approach" to describe the remote sensing approach emphasized here. Other remote sensing methods, some of which have been utilized much longer than the relatively new technology we describe, also have quantitative techniques associated with them. Yet we make no apology for our choice of terms. No other approach has been so fundamentally concerned with the quantitative collection, storage, processing, and application of remote sensing data. The new sensor systems can be calibrated more precisely and operate over greater dynamic and spectral ranges with greater sensitivity than any instrumentation previously employed for this purpose. The integrity of the data is preserved by converting the data to and recording and storing it in digital form (i.e., computer-compatible form) right at the sensor. The computer-implemented analysis methods are inherently quantitative (objective, numerical) and produce results which are quantitative (numerical, statistical). The results are often most appropriately utilized in applications demanding quantitative information, often distilled in a rapid fashion from relatively large volumes of sensor data.

The quantitative approach to remote sensing has evolved rapidly during the past decade, and its documentation is scattered throughout the technical literature of many disciplines. While this multidisciplinary nature of the technology attests to its depth and the breadth of its application, it also makes it harder for the new student to quickly gain a comprehensive understanding of the approach. Prior to the publication of this book, a unified, tutorial treatment of the quantitative approach had not been available.

Remote Sensing: The Quantitative Approach deals with both the theoretical bases and the practical aspects of this approach to remote sensing. The subject is developed through a sequence of chapters that evolves logically from the technology itself. After the multispectral and multitemporal concepts are introduced, a generic description of a quantitative remote sensing system is presented to serve as the framework for the book as a whole. An explanation of the physical theory which underlies data-collection systems leads to an extended description in Chapter 2 of those sensors most appropriate for collecting quantitative data. The statistical theory which supports the various data-analysis steps is presented in Chapter 3, and methods of implementing data-handling and data-analysis tasks are explored in Chapter 4. In Chapters 5 and 6, we see first why the natural spectral conditions of the earth can be described quantitatively and then move to a detailed case study describing an actual large-area agricultural survey which employed the quantitative approach. In the final chapter, multispectral remote sensing, based on both image and multivariate analyses, is viewed in a more complex framework appropriate to the current state of the technology, and future prospects are suggested.

It is our intention to provide the serious reader of this book with a systematic way to develop his understanding of quantitative remote sensing. The material presented here is *both* basic in nature and theoretical. The book is not designed to teach the reader how to use a particular system or how to apply the technology in a specific application but rather to give him both an awareness of some of the options open to him as a remote sensing user or designer and the theoretical understanding necessary for making wise choices.

This book can be used in many ways by people with many different objectives and backgrounds. The level of presentation assumes only a general scientific background, and thus the book can be used by advanced undergraduate or graduate students in a variety of disciplines. It is equally suited for use by professionals in governmental agencies or private industry who need to understand the theoretical bases of quantitative remote sensing, perhaps as a step in evaluating its usefulness in their own work.

As an aid to readers at all levels, each chapter contains statements of educational objectives; these give the reader a quick preview of the material covered and, in the case of group study, provide the teacher or group leader with a convenient aid for focusing the group's attention on those aspects of the technology most relevant to their study. The questions and exercises aid the reader in evaluating his grasp of the concepts presented and serve as departure points for group discussions. These objectives and exercises were prepared by Dr. John C. Lindenlaub, who has long been active in the development and utilization of non-traditional instructional materials in remote sensing and engineering. The references included at the end of each chapter will assist the reader who wishes to continue his study to greater depth in the open literature. The glossary defines technical terms from many disciplines within the context of the quantitative approach to remote sensing of earth resources.

Because of the inherently multidisciplinary nature of quantitative remote sensing, any tutorial presentation of the material can be greatly enhanced by a unified, systematic approach. Ideally such a presentation should also be made in a truly multidisciplinary way, not through the eyes of only the engineer, the computer manager, or the agronomist but as the effective synthesis of the perspectives they all provide. The seven authors and editors of this book have attempted to attain this synthesis. Among them they represent many of the major disciplines associated with remote sensing, and yet, as long-time members of the research and administrative staff of a single laboratory, the Laboratory for Applications of Remote Sensing (LARS) at Purdue University, they have together evolved philosophies and approaches essential to effective interdisciplinary research.

We would like to acknowledge the contributions of several organizations in the development of the remote sensing technology described here: Purdue University for providing the environment conducive to advanced multidisciplinary research; the U.S. Department of Agriculture which promoted the establishment of LARS in 1966; and the National Aeronautics and Space Administration (NASA) for its continued encouragement and support of remote sensing research.

P.H.S.
S.M.D.

THE QUANTITATIVE APPROACH:
CONCEPT AND RATIONALE

David A. Landgrebe

Remote sensing is the science of deriving information about an object from measurements made at a distance from the object, i.e., without actually coming in contact with it. The quantity most frequently measured in present-day remote sensing systems is the electromagnetic energy emanating from the object of interest, and although there are other possibilities (e.g., seismic waves, sonic waves, and gravitational force), our attention in this text is focused on systems which measure electromagnetic energy. Sometimes there is playful debate concerning how far an object must be from the measuring device before it can be considered "remote." But this is merely a debate over semantics; we wish to concern ourselves with methodology. Although the applications to be discussed in this book involve measurements made hundreds or thousands of meters from the object of interest, we can also conceive of applications in which the measuring device might be only a fraction of a meter from the object.

Study objectives
After studying Sec. 1-1, you should be able to:
1. Briefly describe the impact that the digital computer and pattern recognition have had on remote sensing technology.

1-1 BACKGROUND

It is difficult to establish a specific time or event as marking the beginning of remote sensing. Some cite the use of photography from tethered balloons before the American Civil War; others say it began even earlier.[1] However, the development of the techniques involved has been perhaps more intense since early in the twentieth century.

Many of the early techniques used in remote sensing were developed for military reconnaissance during both World Wars I and II, but it was not long before the possibility of applying these methods to the solution of peacetime problems became apparent. Civil engineers were among the first to use remote sensing for nonmilitary purposes when they turned to it early in this century as a means for surveying and mapping; and in the 1930s the U.S. Department of Agriculture began nationwide use of aerial photography for soil survey work. The practice of using a human interpreter to analyze photographs obtained from airborne cameras was well established by the 1940s,[2] and investigations into the use of color and infrared photography increased the effectiveness of the aerial camera as a sensor of the earth's surface.[3]

During the 1950s and early 1960s, the advent of the digital computer and attempts at modeling the processes associated with human intelligence combined to accelerate the development of a data-analysis technology called "pattern recognition."[4,5] During the same period, significant developments were made in the design of sensor systems,[6] particularly in the measurement of energy in the infrared wavelengths. Taken together, these advancements were to give rise to a whole new approach to remote sensing—the quantitative approach—based on the ability to rapidly and precisely measure and process multispectral remote sensing data.

Thus today we find there are two major branches of remote sensing. The branch first mentioned above we refer to as *image oriented* because it capitalizes on the pictorial aspects of the data and utilizes analysis methods which rely heavily on the generation of an image. The second branch is referred to as *numerically oriented* because it results directly from the development of the computer and because it emphasizes the inherently quantitative aspects of the data, treating the data abstractly as a collection of measurements. In this case an image is not thought of as data but rather as a convenient mechanism for viewing the data.

The image-oriented technology is older and more fully developed. The sensors most common to image-oriented systems (photographic cameras) have long been in use, as have the associated analysis techniques (photo interpretation). The technology has proved to be reliable and economical for a wide variety of operational applications. Even so, further development of this technology continues.

By comparison, the technology of numerically oriented systems is still in its infancy, having developed to its present state in a short span of years. A multispectral scanner is often used as the sensor in these systems. This instrument is preferred because it has a greater dynamic range over a larger portion of the electromagnetic spectrum than is available from photographic systems. Although a computer is usually used for analysis in numerically oriented systems, it may also be used in image-oriented systems; likewise, there are many other tools and techniques which are equally applicable to both technologies, e.g., the methods of photogrammetry. For some applications the numerical approach has already achieved a practical level of utility, and, even though it still holds much undeveloped potential, it is now beginning to assume its proper role alongside the older, image-oriented technology.

Today we are acquiring earth observational data from earth-orbiting satellites; because of the wide view possible from satellite altitudes, the speed with which the satellite-borne sensors travel, and the number of spectral bands used, very large quantities of data are being produced. Since in many instances information obtainable through remote sensing has value only if it can be acquired both rapidly and cost effectively, it is quite natural to turn to computers to achieve this economy and speed in extracting reliable information from the large volume of data available.

The purpose of this book is to describe an approach in which modern sensor systems, pattern recognition, computers, and human beings can be used together to synthesize a numerically oriented remote sensing system. In the next section we turn to a brief overview of some of the basic concepts of this approach.

Study objectives
After studying Sec. 1-2, you should be able to:
1. Name and give an example of three major kinds of variation which convey information in remote sensing.
2. Give at least one reason why data values from the same class are not identical but tend to cluster or group themselves around some mean value.

1-2 HOW INFORMATION IS CONVEYED IN REMOTE SENSING DATA

Let us begin by considering some of the fundamentals. Figure 1-1 (between pages 36 and 37) is an air view of a portion of the surface of the earth, and one can immediately gain some information about what is in the scene from the image. Stop for a moment and think about what leads you to the conclusions you can reach with just a glance. We can see many linear features, or straight lines. This suggests human activity in the area.

These features divide the area into fields, and some of the fields are rather uniform, again suggesting human activity—in this case, agriculture. Some fields, notably those in the foreground, have a mottled effect. If there are crops there, apparently the crop canopy has not developed enough to mask the natural variations of the soil types in these fields.

Some of the linear features are roads, and near the roads there are small, angular objects which are buildings on the farmsteads. People not only work but live in this area, and, furthermore, it is possible to estimate the number of people from the number of buildings.

We could go on extracting more and more information from the scene. Looking back now at the information we derived, we realize that what led us to these conclusions were by and large the geometrical forms apparent in the image, i.e., the spatial organization of the energy emanating from the scene and recorded by the photographic film. Thus we may conclude at this point that the spatial organization of the energy contains much information about what is in the scene.

In Fig. 1-2 (between pages 36 and 37) we see the same scene at the same time of year but with something added—color. Now we can see not only the spatial distribution of energy but a significant indication of its spectral distribution as well. Studying this figure in the same way we studied Fig. 1-1, we can derive even more information about what was in the scene and what its condition was. For example, in the color image the tan color of the harvested and unharvested fields of small grain is easier to identify.

Figure 1-3 (between pages 36 and 37) shows the scene once more, but this time the image was formed using color infrared film. This type of film will be discussed in more detail in Chap. 2. For now, though, we will simply state that instead of being sensitive to the energy distribution in the blue, green, and red portions of the spectrum, this film is sensitive to the energy from the near-infrared portions of the spectrum as well, and thus provides new information about the scene. For example, since green vegetation has a relatively high reflectance in the near-infrared wavelengths, it is easier to detect relatively minor changes in a vegetative canopy in this portion of the spectrum than it is in the visible portion of the spectrum.

The color and color infrared images of Figs. 1-4 and 1-5 (between pages 36 and 37) give a clear example of this. Note the trapezoidal field on the right-hand side of each image. This is a corn field in which alternate rows have been planted with different varieties. One variety has become infected by a disease and its condition is much more apparent in Fig. 1-5 because of the infrared sensitivity of the film. The important point to notice is that by having data available from the near-infrared wavelengths, more information is available to us than when we are limited to viewing the scene in only the visible wavelengths.†

So far we have confined our discussion to the regions of the spectrum in which photographic sensors can respond, but other parts of the spectrum contain still more information about the earth scene. Figure 1-6 describes the entire electro-magnetic spectrum. At the top are shown the various types of radiation, displayed as a function of the wavelength. It is the portion of the optical region, shown in detail in the lower portion of the figure, which is of primary interest to us, although other portions of the electromagnetic spectrum such as the microwave region are also useful in remote sensing. Notice that radiation measurements available from photographic sensors include only those in the visible region and a small portion of

† Because remote sensing makes use of spectral information within as well as beyond the visible region of the spectrum, the terminology of spectroscopy (wavelength, power per unit wavelength, etc.) has been chosen in the text rather than that of colorimetry (hue, chroma, etc.).

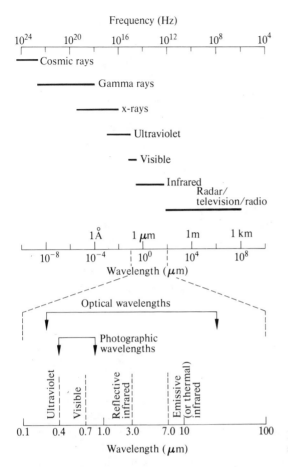

Figure 1-6 The electromagnetic spectrum.

the infrared region. The optical portion of the electromagnetic spectrum will receive more detailed attention early in Chap. 2.

Summarizing to this point, then, we have illustrated the fact that one may derive information about the materials covering the surface of the earth from the *spatial* and *spectral* distributions of energy emanating from those materials. Shortly, we will point out that *temporal* (time) variations in the scene are also useful in the information-extractive process, since much information about what is in an area can be derived by seeing how the area changes with time.

Before doing this, however, let us consider in a bit more detail the manner in which data may be analyzed and information realized about a scene.[7] Figure 1-7 shows a very small portion of an agricultural area imaged in three portions of the spectrum: the visible, the near infrared, and the far infrared. Suppose for the moment that all the data gathered will represent only the four classes of materials contained in this small scene, i.e., corn, alfalfa, bare soil, and stubble. Thus, in analyzing the data gathered from a region of unknown ground cover, we have only four possibilities to choose from. On what can we base our choice?

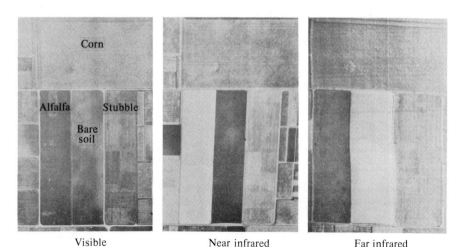

Visible   Near infrared   Far infrared

Figure 1-7 Response of four cover types in three spectral regions.

First, we can carefully inspect the image from the visible region. Notice that the area marked "corn" has a distinctive texture which is not seen in the other three materials. Thus each time we see this texture in an unknown area, we can make a decision in favor of corn over the other classes. Texture, however, is a very difficult image characteristic to quantify. To do so requires a precise definition of what is meant by texture which, if made adequately robust, may be too complex to be utilized efficiently.

There are, however, other approaches to the discrimination problem. Notice, for example, that the response from alfalfa in the three spectral regions shown in the figure changes from very dark to very bright to very dark and that this sequence of responses as a function of wavelength does not occur for any of the other three materials. Thus we have a distinctive way to discriminate alfalfa from the other three materials, and the process relies on a feature more easily quantified than texture, namely, the radiant energy in each of the spectral regions.

Pursuing this line of reasoning further, consider Fig. 1-8. This figure provides a hypothetical but conceptually accurate graph of the relative reflectance as a function of wavelength for three simple ground-cover classes: vegetation, soil, and water. Let us select two wavelengths of interest, marked λ_1 and λ_2 in Fig. 1-8(a), and display the spectral responses in two-dimensional measurement space, as shown in Fig. 1-8(b). In this case, we are showing the response at wavelength λ_1 versus the response at wavelength λ_2. As we saw earlier (Fig. 1-7), distinct response sequences can aid in class identification; they produce points in different portions of the two-dimensional space. The discrimination or identification process can therefore be reduced to determining the class associated with the region in the space where the data points fall.

Notice immediately that we can use response data at more than two wavelengths for the analysis. For example, we have marked a third in Fig. 1-8(a). Use of this wavelength in addition to the others would require a three-dimensional version

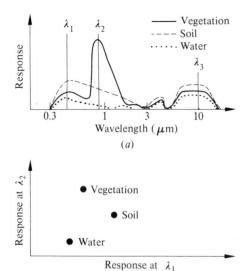

(a)

(b)

Figure 1-8 (a) Hypothetical spectra and (b) their representation in two-dimensional measurement space. (*After Landgrebe.*[7])

of the space shown in Fig. 1-8(b). Indeed, four or more wavelengths could be used and would present no particular difficulty if, for example, one were using a computer in analyzing the data.

There is another characteristic of the reflectance data which must be incorporated into this theoretical example in order to present an adequate basic overview of the quantitative approach. While the spectral response patterns from all healthy, green vegetation tend to be generally the same, there are small variations both within and between vegetative types. The variations, of course, are what make it possible to identify different plant species. In terms of Fig. 1-8(b) this means, however, that the data point for vegetation in two-dimensional space should not really be a point but a cluster of points, which could be described by a statistical distribution. Thus a statistical model for the response of a given material is needed in order to deal properly with, and indeed to take advantage of, the natural variability within classes.

The analysis task therefore becomes one of deciding how to partition the multi-variate space into regions associated with each class so that a data point occurring in any part of the space will be uniquely assigned to a class. The actual situation is illustrated in Fig. 1-9, in which points from three classes of material are plotted in two-dimensional space. Note that the distribution of the points within a class, as illustrated here, is more typical of the actual situation. The question we must raise now is: By what rule should we associate an unknown point, marked U in the figure, with one of the three classes?

There are many algorithms (calculation procedures) which have been proposed in the engineering and statistical literature for answering this question. One of the simplest is illustrated in Fig. 1-10. In this case the conditional mean (class average value) for each class of material has been determined and is marked by the bold-

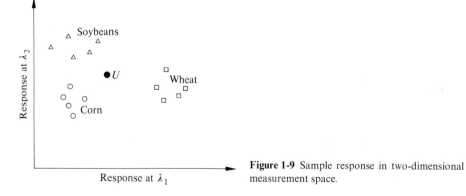

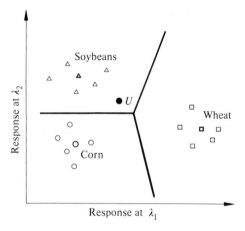

Figure 1-9 Sample response in two-dimensional measurement space.

face symbol for each of the three classes. Next the locus of points equidistant from these conditional means has been drawn. These lines, referred to as *decision boundaries*, then have the effect of partitioning the space into three nonoverlapping regions such that each point in the space is uniquely associated with one of the classes. The unknown sample in Fig. 1-10 is thus associated with the class "soybeans." Although this particular algorithm is useful for illustrating the general approach, often we must use more complex algorithms which result in nonlinear decision boundaries.

The ideal of using a few samples, called *training samples*, to determine decision boundaries in multivariate space is a central one and has become the basis for the quantitative approach to remote sensing data analysis. Note that this approach relies on features which are relatively simple to measure, namely, the response at specific wavelengths. Note also that so far in this approach we have used only spectral variations of the scene. Spatial variations, which are more difficult to quantify, might also be used.

But what about the temporal variations referred to earlier? Let us illustrate the effect of temporal variations in terms of our simple two-dimensional space. Figure

Figure 1-10 Classification rule: minimum distance to means.

1-11 provides an example, in simplified form, using the classes "corn" and "soybeans." In the central part of the United States, these two crops are both planted in the spring, usually as row crops, and are harvested in the fall. Thus, if we are looking down from above during the spring, a field planted in each crop appears essentially as bare soil. Eventually the crop canopies begin to emerge and develop, and, in time, the response points plotted in two-dimensional space would migrate from the bare-soil region toward the green-vegetation region. During the summer months the canopies become complete, and then later the crops begin to ripen and turn brown. Finally the canopy diminishes and a residue of dried leaves remains on the bare soil. Generally, therefore, as the growing season proceeds, the two crops can be expected to progress through a sequence of relatively similar spectral responses. However, even though both are green vegetative crops planted in rows, one develops into a low bushy plant while the other becomes quite tall. This difference in plant geometry results in different shadowing effects and differing amounts of bare soil exposed through the canopy; there is also a slightly different spectral response from the plant leaves themselves. Thus we may expect at various times during the season to find a perhaps small but nevertheless significant difference in the spectral response of corn and soybeans.

In what ways might knowledge of these temporal variations be useful? There are at least three ways. First, a knowledge of the temporal variations helps in determining the optimal time for deriving information relative to the classes of interest. For example, based on the situation depicted in Fig. 1-11, the best time to gather data is 30 days after the season begins, since this appears to be the time of maximum separation of spectral response for corn and soybeans.

Second, in many problems involving the observation of earth-cover materials, the requirement is simply for a time history of what takes place. For example, observing an urban and near-urban area for a period of years would allow one to have an accurate record of how the urban area developed and grew. This information could be used in making projections about future growth and development to be expected.

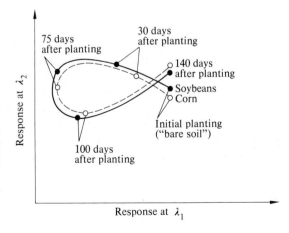

Figure 1-11 Temporal changes in two-dimensional measurement space. (*After Landgrebe.*[7])

The third use of temporal information is possible because of the multivariate approach to data analysis. Suppose, for example, we have no option as to when the data are to be gathered, and that data from 75 and 100 days into the growing season are all that are available. The spectral separation of these two crops at these two times is minimal; however, by replotting these two data sets in four-dimensional space (the response at wavelengths λ_1 and λ_2 at 75 days as dimensions 1 and 2, and at the same wavelengths at 100 days as dimensions 3 and 4), it may be possible to make at least partially additive the small separations that exist at these two times so that a greater net separation between the two classes exists and greater classification accuracy is possible.

Thus temporal variation in the scene along with spectral and spatial variations are all information-bearing. The task then is to further study how we might devise suitable means for utilizing all of them. In the next section we probe deeper into the multispectral concept to aid us in this process.

Study objectives

After studying Secs. 1-3 and 1-4, you should be able to:

1. Illustrate that by presenting data from two spectral bands in multivariate space, discrimination between two spectrally similar materials may be possible even though spectral differences in either band alone are too subtle for class differentiation.
2. Sketch a block diagram of a remote sensing system and briefly describe each component in the diagram.

1-3 INFORMATION FROM MULTISPECTRAL DATA: A CLOSER LOOK

The differences between an image-oriented approach to data analysis and a numerically oriented approach are, in some respects, rather subtle. In order to deepen our understanding of the differences, it may be helpful at this point to study a further example of data classification. In the process, we will introduce another aspect of the multispectral approach, namely, spectral band selection.

For these purposes, then, consider the following example. Figure 1-12 shows data from 10 spectral bands as reproduced on a video display.† A conventional air photo of the same area is included with letter symbols superimposed on the photo to indicate the crop present in each agricultural field. Suppose we wish to analyze the multispectral data to identify the crops in the area. One's first impression is that images in 10 spectral bands result in a very large amount of data. Subtle variations are present in each spectral band which seem to be information-bearing, but, as human interpreters, it is difficult for us to see how to relate data from all these bands.

---

† In practice instruments cannot be built which will measure energy at a single wavelength. Instead, they measure energy in narrow ranges or bands of wavelengths.

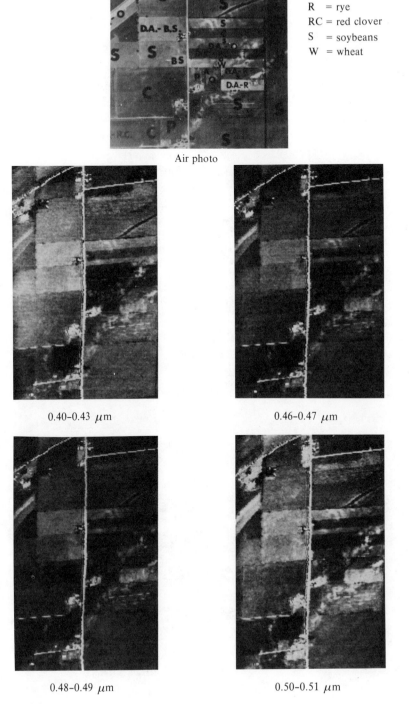

A = alfalfa
BS = bare soil
C = corn
DA = diverted
O = oats
P = pasture
R = rye
RC = red clover
S = soybeans
W = wheat

Air photo

0.40–0.43 μm

0.46–0.47 μm

0.48–0.49 μm

0.50–0.51 μm

Figure 1-12 Conventional aerial photograph and video display of individual spectral bands. (Continued on page 12.)

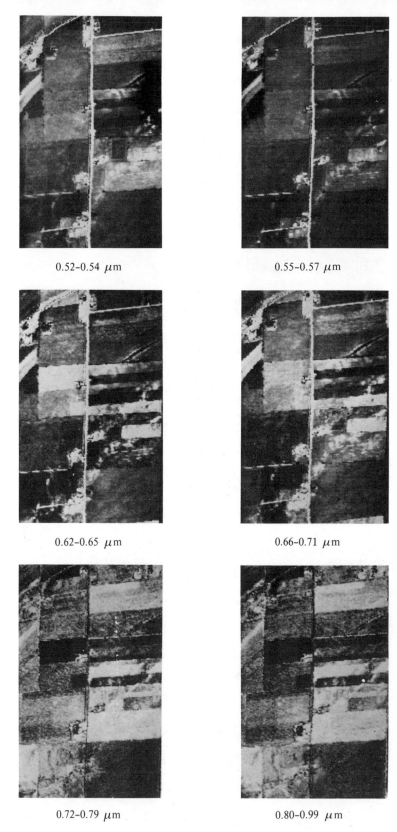

0.52–0.54 μm

0.55–0.57 μm

0.62–0.65 μm

0.66–0.71 μm

0.72–0.79 μm

0.80–0.99 μm

Figure 1-12 (concluded).

For our present purposes, let us therefore reduce the complexity. Refer to Fig. 1-13. Suppose that we have proceeded to the point in the analysis process of having selected two spectral bands, those shown in the lower half of Fig. 1-13, and we are now considering whether there would be additional benefits in using the data in the third band displayed on the upper right. A quick comparison of the data on the upper right with that on the lower left indicates that there appears to be very little difference in these spectral bands. Thus the value of including the third band

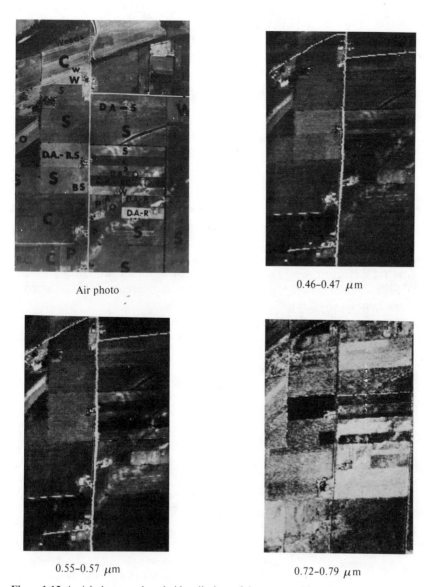

Air photo

0.46–0.47 μm

0.55–0.57 μm

0.72–0.79 μm

Figure 1-13 Aerial photograph and video displays of three spectral bands.

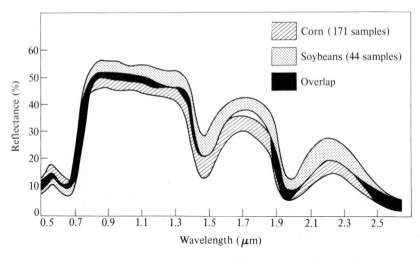

Figure 1-14 Laboratory reflectance measurements of corn and soybean leaves.[8]

appears small. However, remember that we have reached this conclusion from looking at the data in image form, but we plan to analyze it numerically.

To further appreciate why we are emphasizing numerical analysis, let us consider the multispectral analysis problem in even finer detail. Shown in Fig. 1-14 are the results from measuring the percentage of reflectance of a large number of corn and soybean leaves over the spectral range of 0.5 to 2.5 micrometers (μm).[8] The results for 171 corn samples and 44 soybean samples are shown by the patterned areas. The black areas indicate where the measurements from the corn and soybean samples overlapped.

The question is: Given this data, would it be possible to discriminate between soybeans and corn on a spectral basis alone? The answer, of course, is "yes." To do

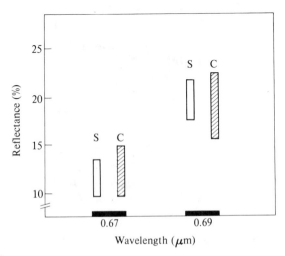

Figure 1-15 Data distribution in two spectral bands for corn and soybeams.

so, one would only need data from the 1.7-μm region. Since there is no overlap at this wavelength, all data points above a certain reflectance percentage would be classed as soybeans and those below as corn. Discrimination on this basis is sometimes referred to as "level slicing."

However, let us make the problem more difficult by assuming that we only have data available in the region near 0.7 μm where there is maximum overlap between the two classes. Figure 1-15 shows data from narrow bands centered at 0.67- and 0.69-μm wavelengths in greater detail. It is apparent that for these two classes there is a high degree of overlap in both spectral bands. Indeed, if the data were displayed as images in these two bands, corn and soybeans would look very similar. Suppose we now ask: Using these two bands alone, is it possible to discriminate between corn and soybeans?

Perhaps surprisingly, the answer to this question is still "yes." Figure 1-16 shows the actual data replotted in the form we used in Fig. 1-8(b). In this case, the data are plotted in multivariate space so that the actual distribution of the two classes can be observed. When the two data sets are replotted on the same set of coordinate axes, as shown in Fig. 1-17, it is clearly possible to separate the classes almost completely, even with a linear decision boundary, despite the high degree of overlap of the classes still apparent in each of the individual spectral bands.

The significant difference between the data presentation in Fig. 1-15 (where discrimination is not possible) and that in Fig. 1-17 (where it is possible) is that the latter presentation is in multivariate space whereas that in Fig. 1-15 may be referred to as a "multiple univariate" presentation. The information present in Fig. 1-17 but lacking in Fig. 1-15 is the relationship between the data in the two spectral bands. Correlated data tend to be distributed in a band along a line whereas uncorrelated data would be circularly distributed in such a diagram. In the present case both classes show a high degree of correlation between the spectral

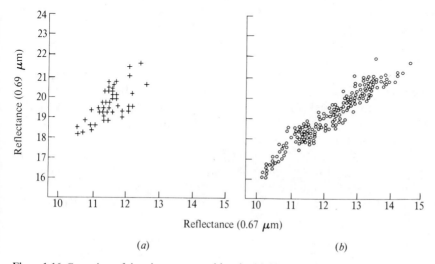

(a)   (b)

Figure 1-16 Crossplots of data in two spectral bands: (a) 44 soybean samples, (b) 171 corn samples.

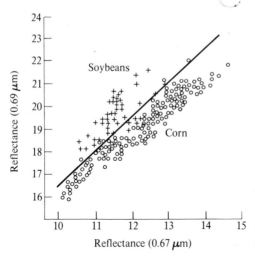

Figure 1-17 Overlay of corn and soybean crossplots.

bands. This correlation plus a relatively minor difference in the mean values of the data in the two bands makes the discrimination possible. † This interrelationship of the spectral bands is not apparent when we view the data in image form.

1-4 A SYSTEM PLAN FOR THE QUANTITATIVE APPROACH

To provide a perspective for the following chapters, it will be helpful at this point to introduce a rather generalized system plan which shows the interrelationships among the three major components of the remote sensing system: the earth's surface, the sensor, and the data-processing system. Figure 1-18 provides such a generalized system plan. There must be, of course, a sensor system which views the portion of the earth's surface of interest. Types of sensors which have been used include photographic cameras, television, and multispectral scanners. Sensors are treated in the next chapter.

Following the sensor there is usually needed some type of on-board processing, plus telemetry. Examples of on-board processing might include calibrating the data radiometrically based upon a standard, associating a geographic reference with the sensor data stream, and reformatting the data, perhaps utilizing a data-compression algorithm.

Once on the ground the data would pass through additional stages of pre-processing. Examples of this might be the geometric manipulation of the data to improve its cartographic quality and the addition of a grid or other geographic reference system so as to facilitate the association of remotely sensed data with its

---

† Note specifically in this case (and perhaps in contrast with one's intuition) that the degree of correlation present in these two bands aided in making the classes separable. Generally, it is an over-simplification to think of correlation as an indication of redundancy in the data.

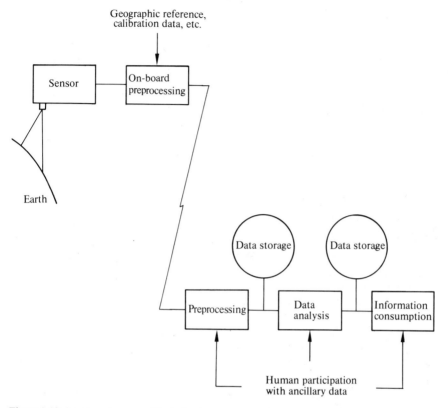

Figure 1-18 An example system block diagram.

corresponding location on the earth's surface. Possible preprocessing operations are given further attention in Chap. 4.

Of course a very important step is the analysis of the data, for it is through this process that "data" becomes "information." We have already pointed to the fact that there are two basic approaches to remote sensing, image oriented and numerically oriented, and it is at this point in the system, the analysis step, where the difference between these two is most apparent. Even within the numerically oriented technology there is a diversity of possible analysis approaches. The simplest involve univariate methods applicable to a single spectral band of data (e.g., level slicing). Other methods are available for the multiband case, both where data from the various bands or other data features are utilized simultaneously (multivariate methods) and where they are utilized sequentially (multiple-univariate methods). Analysis methods are discussed in Chap. 3.

Data storage may occur at several points throughout the system, for reasons of economy and reliability, but it needs to be available minimally at the two points indicated in Fig. 1-18. It is needed just before data analysis because, for example, it sometimes will be desirable to analyze a data set combining data gathered at different times, i.e., a multitemporal data set. Data storage will also be required after analysis and prior to information consumption, primarily to buffer the information flow so that analysis results may be used as needed rather than as produced. Similarly postanalysis data storage will be needed when results of different analyses must be compared, such as in studies monitoring land-use change. Data-system considerations are discussed in Chap. 4.

It is very important to keep in mind that although such numerically oriented systems tend to be computer oriented, they should not be considered automatic, relegating human involvement to only monitoring and maintenance functions. Indeed, interactive participation in both preprocessing and analysis phases has proved in many cases to be indispensable for meeting the accuracy and cost constraints placed upon such systems. Human involvement is, of course, an absolute requirement in the information-consumption step as well as in the use of ancillary data. Effective use of available ancillary data, effective participation in the system processing, and effective understanding and use of results require an understanding of the portion of the system which is in front of the sensor, i.e., the scene. This very complex subject is introduced in Chap. 5.

It is helpful for the information consumer to be regarded not simply as a user of the information-system output but really as a part of the overall system. This viewpoint is helpful because it emphasizes the importance of user considerations in system design. The influence of user requirements is reflected through the system all the way back to the design of the sensor in just the same way as the requirements of the analysis algorithms; both must be taken just as seriously into account by the system designer and operator. The applications of Chap. 6 illustrate this fact.

Chapter 7 provides a concluding overall look at the matter of extracting information from remotely sensed data by numerically oriented means.

PROBLEMS

1-1 What impact did the digital computer and pattern recognition have on remote sensing?

1-2 What are the three kinds of variations through which information is conveyed in remote sensing systems? Give an example of each.

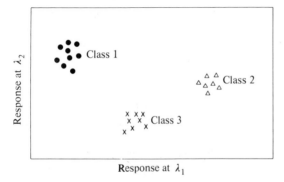

Response at λ_1 **Figure PI-3**

1-3 In Fig. Pl-3, λ_1 represents relative reflectance in wavelength band 1 and λ_2 represents relative reflectance in wavelength band 2. Data known to represent classes 1, 2, and 3 have been plotted in the λ_1, λ_2 space.

(a) Explain why there is variability in the data associated with any one of the classes.

(b) Based on the given data, draw in decision boundaries which divide the λ_1, λ_2 space into non-overlapping regions.

1-4 For each of two cover types, C_1 and C_2, five measurements are made in each of two spectral bands. The resultant data are tabulated in Table P1-4.

Table P1-4

Cover type C_1 data values		Cover type C_2 data values	
Band 1	Band 2	Band 1	Band 2
1	1.5	1	0.8
1.5	2	2.5	2.1
2	2.5	3	2
3	3.5	2	1
3.5	3.6	4	3.5

One way of presenting the data in graphical form is shown in Fig. P1-4. This pictorial presentation of the data would suggest that the two cover types are indistinguishable in either band. (An image of the two cover types would indicate nearly equal gray tone values.) By displaying the data in multivariate space, show that the two sets of data may be separated by a linear (straight line) decision boundary.

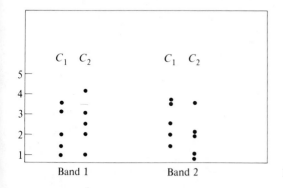

Figure Pl-4

1-5 Prepare a sketch and notes to help you explain the major components in a remote sensing system.

REFERENCES

1. Fischer, W. A.: History of Remote Sensing, in R. G. Reeves (Ed.), "Manual of Remote Sensing," American Society of Photogrammetry, Falls Church, Va., 1975.
2. Colwell, R. N. (Ed.): "Manual of Photographic Interpretation," American Society of Photogrammetry, Washington, D.C., 1960.

3. Smith, J. T., Jr. (Ed.): "Manual of Color Aerial Photography," American Society of Photogrammetry, Falls Church, Va., 1968.
4. Nilsson, N. J.: "Learning Machines," McGraw-Hill Book Company, New York, 1965.
5. Fukunaga, K.: "Introduction to Statistical Pattern Recognition," Academic Press, New York, 1972.
6. National Aeronautics and Space Administration: "Advanced Scanners and Imaging Systems for Earth Observations," NASA, Washington, D.C., 1973.
7. Landgrebe, D. A.: Systems Approach to the Use of Remote Sensing, *Proc. International Workshop on Earth Resources Survey Systems.* NASA, Ann Arbor, Mich., NASA Special Publication SP-283, vol. I, pp. 139–154, 1971.
8. Laboratory for Agricultural Remote Sensing: "Remote Multispectral Sensing in Agriculture," vol. 3, Research Bulletin no. 844, Agricultural Experiment Station, Purdue University, West Lafayette, Ind., 1968.

RADIATION AND INSTRUMENTATION IN REMOTE SENSING

LeRoy F. Silva

Following the model of Chap. 1, it is convenient to use in this chapter, as well, a systems approach when discussing remote sensing technology. Here we will discuss the physical aspects of remote sensing data acquisition and will utilize the concept of energy flow through a system and the conversion of that energy into data for subsequent processing into information.

The remote sensing data-acquisition system is considered to have four basic parts: the radiation source, the atmospheric path, the target, and the sensor.† In a passive remote sensing system the primary radiation source is the sun, whose energy is spectrally distributed throughout the electromagnetic spectrum. The energy propagates through the atmosphere, and its intensity and spectral distribution are modified by the atmosphere. The energy then interacts with the target and is reflected, transmitted, and/or absorbed by it. A portion of the energy absorbed in one region of the spectrum may be emitted in another region of the spectrum. The reflected/emitted energy then passes back through the atmosphere and again is subjected to spectral and intensity modifications. Finally, the energy reaches the sensor where it is measured and converted into data for subsequent processing.

† In Chap. 1 the *entire* remote sensing system was modeled in three parts: the earth's surface, the sensor, and the data-processing system. Here we use the system concept to model only the data-acquisition portion of the remote sensing system.

The way in which the target reflects and/or absorbs and emits incident energy (a modification of the intensity and spectral distribution of the incident energy) is the signal that is to be detected and placed in a quantitative format. In this book, it is generally assumed that the format is digital data suitable for processing on a computer. The processing of the data is done in such a way as to enhance the signal. The atmospheric effects appear as noise in the sensor and digital data-conversion systems; the entire data-acquisition system is carefully designed to maximize the signal-to-noise ratio in order to minimize the amount of enhancement required in subsequent data processing.

In this chapter a discussion of solar radiation and its spectral distribution will be followed by a description of the electromagnetic spectrum oriented specifically toward remote sensing technology. A review of contemporary radiation terminology is presented, followed by a simplified treatment of blackbody radiation, a topic necessary for the understanding of both solar and terrestrial emitted radiation.

Atmospheric effects and their impact on spectral band selection are considered. A simplified optics review is followed by a careful discussion of target reflectance. Some basic principles of remote sensing instrumentation and a discussion of radiation detectors are followed by examples of several general remote sensing instrumentation systems. The chapter concludes with a discussion of data flow in data-acquisition systems and examples of operational multispectral scanners.

How to Read This Chapter

The user of a remote sensing system frequently is able to process data for relatively simple applications without serious concern for radiation and instrumentation principles. However, when working on more difficult problems, e.g., crop-stress determination or detailed urban land-use analysis, the characteristics of the sensor and of the reflectance from the natural material under consideration must affect the approach to the problem. Therefore, it is necessary for the serious remote sensing analyst to have a clear understanding of radiative principles and of the sensors used. Because of the very broad range of material relevant to these subjects, this chapter, even though providing only an introductory treatment, is necessarily rather long. The reader who is new to remote sensing would do well to adopt the following strategy in order to learn the material effectively.

First, read this chapter lightly, paying close attention only to the objectives and the opening few paragraphs of each major section. Then reread and study one section at a time, interspersed with studying the other chapters in the text. In addition to the variety of reading provided by this approach, a further benefit will be the opportunity to see how the radiation and instrumentation concepts are applied and in what respects they are important.

Study objectives
After studying Secs. 2-1, 2-2, and 2-3 you should be able to:
    1. Sketch the important qualitative features of the extraterrestrial solar

spectrum and label (in micrometers) the wavelength scale. The intensity scale need only show relative intensity.

2. State the wavelength ranges for:
   the optical portion of the spectrum,
   the visible region,
   the near-, middle-, and far-infrared regions,
   the "reflective" portion of the spectrum,
   the "emissive" portion of the spectrum.

3. Give a physical interpretation of a blackbody.

4. Use the Stefan-Boltzmann law and equation for λ_{max} to sketch the approximate shape and magnitude of blackbody radiation curves as a function of temperature.

5. Predict whether a radiation-sensing instrument could detect an object of known temperature T_t in the presence of background radiation having an equivalent temperature T_b.

6. Name the chief mechanisms by which the atmosphere interacts with electromagnetic radiation and describe what happens to the incident energy in each case.

7. Define "atmospheric window."

8. Sketch thin-lens optical systems illustrating the concepts of optical stops, speed of lens, field of view, and the thin-lens formula.

9. Give a physical interpretation of each of the reflectance functions introduced and describe how they are interrelated.

2-1 RADIATION AND RADIATION SOURCES

Remote sensing systems in current use are primarily passive; that is, the sensor merely receives energy from a target that has been illuminated by an external radiation source, usually the sun. An active remote sensing system generates the radiation within the system (e.g., radar). This chapter will deal primarily with passive systems.

Solar Radiation

Figure 2-1 is a graph of the solar spectral irradiance falling upon the earth, plotted as a function of wavelength. The graph of the *terrestrial* irradiance illustrates the atmospheric effects on the solar radiation on its passage through one air mass to the earth's surface. The atmosphere's constituent gases introduce complex "structure" to the solar spectrum. The extraterrestrial solar spectrum is, by contrast, very smooth. In fact, it strongly resembles the spectrum produced by an ideal blackbody radiator at a temperature of approximately 6000 K. Interpretation of the solar spectrum in terms of an ideal blackbody radiator is convenient in remote sensing, and the spectra of targets at a temperature of approximately 300 K (the earth's surface temperature) are also of major interest. It is appropriate to discuss some principles of radiation in general, followed by a discussion of blackbody radiation in particular.

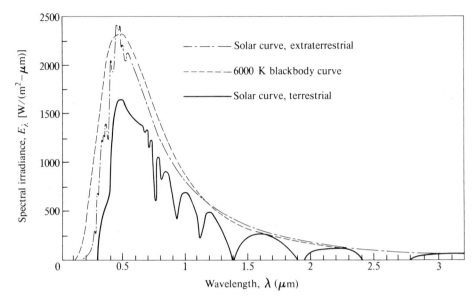

Figure 2-1 Solar spectral irradiance at the earth, through one air mass at surface (Source: *Handbook of Geophysics*,[1] © copyright 1961, The Macmillan Company. Used with permission.)

Radiation Terminology

The units normally used to measure light are *photometric* units. This system of units is based on the assumption that the human eye is the ultimate sensor of the radiation, and thus the sensitivity of the human eye is one basis for the formulation of these units. In remote sensing, however, sensors other than the human eye are used to detect radiation. The units most useful for measuring radiation in remote sensing are the *radiometric* units, which have been used for many years in the calibration of radiation standards. While many more units have been defined than are described here, we will discuss only those which are most frequently used in remote sensing. Table 2-1 lists those units.

The entries in Table 2-1 deserve some comment. The symbols in the left-hand column are those most recently recommended; the old symbols are also included in the table because of their prevalence in the literature. Notice that the units and defining expressions for both radiant exitance and irradiance are identical. The only difference between these two radiometric terms is that *radiant exitance* refers to radiation that is leaving a surface or object of interest, whereas *irradiance* refers to radiation that is incident on an object or surface of interest.

Radiance is a geometric radiation quantity that allows us to describe how radiation is distributed in space. To account for view-angle effects, a cosine term appears in the denominator of the defining expression. If an object which does not fill the field of view of an optical system is viewed at normal incidence, it will appear to the optical system to have a certain area. As the optical instrument is moved away in an angular fashion from the normal to the surface, the apparent area of the

Table 2-1 Radiometric terms (*After Richmond et al.*[2])

Symbol	Description	Defining expression	Units	Units (Abbreviation)	Old symbol
Q	Radiant energy		Joule	J	U
Φ	Radiant flux (power)	$\dfrac{dQ}{dt}$	Watts	W	P
M	Radiant exitance	$\dfrac{d\Phi}{dA}$ (out)	Watts per square meter	W/m^2	W
E	Irradiance	$\dfrac{d\Phi}{dA}$ (in)	Watts per square meter	W/m^2	H
L	Radiance	$\dfrac{d^2\Phi}{\cos\theta\, dA\, d\omega}$ (out)	Watts per square meter per steradian	$W/(m^2 - sr)$	N
M_λ	Spectral radiant exitance	$\dfrac{dM}{d\lambda}$ (out)	Watts per square meter per micrometer	$W/(m^2 - \mu m)$	W_λ
E_λ	Spectral irradiance	$\dfrac{dE}{d\lambda}$ (in)	Watts per square meter per micrometer	$W/(m^2 - \mu m)$	H_λ
L_λ	Spectral radiance	$\dfrac{dL}{d\lambda}$	Watts per square meter per steradian per micrometer	$W/(m^2 - sr - \mu m)$	N_λ

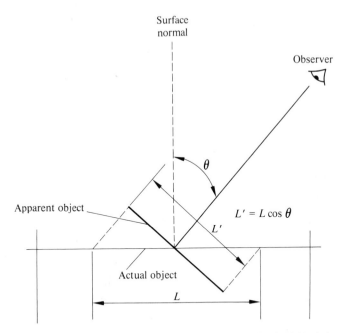

Figure 2-2 Projected area effects when object does not fill the field of view.

surface decreases according to a cosine law. This effect is illustrated in Fig. 2-2; the angular measures used will be defined in Fig. 2-3.

The last three terms in the table are spectral terms. They differ from their companion terms earlier in the table only in that they describe how the energy is distributed with respect to wavelength across the electromagnetic spectrum. In all other respects they are identical to the analogous terms described earlier in the table.

In order to consider some of the geometric effects that occur in radiometry, it is necessary to define a few geometric terms which are convenient in such a discussion. Standard radian measure will be used to describe plane angles and standard solid-angle measure will be used to describe three-dimensional angles. The geometry of

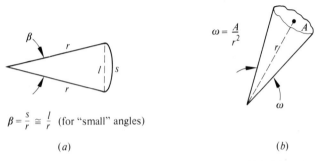

Figure 2-3 Angle definitions. (a) Plane-angle measure. (b) Solid-angle measure.

these two angular measures is illustrated in Fig. 2-3. To measure a plane angle in radians, draw a circle with center at the vertex of the angle. The size of the plane angle β, in radians, is the length of the arc subtended by the angle divided by the radius of the circle. For small angles, the length of the chord approximates the length of the arc. The solid-angle measure is the quotient of the area of the face of the cone divided by the square of the radius.

Blackbody Radiation

Conceptually, a blackbody is an ideal radiator of thermal radiant energy. An ideal radiator is also an ideal absorber. Planck used a model of a blackbody that consisted of a large sphere with a very small hole cut in the side. If a quantity of radiation entered the hole, it most likely would not come back through the hole since the area of the hole was very small compared to that of the sphere. Rather, it would bounce around inside the sphere, heating up the sphere until it was absorbed. Therefore, the hole itself acts like an ideal blackbody—a complete absorber. Planck then applied his famous criterion of discrete energy levels to the thermal radiation field within the sphere.[3] The mathematical formulation that resulted is now known as Planck's radiation law, expressed by the equation:

$$M_\lambda = \frac{\varepsilon c_1}{\lambda^5(e^{c_2/\lambda T} - 1)} \qquad W/(m^2 - \mu m) \qquad (2\text{-}1)$$

where M_λ = spectral radiant exitance, $W/(m^2 - \mu m)$
   ε = emittance (emissivity), dimensionless
   c_1 = first radiation constant, $3.7413 \times 10^8 \, [W - (\mu m)^4]/m^2$
   λ = radiation wavelength, μm
   c_2 = second radiation constant, $1.4388 \times 10^4 \, \mu m - K$
   T = absolute radiant temperature, K

Alternatively, the law can be formulated in terms of the radiation frequency v by applying the following relationships:

$$M_\lambda \, d\lambda = -M_v \, dv$$

$$\lambda = \frac{c}{v}$$

$$\frac{d\lambda}{dv} = -\frac{c}{v^2}$$

where c is the velocity of light in appropriate units. When the substitutions are made, Planck's law becomes

$$M_v = \frac{\varepsilon c_1 v^3}{c^4(e^{c_2 v/cT} - 1)} \qquad W/(m^2 - Hz) \qquad (2\text{-}2)$$

where c = velocity of light, $2.9996 \times 10^{14} \, \mu m/s$
   v = radiation frequency, Hz

The other terms in the equation are as before.

In remote sensing technology the wavelength form of Planck's law is usually used. It is convenient, however, to use the frequency form of the law to obtain another useful relationship. If we integrate M_v over all frequencies, the radiant exitance M will be obtained for a blackbody. That is,

$$M = \int_0^\infty M_v \, dv = \int_0^\infty \frac{\varepsilon c_1 v^3 \, dv}{c^4(e^{c_2 v/cT} - 1)} \qquad W/m^2$$

Using the substitutions

$$x = \frac{c_2 v}{cT} \qquad \text{and} \qquad dx = \frac{c_2}{cT} \, dv$$

we obtain

$$M = \frac{\varepsilon c_1 T^4}{c_2^4} \int_0^\infty \frac{x^3 \, dx}{(e^x - 1)}$$

The integral on the right is the Riemann zeta function[4] for $n = 3$ and has a value of $\pi^4/15$. Therefore,

$$M = \frac{\varepsilon c_1 \pi^4}{15 c_2^4} T^4 = \varepsilon \sigma T^4 \qquad W/m^2 \tag{2-3}$$

where σ is the Stefan-Boltzmann radiation constant:

$$\sigma = 5.6693 \times 10^{-8} \, W/(m^2 - K^4)$$

This equation is known as the Stefan-Boltzmann radiation law. The quantity ε appearing in the above equations will be discussed below.

If M_λ is differentiated with respect to wavelength, the derivative set equal to 0, and the resulting equation solved for λ_{max}, an expression for the wavelength at which M_λ is a maximum is obtained. The result is

$$\lambda_{max} = \frac{2898}{T} \qquad \mu m \tag{2-4}$$

where T is in K.

Planck's law for several temperatures is plotted in Fig. 2-4. Note that as temperature increases, the peak of M_λ moves to shorter wavelengths and the area under the curve increases, as predicted by Eqs. (2-3) and (2-4).

Prior to the formulation of Planck's law, two other laws of radiation had been derived from classical physics. They are the Rayleigh-Jeans radiation law and the Wien radiation law. These laws are approximations of Planck's law for certain portions of the spectrum. For long wavelengths and high temperatures the Rayleigh-Jeans law approximates experimental data; at shorter wavelengths and low temperatures Wien's law more closely approximates physical observations.

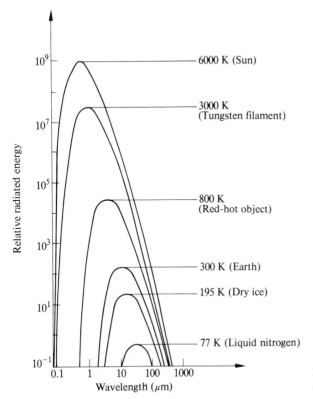

Figure 2-4 Blackbody radiation at various temperatures.

These laws are:

$$(\text{Rayleigh-Jeans}) \quad M_\lambda = \frac{K_1 T}{\lambda^4} \qquad W/(m^2 - \mu m)$$

$$(\text{Wien}) \qquad M_\lambda = \frac{K_2}{\lambda^5 e^{hc/\lambda T}} \qquad W/(m^2 - \mu m)$$

where K_1 and K_2 are unit-fitting constants and h is Planck's constant. Planck's law is derived from quantum mechanics as opposed to classical physics. It exactly explains experimental observations over all wavelengths (or frequencies); nevertheless the Rayleigh-Jeans and Wien laws are sometimes used because of their simpler form in the particular portions of the spectrum in which they are valid approximations. The Rayleigh-Jeans law may be used when the product of wavelength and temperature exceeds approximately 10^5 $\mu m - K$; Wien's law may be used when the product of wavelength and temperature is less than 3×10^3 $\mu m - K$.

All of the radiation laws assume that the radiator is an ideal blackbody—a perfect radiator. A measure of an object's approach to being an ideal blackbody is quantified by including in the Stefan-Boltzmann radiation law (or the Planck radiation law) a constant multiplier known as emittance ε. Emittance is usually referred to as emissivity in the case of opaque targets. A perfect blackbody has an emittance equal to 1, whereas an object which is not a perfect radiator would have an emittance less than 1. An ideal radiator is also a perfect absorber, and so this means that an object with an emittance of 0 would be a perfect reflector with a reflectance equal to 1.

Consider for the moment an object at room temperature, approximately 300 K. The spectral radiant exitance of such an object would appear as in Fig. 2-5. Notice that the graph has a peak wavelength of approximately 10 μm, and most of the radiation occurs between 7 and 15 μm, the so-called *thermal* region. In this region of the spectrum, water is opaque; i.e., its transmittance is equal to 0. In general, any object will reflect and absorb and also transmit radiation. This fact is summarized in the following equation:

$$\rho + \alpha + \tau = 1$$

where $\rho = \dfrac{\text{reflected radiation}}{\text{incident radiation}}$

$\alpha = \dfrac{\text{absorbed radiation}}{\text{incident radiation}}$

$\tau = \dfrac{\text{transmitted radiation}}{\text{incident radiation}}$

In that portion of the spectrum where water-bearing objects are opaque, τ will be equal to 0. Then the equation becomes

$$\rho + \alpha = 1 \qquad \text{or} \qquad \alpha = 1 - \rho$$

Applying the concept of an ideal absorber being an ideal radiator, we then identify absorptance with emittance and we obtain

$$\varepsilon = 1 - \rho$$

This is an expression of the Kirchhoff radiation law. It should be noted that this relationship can be applied only in that portion of the spectrum where the materials in question are opaque, which frequently occurs in the thermal portion of the

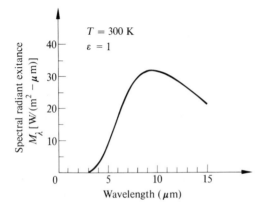

Figure 2-5 Emission from a blackbody at 300 K.

spectrum. In the shorter wavelength region where solar energy predominates, we must, however, consider transmission of radiation through natural materials, since water is partially transparent in the shorter wavelengths.

If one has enough knowledge about a target of interest, one may be able to interpret radiation in the 7- to 15-μm band in terms of temperature; however, care must be taken to account for emittances different from unity and also to account for the fact that in examining the surface of an object one may or may not have measurements indicative of the bulk temperature of an object. For example, on a day of low humidity, the surface temperature of a body of water may be considerably lower than its bulk or contact temperature due to evaporative cooling effects at the surface; however, the observable radiant temperature of the water would be that of the surface of the waterbody.

Targets departing from the ideal blackbody status can be described in a number of ways. For example, a graybody has a spectral radiant exitance curve of the same functional form as an ideal blackbody radiator curve, but it is a constant factor lower than the ideal curve for every wavelength. It is said to have an emittance less than 1 but greater than 0, and all of the radiation laws apply when this constant fraction is factored into the appropriate mathematical relation. Occasionally an object has a spectral radiant exitance curve with a different functional form than Planck's law; in this case it is neither a blackbody nor a graybody, and it is convenient to associate the spectral radiant exitance on a wavelength-by-wavelength basis with a blackbody whose radiant temperature is equal to the contact temperature of the object of interest. If one relates the radiation at each wavelength to that of the ideal radiator, it is then possible to assign an emittance value to the object of interest at each wavelength so that in this case emittance is a function of wavelength. This allows us to summarize the thermal characteristics of materials in the following fashion:

Blackbody	$\varepsilon = 1$
Graybody	$0 < \varepsilon < 1$
Perfect reflector	$\varepsilon = 0$
All others	$\varepsilon = f(\lambda)$

Alternatively, we can account for the difference in spectral radiant exitance between an ideal blackbody and a natural target by assuming that the target has an emittance equal to 1, but that its temperature is a function of wavelength. That is, on a wavelength-by-wavelength basis, the spectral radiant exitance of the actual target is compared to a blackbody which would have the same spectral radiant exitance at that temperature. Since the natural material is assumed to have an emittance equal to 1, then an equivalent radiant temperature can be assigned to the natural target on a wavelength-by-wavelength basis. If the target were an ideal blackbody itself, it would have a constant temperature as a function of wavelength precisely equal to that of the corresponding blackbody. If the target were a graybody, it would have a constant effective radiant temperature as a function of wavelength, but this temperature would be some factor less than the corresponding reference blackbody. If the target were not a blackbody or graybody, then the radiant temperature of the target would be a function of wavelength, with all temperatures being equal to or less than the reference blackbody.

An interesting problem arises when one is attempting to measure the effective radiant temperature of an object, especially when the object is not a perfect blackbody and has a contact temperature approximately equal to its surroundings. Consider the situation diagrammed in Fig. 2-6. Assume that the target, at a temperature of 25°C, is a graybody with an emittance of 0.8. Assume further that the target does not transmit any radiation and therefore obeys the Kirchhoff radiation law, $\varepsilon = 1 - \rho$. The background radiation from the surroundings will reflect off the target, which has a reflectivity $\rho = 1 - \varepsilon = 1 - 0.8 = 0.2$, into the sensor; but the sensor cannot tell the difference between the radiation being emitted by the target and the background radiation which has been reflected by the target. Since the background radiation and the target radiation come from objects that are very close in absolute radiant temperature, it is impossible to correctly discriminate the target's thermal characteristics. However, if the target were at a temperature considerably above the ambient temperature, e.g., at 100°C, the Stefan-Boltzmann radiation law shows that the radiant exitance of the target would then overwhelm the radiation coming from the background, since the radiant exitance varies as the fourth power of temperature. The sensor could then adequately discriminate between the target's radiant exitance and the background radiant exitance. Alternatively, if one made the measurement of the target when its temperature was at 25°C and were careful to ensure that the background radiant exitance came from sources that were at radiant temperatures considerably

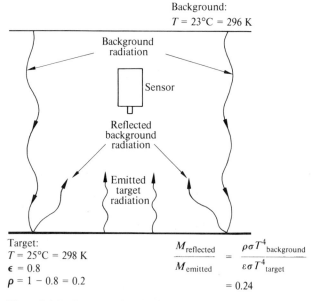

Figure 2-6 Background effects in thermal remote sensing. Notice that the target radiation is actually less than the background radiation.

below that of the target, then the sensor would be capable of discriminating between the background radiant exitance and the target radiant exitance.

These background effects must be considered carefully when calibrating instruments which measure radiation in the far-infrared portion of the spectrum. They must also be considered when making measurements on natural targets in field situations. For example, when the sky is heavily clouded, it is usually not possible to make measurements of thermal radiation on a natural scene, such as a corn field; the radiation from the clouds may be detected by the sensor. It is usually necessary to wait for a relatively clear sky since the effective radiant temperature of a clear sky is considerably below the earth's surface temperature. These problems, along with the associated calibration problems, will be discussed later in this chapter.

The Electromagnetic Spectrum

Electromagnetic energy, you will recall from Chap. 1 (Fig. 1-6), is commonly considered to span the spectrum of wavelengths from 10^{-10} μm, the cosmic rays, up to 10^{10} μm, the broadcast wavelengths. The wavelengths which are of greatest interest to us in remote sensing are the *optical* wavelengths, which extend from 0.30 to 15 μm. At these wavelengths, electromagnetic energy can be reflected and refracted with solid materials, like mirrors or lenses, capable of being manufactured to precision tolerances.

Figure 2-7 depicts the optical portion of the spectrum. The region between 0.38 and 3.0 μm is frequently referred to as the *reflective* portion of the spectrum. Energy sensed in these wavelengths is primarily radiation originating from the sun and reflected by objects on the earth.

The reflective portion of the spectrum is further divided into the visible wavelengths and the reflective-infrared wavelengths. Since the human eye responds to radiation between wavelengths of approximately 0.38 and 0.72 μm, these wavelengths are referred to as the *visible* wavelengths. The region between 0.72 and

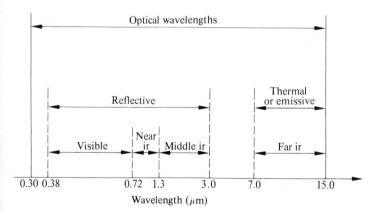

Figure 2-7 The optical spectrum.

3.0 μm is referred to as the *reflective-infrared* portion of the spectrum, which in turn is further subdivided into the *near-infrared* (0.72 to 1.3 μm) and the *middle-infrared* (1.3 to 3.0 μm) wavelength regions. In fact, these latter two designations are preferred to the designation "reflective infrared" for reasons we will discuss shortly.

A special term is not usually applied to the wavelength region from 3.0 to 7.0 μm; atmospheric effects greatly complicate interpretation of the radiation data in this region and, in fact, limit the usefulness of these wavelengths for satellite remote sensing applications.

Electromagnetic energy in wavelengths from 7.0 to 15 μm is in the *far-infrared* region of the spectrum. The terms "emissive" and "thermal" are sometimes used to designate this portion of the spectrum, but, while we need to be aware of the use of these terms in the literature, they are really somewhat inappropriately applied, since both reflection and solar radiation also occur in this wavelength region as well as emission.

Figures 2-1 and 2-5 exhibit the relationships between reflected and emitted energy at the earth's surface. All matter at temperatures above absolute zero radiates electromagnetic energy, but the amount of energy radiated and the predominant wavelength bands vary according to the nature of the material. Radiation from the sun, when viewed from outside the atmosphere, is greatest between the wavelengths of approximately 0.35 and 3.0 μm, the reflective region of the spectrum, but the sun also emits energy throughout the entire electromagnetic spectrum, as shown by the curve. The sun's radiation can be said to resemble that of a blackbody, as discussed previously, at 6000 K. An object at room temperature (300 K) tends to concentrate the bulk of its radiation in the wavelengths from 7.0 to 15 μm (see Fig. 2-5), but its radiation also extends throughout the entire spectrum. At no point, however, is the energy emitted by a 300 K blackbody as great as that emitted by a 6000 K blackbody, even in the 7.0- to 15-μm range. But since we are looking at objects on the earth's surface, and since the great distance between us and the sun makes it appear so much smaller than it actually is, the energy apparent to us in the 7.0- to 15-μm wavelength region is predominantly that radiated from nearby objects rather than from the sun. For this reason, the terms "reflective infrared," "thermal infrared," and "emissive infrared" are imprecise. Radiation from both the sun and the earth exists in the 0.90- to 3.0-μm region, and both solar radiation and reflection also exist in the 7.0- to 15-μm region. Summarizing, then, the preferred terminology for remote sensing, and the one used throughout the book, is as follows:

Visible	0.38–0.72 μm
Near infrared	0.72–1.30 μm
Middle infrared	1.30–3.00 μm
Far infrared	7.0 –15.0 μm

2-2 ATMOSPHERIC EFFECTS

In remote sensing applications involving the acquisition of data from either aircraft or spacecraft sensor platforms, the atmosphere between the sensor and the target and between the radiation source and the target has an effect on the data. Generally, unless the atmospheric effect is constant over the frame of data, corrections for the effect may have a significant impact on the final analysis of the frame.[5] The emphasis in this section will be on the impact of the atmospheric effects on remote sensing data quality and on spectral band selection.

The atmosphere may affect remote sensing data in two ways, through scattering and absorption of energy. *Scattering* occurs when radiation is reflected or refracted by particles in the atmosphere which may range from molecules of the constituent gases to dust particles and large water droplets. The usual assumption is that scattered radiation, whether coming from the sun (downwelling) or reflected from the earth's surface (upwelling), is not attenuated but rather is redirected. This redirection is frequently wavelength dependent. The radiation that is not scattered is *absorbed* by the atmosphere, usually also in a wavelength-dependent manner, and the atmosphere is heated by the absorbed radiation. The meteorological characteristics of the atmosphere strongly affect the relative dominance of the scattering and absorption mechanisms.

Pure scattering is said to occur in the absence of all absorption; there is no loss of energy—only a redirection of energy. In the remote sensing context, scattering causes some of the energy to be directed outside the field of view of the sensor. If the field of view is very large, some scattered radiation will still be accepted; however, if the field of view is small, virtually all the scattered radiation will be rejected. In the latter case, scattering produces an apparent attenuation or dimming of the image, whereas in the former case scattering increases the signal being received by the instruments due to the additional radiation entering the instrument aperture. In either case, however, the scattering degrades the quality of the received data.

Theoretically, scattering can be divided into three different categories, depending upon the relationship between the wavelength of the radiation being scattered and the size of the particles causing the scattering. The scattering mechanisms are Rayleigh scattering, Mie scattering, and nonselective scattering. *Rayleigh scattering* occurs when the radiation wavelength is much larger than the size of the scattering particles. In Rayleigh scattering, the volume-scattering coefficient σ_λ is given by[6]

$$\sigma_\lambda = \frac{4\pi^2 N V^2}{\lambda^4} \frac{(n^2 - n_0^2)^2}{(n^2 + n_0^2)^2} \qquad \text{cm}^{-1} \qquad (2\text{-}5)$$

where N = number of particles per cm^3
$\quad V$ = volume of scattering particles (cm^3)
$\quad \lambda$ = radiation wavelength (cm)
$\quad n$ = refractive index of particles
$\quad n_0$ = refractive index of medium

Using this equation, the scattering coefficient of spherical water droplets in air, for which $n_0 = 0$ and $n = 1.33$ for the entire visible and infrared spectrum (except in the vicinity of the absorption bands), reduces to $\sigma_\lambda = 0.827 \, A^3 N/\lambda^4$. Here A is the cross-sectional area of the scattering droplet. In this case, of course, the droplets are assumed to be of molecular size and therefore much smaller than the wavelength of radiation in the optical spectrum. To find the total scattering, Eq. (2-5) must be integrated over the range of wavelengths and cross-sectional areas encountered in any given situation subject to the previously cited restriction involving the wavelength relative to the size of the particles.

Rayleigh scattering causes the sky to appear blue. Since the scattering coefficient is proportional to the inverse fourth power of wavelength, radiation in the shorter blue wavelengths is scattered toward the ground much more strongly than radiation in the red wavelengths. The red of the sunset is also caused by Rayleigh scattering. As the sun approaches the horizon and its rays follow a longer path through the atmosphere, the shorter wavelength radiation is scattered, leaving only the radiation in longer wavelengths, red and orange, to reach our eyes. Because of Rayleigh scattering, multispectral data from the blue portion of the spectrum is of relatively limited usefulness. In the case of aerial photography special filters are used to filter out scattered blue radiation due to haze present in the atmosphere.

The second scattering mechanism to be considered here is *Mie scattering*. This scattering occurs when the radiation wavelength is comparable to the size of the scattering particles. In this case, the scattering coefficient is an area coefficient defined as the ratio of the incident wavefront affected by the particle to the cross-sectional area of the particle itself. For the most universal situation, in which there is a continuous particle-size distribution, the scattering coefficient is given by the following relationship:[7]

$$\sigma_\lambda = 10^5 \, \pi \int_{a_1}^{a_2} N(a) \, K(a, n) \, a^2 \, da \qquad (2\text{-}6)$$

where
σ_λ = scattering coefficient at wavelength λ, km^{-1}
$N(a)$ = number of particles in interval a to $a + da$ (cm^{-3})
$K(a, n)$ = scattering coefficient (cross section) (cm^{-1})
a = radius of spherical particles (cm)
n = index of refraction of particles

Mie scattering may or may not be strongly wavelength dependent, depending upon the wavelength characteristics of the scattering-area coefficient. In remote sensing, Mie scattering usually manifests itself as a general deterioration of multispectral images across the optical spectrum under conditions of heavy atmospheric haze. Under these conditions the water droplets involved are considerably larger than those discussed in connection with Rayleigh scattering.

The final scattering mechanism to be considered is *nonselective scattering*. This occurs when the scattering particle size is much larger than the radiation wavelength. The total effect of large-particle scattering is the sum of the contributions from the three processes involved in the interaction of the radiation with the particle, i.e., reflection from the surface of the particle with no penetration, passage of the radiation through the particle with or without internal reflections, and refraction at the edge of the particle. Nonselective scattering usually occurs when the atmosphere is heavily dust-laden and results in a severe attenuation of

Figure 1–1 Black-and-white aerial photograph (oblique view).

Figure 1–2 Color aerial photograph (oblique view).

Figure 1–3 Color infrared aerial photograph (oblique view).

Figure 1–4 Color aerial photograph (vertical view).

Figure 1–5 Color infrared aerial photograph (vertical view).

Figure 2–29 A CVF instrument being calibrated in the field. Two units are employed permitting wavelength coverage from 0.4 to 15 μm.

Figure 2–30 A CVF system being used in the field.

(*a*)

(*b*)

Figure 5–3 Coleus leaf showing differences in pigmentation on color and color infrared film.[4] Compare to the spectral reflectance curves shown in Fig. 5–2(*a*). (*a*) Color photo. (*b*) Color infrared photo.

(a)

(b)

Figure 5–4 Pigmentation effects as seen on color and color infrared photographs of a small maple tree.[4] Compare to the spectral reflectance curves of these leaves shown in Fig. 5–2(b). (a) Color photo. (b) Color infrared photo.

(a)

(b)

Figure 5–30(a) Variation in the amount of ground cover, due to differences in soil type and soil moisture conditions. This type of variation might cause problems in identifying the entire field as being soybeans, but would be very useful in obtaining more accurate yield predictions and in soil-mapping programs. (b) Same area showing bare soil at planting time. This photo was taken 9 weeks earlier than photo (a). Together, these photos provide a good example of the effect of soil on vegetation growth.

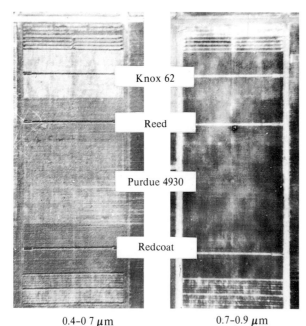

Knox 62

Reed

Purdue 4930

Redcoat

0.4-0 7 μm   0.7-0.9 μm

Figure 5–31 Differences in reflectance due to differences in maturity of wheat.[44] In this case, the basic cause is a difference in varieties. Spectral variation such as this would normally be evident for only a relatively short period of time late in the growing season.

Figure 5–32 Spectral variation due to harvesting of alfalfa.[6] The harvesting has caused many stems, dead leaves near the bottom of the plants, soil, and the more reflective lower sides of leaves to be exposed, thereby creating this distinct difference in color. On a color infrared photo of the same field, the unharvested area appears red whereas the harvested area is pink.

Figure 5–33 Geometric configuration causes differences in spectral response, such as seen in this wheat field where the mature wheat has been knocked down or "lodged."[45] Sometimes hailstorms cause extensive damage by knocking down the stalks of wheat so that the combines used to harvest the wheat cannot pick up the lodged stalks. This type of spectral variation is of considerable interest to insurance companies involved in settling storm damage claims.

Figure 5–34 Spectral differences due to disease.[46] In this case, the area at letter A is healthy corn, whereas that at letter B has a severe infestation of Southern corn leaf blight. At letter C is a field of corn that has been killed by the disease. Spectral differences such as these have proven effective for improving the accuracy of both state and national crop-yield predictions.

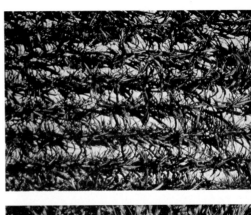

Figure 5–35 Spectral differences due to disease.[47] This striking photo shows plots of healthy potatoes (red) and plots that have been infected with a disease called potato late blight, which has caused a distinct decrease in reflectance such that the affected plots appear black on this color infrared photo. Color infrared film is being operationally utilized in potato late blight surveys.

Figure 5–36 Moisture stress effects in corn Notice how the stressed corn in the top pictu has a distinct difference in crop geometry (curl leaves extending straight up from the stal thereby causing a difference in the amount of s that is exposed. The same area is shown in t lower picture as seen the following day after good rain had relieved the moisture str conditions.

Figure 5–37 Environmental variables, as shown on this Gemini photo depicting the path of rainfall in northern Texas.[48] This photo was taken only 3 to 4 hours after the rainfall, and the decreased reflectance due to the increased soil moisture clearly delineates the area where the rain fell.

Figure 6–8 A color infrared photograph of a portion of the area shown in Fig. 6–7.

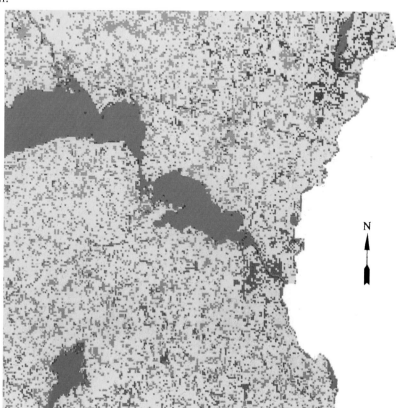

Figure 6–11 Color-coded classification of Winnebago County, Wisconsin, showing level 1 land-use classes (blue=water; red=urban; green=forest; yellow=agriculture).

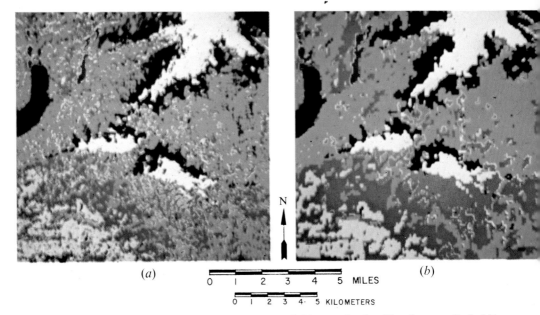

(a)   N   (b)

0 1 2 3 4 5 MILES

0 1 2 3 4· 5 KILOMETERS

Figure 6–14 Comparison of (a) point-by-point and (b) sample classification results [white= snow; black (far left)=water; black (near snow)=Engelmann spruce and subalpine fir; blue=Douglas and white fir; green=ponderosa pine; red=hay and pasture; orange-brown=aspen; yellow=gambel oak].

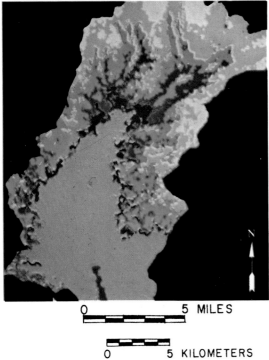

0 5 MILES

0 5 KILOMETERS

Figure 6–19 Snow-cover map of the Lemon Reservoir (dark blue=water; white=95% snow, 5% forest; yellow=85% snow, 15% forest; light blue=65% snow, 35% forest; dark green=30% snow, 70% forest; red=10% snow, 90% forest; light green=other).

Figure 6–22 Thermal map of water only, obtained from the layered classification procedure (yellow and green=20.4 to 21.2°C; red-orange =21.2 to 24.8°C; pink and red =24.8 to 28.4°C).

the received data. However, the occurrence of this scattering mechanism frequently is a clue to the existence of large particulate matter in the atmosphere above the scene of interest, and this in itself is sometimes useful data.

In atmospheric scattering the radiation is reflected or refracted by the particles in the atmosphere. Absorption, on the other hand, involves energy transfer from the impinging radiation into the molecular motion of the intervening atmosphere. The constituent molecules of the atmosphere can absorb energy either by a translational vibration or by a rotational mechanism. The molecular structure of the atmospheric gases is complex enough that several resonant frequencies of absorption are possible for a particular constituent gas. For example, carbon dioxide absorbs radiation at several wavelengths. In the vicinity of these critical wavelengths, the absorption mechanism of the atmosphere predominates over scattering. Elsewhere, the scattering mechanisms discussed previously are the dominant factors affecting atmospheric transmission.

To illustrate, a plot of the percentage of transmission versus wavelength for an atmospheric path is shown in Fig. 2-8. This plot does not include the effects of scattering. The peaks on the graph indicate those regions of the spectrum where radiation is passed through the atmosphere with relatively little attenuation. These spectral regions are called *atmospheric windows*.

Figure 2.8 also does not show the details of the absorption in the spectral region below 2.5 μm. There are narrow but strong bands of absorption at approximately 1.4 μm and 1.9 μm wavelength due to water vapor in the atmosphere. Water in natural materials similarly produces low reflectances (i.e., absorptions) in these wavelength bands. Atmospheric molecular (Rayleigh) scattering and ozone layer absorption severely attenuate radiation at wavelengths below 0.3 μm. This leads to an identification of atmospheric windows in the spectral regions 0.3–1.3, 1.5–1.8, and 2.0–2.6 μm. Combining this information with that of Fig. 2.8, the following list of atmospheric windows can be constructed.

Atmospheric window	Spectral region (μm)
1	0.3–1.3
2	1.5–1.8
3	2.0–2.6
4	3.0–3.6
5	4.2–5.0
6	7.0–15.0

The spectral bands of remote sensing data acquisition systems are usually selected from windows 1, 2, 3 and 6, since reflective and emissive effects are clearly separated in these spectral regions (see Sec. 5-2 for a discussion of vegetative spectral response and the consequent impact of atmospheric windows).

The data presented here assume a horizontal atmospheric transmission path with constant temperature, humidity, and pressure. For slant-path transmission, very little quantitative data are available. Since gas pressure, temperature, and composition vary along the path, the transmission must generally be predicted on the basis of theoretical calculations. To determine the atmospheric transmissivity

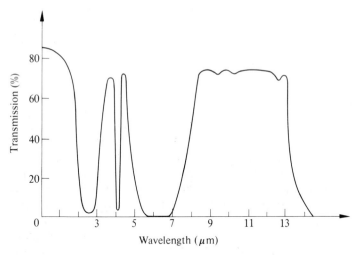

Figure 2-8 Percent of radiation passed through an atmospheric path 2000 meters long (H_2O partial pressure 17 mm) (Source: *Handbook of Geophysics,*[1] © copyright 1961, The Macmillan Company. Used with permission.)

for a given slant path, one must first determine the equivalent sea-level path for the absorber of interest and then find the absorption which is characteristic of the wavelength interval of interest. The final step, then, is to establish the functional relationship between transmissivity and the absorption which best suits the slant path under consideration. Slant-path calculations are important in those cases where the total field of view of the sensor being used to generate the data is wide enough for the path length near the edge of the total field of view to be considerably longer than the path length at nadir (straight down).

2-3 REFLECTANCE IN REMOTE SENSING

The manner in which solar radiation interacts with a target determines the reflectance of that target. Many details of the target-energy interactions are discussed in Chap. 5. In this section the geometric aspects of reflectance will be discussed, especially with regard to the interaction of source, target, and sensor. A careful discussion of reflectance involves consideration of the optical characteristics of the sensor as well as the target. Therefore, this section begins with a review of some elementary optics to facilitate the subsequent discussion of reflectance. The discussion on reflectance closely follows that of DeWitt and Robinson.[8]

Elementary Optics Review

In describing the optical components of a remote sensor, it is necessary to speak of the *optical stops* in the critical image-forming portions of the system. The location and size of these optical stops affect the flow of radiation through the optical portion of the sensor; therefore, it is necessary to acquire a basic understanding of stops and their effect on data characteristics and quality.

Although remote sensing optical systems may be quite complex, we can present the basic ideas of optical stops with a very simple example involving a thin lens. A *thin lens* is one in which the focal length is much greater than the lens thickness; it is a refractive element. Usually a remote sensing optical system will contain both refractive and reflective elements; however, the basic concepts can be more easily understood from an example of a thin-lens refracting system.

Consider the thin lens shown in Fig. 2-9. Any imperfections in the lens are ignored, and a simple ray-tracing diagram is used to locate the *image* of the arrow

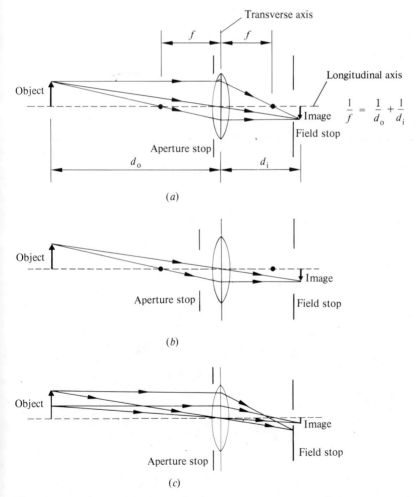

Figure 2-9 Optical stop definitions. (*a*) Aperture and field stops open. (*b*) Aperture stop closed down, fewer rays reach image plane. (*c*) Field stop closed down, portion of image cut off.

(on the right), which represents the *object* located on the left-hand side of the diagram. In the ray-tracing method used here, any ray parallel to the *longitudinal* axis of the lens and passing through the transverse axis of the lens is refracted through the *focal point*, and any ray which passes through the *center* of the lens is not refracted at all. The tip of the image can be located at the intersection of two rays, as shown. The relationship between the focal length f and the object and image distances, d_o and d_i, is shown in the thin-lens equation that accompanies Fig. 2-9(a). Notice that any ray which originates from the tip of the object will be refracted at such an angle when it passes through the transverse axis of the lens so as to terminate at the tip of the image arrow. When an aperture is placed very close to the transverse axis of the lens and is closed (i.e., "stopped down"), some of the rays originating from the tip of the object arrow are blocked by the *aperture stop*, as shown in Fig. 2-9(b). The entire image still is present, but it is dimmer since fewer rays reach the image plane. In general, as one shuts down the aperture stop in a simple thin-lens system, the image gets dimmer.

If a diaphram is placed directly in front of the image and is closed down, *all* of the rays which terminate on the tip of the image arrow are blocked, as shown in Fig. 2-9(c), and part of the image is cut off. Thus the stop which is located directly in front of the image plane is a *field stop*, i.e., a stop which affects the field of view, the portion of the object that is actually imaged by the lens-and-stop system.

The aperture stop determines the light-gathering ability or "speed" of the lens. As the aperture stop is closed down, the angular deflection of the rays at the edge of the stop decreases as they pass through the lens; this can be seen by examining

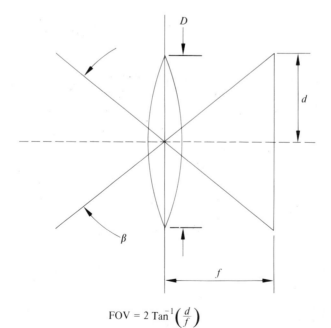

$$FOV = 2\,Tan^{-1}\left(\frac{d}{f}\right)$$

Figure 2-10 Field of view of a simple lens.

Fig. 2-9(a) and (b). For a given focal length the angular deflection of the rays near the edge of the aperture stop is greatest, and this deflection increases as the stop is opened. Therefore, one can relate the magnitude of the angular deflection of the rays at the aperture stop to the light-gathering ability of the lens/aperture-stop combination. Notice that if the focal length of the lens is shortened while the aperture-stop diameter is held constant, then the angular deflection of the rays near the edge of, the aperture stop increases. Thus, of two lenses having identical diameters but different focal lengths, the lens with the shorter focal length has the greater light-gathering power. A convenient way to express this fact is to form the ratio f/D where f is the focal length of the lens and D is its diameter. The smaller this ratio, the greater the light-gathering power (speed) of the lens.

The geometry of the field-of-view (FOV) calculation is shown in Fig. 2-10:

$$\beta = \text{FOV} = 2 \, \text{Tan}^{-1}\left(\frac{d}{f}\right) \qquad ° \text{ or rad} \qquad (2\text{-}7)$$

where d is one-half the diameter of the field stop (located at the focal plane) and f is the focal length of the lens. Note that D, the lens diameter, does not affect the FOV of the lens. As an example, consider the lens system of a typical 35-mm camera with a 50-mm focal-length lens. The diagonal distance across the 35-mm film (the effective field stop) is 41.6 mm so that $d = 20.8$ mm. Since $f = 50$ mm, then

$$\text{FOV} = 2 \, \text{Tan}^{-1}\left(\frac{20.8}{50}\right)$$

$$= 2 \times 22.5°$$

$$= 45°$$

If the field of view is small enough that the approximation $\tan x = x$ applies, then Eq. (2-7) can be modified to yield

$$\tan \beta \cong \beta = \frac{2d}{f} \qquad (2\text{-}8)$$

The distance, $2d$, represents the limiting field-stop dimension. In remote sensing systems the object distance d_o is much greater than the image distance so that the thin-lens formula in Fig. 2-9(a) reduces to

$$\frac{1}{f} = \frac{1}{d_o} + \frac{1}{d_i} \cong \frac{1}{d_i}$$

That is, the image is formed approximately at the focal plane (which contains the focal point) of the lens. The field stop is located at this focal plane. Usually the detector dimensions establish the field stop and determine the field of view of the sensor according to Eq. (2-8), since, as we shall see, β is typically less than 1 degree. The aperture stop is generally located very near the lens itself. In multispectral scanners discussed in Sec. 2-6, the diameter of the primary lens (or mirror) determines the size of the aperture stop.

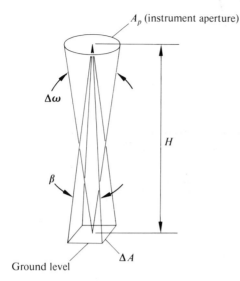

A_p (instrument aperture)

$\Delta\omega$

H

β

Ground level

ΔA

Figure 2-11 Relationship between aperture and field of view.

Reflectance in General

Much of the data acquired in a remote sensing system results from the measurement of reflected solar radiation. Here we shall first discuss reflectance from the mathematical viewpoint and then examine how it affects the remote sensing problem. Next, we will consider how the reflectance that is observed in a field situation is correlated with that observed in a laboratory.

First we must examine the effect that field of view and instrument aperture have upon the measurement of radiation. If a measuring instrument is located a distance H above a surface from which the reflected radiation is to be measured (Fig. 2-11), the area observed by the instrument, ΔA, is equal to $(\beta H)^2$, where β is the plane angle of the instantaneous field of view. Obviously, as H increases, the area defined by the field of view increases. On the other hand, the instrument aperture, with area A_p, subtends a solid angle $\Delta\omega$ at the surface. The relationship between $\Delta\omega$ and H is given by $\Delta\omega = A_p/H^2$. Thus, if the aperture A_p and field of view β are assumed fixed, then as the altitude H increases, A increases and $\Delta\omega$ decreases. It can easily be shown that the power into the sensor is proportional to the product $\Delta\omega \cdot \Delta A = (A_p/H^2)(H^2\beta^2) = A_p\beta^2$. Therefore, for a uniformly illuminated target which "fills ΔA," *the power into the sensor will be independent of H* (ignoring atmospheric transmission effects).

Bidirectional Reflectance Distribution Function (BRDF)

The most fundamental property describing the reflection characteristics of a surface is the bidirectional reflectance distribution function (BRDF), usually denoted by the symbol f, and defined as

$$f(\theta, \phi; \theta', \phi') = \frac{dL'(\theta', \phi')}{dE(\theta, \phi)} \qquad \mathrm{sr}^{-1}$$

(The unprimed angles are those describing the location of the radiation source and the primed angles refer to the location of the sensor, as in Fig. 2-12. The quantities ω and ω' will appear in an expression only when their magnitudes are a significant factor in the reflectance property. Generally ω and ω' are small enough in practical remote sensing geometric arrangements that the property or quantity in question does not vary appreciably as a function of these solid angles.) The BRDF is the quotient dL/dE with the units of inverse steradians, and is a function of the incident and reflected directions expressed by $\theta, \phi; \theta', \phi'$. Here the component angles of source or sensor are separated by a comma; the sets of angles describing incident and reflected flux are separated by a semicolon. In the literature $f(\theta, \phi; \theta', \phi')$ sometimes is denoted as $\rho'(\theta, \phi; \theta', \phi')$. The term dL [units of $W/(m^2 - sr)$] is the reflected radiance in the direction θ', ϕ', produced by the incident irradiance dE (units of W/m^2) of a well-collimated beam (parallel rays) from the direction θ, ϕ. Using the radiation geometry of Fig. 2-13, the irradiance is

$$dE(\theta, \phi) = L(\theta, \phi) \cos \theta \, d\omega$$

$$= L(\theta, \phi) \cos \theta \sin \theta \, d\theta \, d\phi \qquad W/m^2 \qquad (2\text{-}9)$$

where $L(\theta, \phi)$ is the incident radiance from the direction θ, ϕ through the solid angle $d\omega = \sin \theta \, d\theta \, d\phi$. The measurement of the BRDF is quite complex since accurate measurement of $dE(\theta, \phi)$ at the target surface is generally inconvenient.

Measurement of $L'(\theta', \phi')$, however, is straightforward. The sensor is located at the (θ', ϕ') of interest. The radiant flux entering the instrument, following the

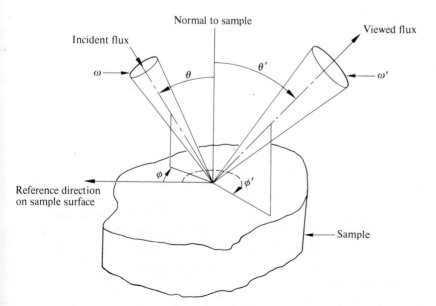

Figure 2-12 Geometric parameters describing reflection from a surface: $\theta = $ zenith angle, $\phi = $ azimuthal angle, $\omega = $ beam solid angle; A prime on a symbol refers to viewing (reflected) conditions.

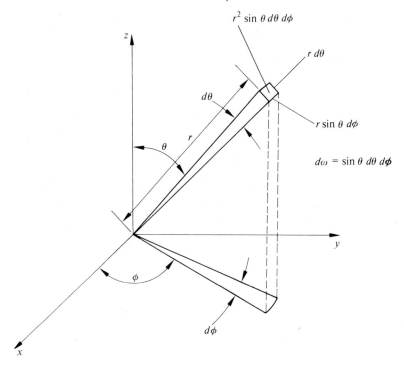

Figure 2-13 Radiation geometry.

terminology of Fig. 2-12, is given by

$$\Phi = L'(\theta', \phi') H^2 \beta^2 (\cos \theta') \frac{A_p}{H^2}$$

$$= L'(\theta', \phi') \beta^2 A_p \cos \theta' \qquad W$$

Therefore,

$$L'(\theta', \phi') = \frac{\Phi}{\beta^2 A_p \cos \theta'}$$

$$= \frac{k V_s}{\beta^2 A_p \cos \theta'} \qquad W/(m^2 - sr) \qquad (2\text{-}10)$$

where V_s is the output voltage and k the calibration constant of the sensor. In the limit of small $\Delta\omega$ and ΔA these quantities can be interpreted as differential quantities $d\omega$ and dA, and Eq. (2-10) yields $dL'(\theta', \phi')$ as well.

Bidirectional Reflectance

Another frequently encountered reflectance quantity is bidirectional reflectance, which is expressed as

$$d\rho(\theta, \phi; \theta', \phi') = \frac{dL'(\theta', \phi') \cos \theta' \, d\omega'}{L(\theta, \phi) \cos \theta \, d\omega}$$

$$= f(\theta, \phi; \theta', \phi') \cos \theta' \, d\omega' \qquad \text{(dimensionless)} \qquad (2\text{-}11)$$

Although frequently measured and used, $d\rho$ does not indicate a unique property of the sample but is dependent on the instrument configuration used to make the measurement, since the value of $d\rho$ depends on $\cos \theta' \, d\omega'$, the projected solid angle of reflection. The incident flux is determined by source characteristics and the solid angle of irradiation $d\omega$. The flux reflected in any direction over the hemisphere that will be sensed by the instrument is determined by $d\omega'$, the solid angle of reflected radiance, which may be quite different from $d\omega$. This may cause confusion to the user of the data from such measurements.

Reflectance Reciprocity

A property of BRDF which is useful in analyzing the various types of reflectance is the Helmholtz reciprocity theorem:[9]

$$f(\theta, \phi; \theta', \phi') = f(\theta', \phi'; \theta, \phi) \qquad (2\text{-}12)$$

Physically this means that the location of the radiation source and sensor may be interchanged without affecting the value of the BRDF that is measured. Using the geometry of Fig. 2-12 where subscripts indicate specific values of the parameters $\theta, \phi,$ and ω (with the restriction that the radiance does not vary over the solid angle of incidence in either case), we can write, using Eq. (2-11),

$$\rho(\theta_1, \phi_1, \omega_1; \theta_2, \phi_2, \omega_2) = f(\theta_1, \phi_1, \theta_2, \phi_2) \cos \theta_2 \omega_2 \qquad (2\text{-}13)$$

or, interchanging the subscripts and rewriting,

$$\rho(\theta_2, \phi_2, \omega_2; \theta_1, \phi_1, \omega_1) = f(\theta_2, \phi_2, \theta_1, \phi_1) \cos \theta_1 \omega_1 \qquad (2\text{-}14)$$

Applying Eq. (2-12) to the BRDF terms in Eq. (2-13) and (2-14), we obtain

$$\frac{\rho(\theta_1, \phi_1, \omega_1; \theta_2, \phi_2, \omega_2)}{\cos \theta_2 \omega_2} = \frac{\rho(\theta_2, \phi_2, \omega_2; \theta_1, \phi_1, \omega_1)}{\cos \theta_1 \omega_1} \qquad (2\text{-}15)$$

That is, the reciprocity of bidirectional reflectance is not as direct as that of BRDF and must be modified by considering the projected solid angle of both source and sensor. Equation (2-15) will be used later to establish additional reciprocity relationships.

Bidirectional Reflectance Factor (BRF)

The BRDF, although completely descriptive of a reflecting surface, is difficult to measure. An alternative quantity, the bidirectional reflectance factor (BRF), proves to be straightforward to measure and, under certain assumptions, can be directly related to BRDF. The measurement method used to obtain BRF is based upon the utilization of a perfectly diffuse, completely reflecting surface as a reference.

Reflectance factor is defined as the ratio of flux reflected by a target under specified conditions of irradiation and viewing to that reflected by an ideal, completely reflecting, perfectly diffuse surface, identically irradiated and viewed. "Perfectly diffuse" means that the surface reflects equally in all directions (sometimes called a Lambertian reflector). The radiance of a uniformly illuminated Lambertian surface of infinite extent is constant for any viewing angle, θ'. If a uniformly illuminated Lambertian surface is small enough so as not to fill the field of view of an observing sensor, the radiance measured by the sensor will be proportional to the cosine of the viewing angle (see Fig. 2-2). "Completely reflecting" means that all of the flux falling on the surface is reflected from the surface. That is, M equals E for a perfectly reflecting surface. Let us use this information to calculate the BRDF of such an ideal surface.

Express the radiant exitance from the surface in terms of $L'(\theta', \phi')$ by integrating the projected reflected radiance $L'(\theta', \phi') \cos \theta'$ over a hemisphere:

$$M = \int_{\text{hemi}} L'(\theta', \phi') \cos \theta' \, d\omega'$$

The integration takes place over the viewing angles (primed variables). Inserting $d\omega' = \sin \theta' \, d\theta' \, d\phi'$, the integral becomes

$$M = \int_0^{2\pi} \int_0^{\pi/2} L'(\theta', \phi') \cos \theta' \sin \theta' \, d\theta' \, d\phi'$$

Since the magnitude of $L'(\theta', \phi')$ is independent of θ', ϕ' for a perfectly diffuse surface, we have

$$M = L' \int_0^{2\pi} \int_0^{\pi/2} \cos \theta' \sin \theta' \, d\theta' \, d\phi' = \pi L'$$

and thus

$$L' = \frac{M}{\pi} \tag{2-16}$$

or

$$dL' = \frac{dM}{\pi}$$

Equation (2-16) was derived for a perfectly diffuse reflecting surface, but the result is equally applicable to a Lambertian radiating surface. As a result Eq. (2-1) can be modified to read

$$L_\lambda = \frac{\varepsilon c_1'}{\lambda^5 (e^{c_2/\lambda T} - 1)}$$

where $c_1' = c_1/\pi = 1.1908 \times 10^8$ W/(m^2 $-$ μm $-$ sr). This relation gives the spectral radiance for a Lambertian blackbody radiator. The rest of the terms in the equation for L_λ are the same as those in Eq. (2-1).

Now $M = E$ for the perfectly reflecting surface, and, using the definition for BRDF,

$$f_p(\theta, \phi; \theta', \phi') = \frac{dL'}{dE} = \frac{dM/\pi}{dM} = \frac{1}{\pi} \tag{2-17}$$

where the subscript p refers to the ideal (perfect) diffuser. Using this result the ratio of the reflected radiance (numerator of the BRDF) of a target to the reflected radiance of a perfectly reflecting perfect diffuser can be determined (the T subscript refers to the target):

$$\frac{dL'_T}{dL'_p} = \frac{f_T \, dE}{f_p \, dE} = \frac{f_T}{1/\pi} = R$$

that is,

$$R(\theta, \phi; \theta', \phi') = \pi f_T(\theta, \phi; \theta', \phi') \tag{2-18}$$

Under the assumption of an ideal reference surface, the easily measured BRF is directly relatable to the BRDF. Usually, the reference target is fabricated from barium sulfate ($BaSO_4$). Properly prepared barium sulfate (and also magnesium oxide, MgO) reference surfaces closely approximate perfect diffusers for $\theta' \leq 45°$, but depart from the completely reflecting assumption at some wavelengths. The left-hand side of Eq. (2-18) can be obtained from the measured BRF, here referred to as $R'(\theta, \phi; \theta', \phi')$, by using the relation:

$$R(\theta, \phi; \theta', \phi') = \rho_s R'(\theta, \phi; \theta', \phi') \tag{2-19}$$

where ρ_s is the known spectral reflectance of the reference surface with the values given in Table 2-2.

Now we shall relate the reflectance terms described so far to some special cases. For simplicity, the notation for azimuthal variation (ϕ and ϕ') will be suppressed. Applying Eqs. (2-11) and (2-18), we have

$$\rho(\theta, \omega; \theta', \omega') = f(\theta, \theta') \cos \theta' \omega'$$

$$= \frac{1}{\pi} R(\theta; \theta') \cos \theta' \omega'$$

$$= \frac{\omega' \cos \theta'}{\pi} R(\theta, \omega; \theta', \omega') \tag{2-20}$$

Equation (2-20) is called the *bidirectional relation* and illustrates the relationship between bidirectional reflectance and BRF. Specializing Eq. (2-20) to two particular values of θ and ω, namely, θ_1, θ_2 and ω_1, ω_2, then

$$R(\theta_1, \omega_1; \theta_2, \omega_2) = \pi \frac{\rho(\theta_1, \omega_1; \theta_2, \omega_2)}{\cos \theta_2 \omega_2}$$

and

$$R(\theta_2, \omega_2; \theta_1, \omega_1) = \pi \frac{\rho(\theta_2, \omega_2; \theta_1, \omega_1)}{\cos \theta_1 \omega_1} \tag{2-21}$$

Table 2-2 Absolute spectral reflectance reference data for smoked magnesium oxide (MgO) and barium sulfate (BaSO₄) paint. (*After Grum and Luckey*[10])

Wave-length, μm	BaSO$_4$		MgO	
	Pressed powder	Coating of paint (1.0 mm)	Fresh	Aged (2 months)
0.225	0.925	0.905	0.865	0.672
0.250	0.939	0.932	0.926	0.746
0.275	0.979	0.962	0.951	0.775
0.300	0.987	0.962	0.965	0.799
0.325	0.980	0.983	0.967	0.832
0.350	0.985	0.978	0.970	0.869
0.375	0.992	0.982	0.978	0.887
0.400	0.995	0.980	0.982	0.906
0.420	0.999	0.986	0.984	0.992
0.440	0.999	0.980	0.986	0.933
0.460	0.999	0.993	0.987	0.942
0.480	0.999	0.993	0.987	0.951
0.500	0.998	0.992	0.986	0.957
0.520	0.998	0.992	0.986	0.963
0.540	0.998	0.993	0.986	
0.560	0.998	0.992	0.983	
0.580	0.998	0.992	0.982	
0.600	0.998	0.992	0.981	
0.620	0.998	0.992	0.981	
0.640	0.998	0.992	0.981	
0.660	0.998	0.992	0.981	
0.680	0.998	0.992	0.981	
0.700	0.997	0.992	0.981	
0.800	0.996	0.992	0.987	
0.900	0.995	0.989	0.972	
1.000	0.991	0.983	0.966	
1.100	0.992	0.982	0.965	
1.200	0.983	0.973	0.957	
1.300	0.980	0.966	0.956	
1.400	0.932	0.895	0.939	
1.500	0.929	0.896	0.922	
1.600	0.944	0.920	0.934	
1.700	0.939	0.913	0.935	
1.800	0.915	0.892	0.921	
1.900	0.829	0.770	0.881	
2.000	0.811	0.754	0.854	
2.100	0.842	0.765	0.870	
2.200	0.842	0.792	0.861	
2.300	0.816	0.735	0.847	
2.400	0.768	0.713	0.814	
2.500	0.703	0.650	0.773	

Applying Eq. (2-15) to Eq. (2-21), then

$$R(\theta_1, \omega_1; \theta_2, \omega_2) = R(\theta_2, \omega_2; \theta_1, \omega_1) \tag{2-22}$$

Equation (2-22) illustrates the reciprocity of reflectance factors. If the reflectance factor is measured over several viewing angles over a hemisphere and the results numerically integrated, the result is denoted by $R(\theta_1, \omega_1; 2\pi)$. Applying Eq. (2-22), we can write

$$R(\theta_1, \omega_1; 2\pi) = R(2\pi; \theta_1, \omega_1) \tag{2-23}$$

That is, reciprocity also applies to hemispherical measurements. Now we shall apply Eq. (2-18) to Eq. (2-11) and integrate both sides over a hemisphere of viewing angles:

$$\int_{\text{hemi}} d\rho(\theta, \phi; \theta', \phi') = \frac{1}{\pi} \int_{\text{hemi}} R(\theta, \phi; \theta', \phi') \cos \theta' \, d\omega'$$

Integrating

$$\rho(\theta, \phi; 2\pi) = \frac{1}{\pi} R(\theta, \phi; 2\pi) \int_{\text{hemi}} \cos \theta' \, d\omega'$$

Therefore,

$$\rho(\theta, \omega; 2\pi) = R(\theta, \omega; 2\pi) \tag{2-24}$$

The term $\rho(\theta, \omega; 2\pi)$ is the *directional reflectance* and is in one-to-one correspondence to the directional reflectance factor $R(\theta, \omega; 2\pi)$. If Eq. (2-20) is integrated over the hemisphere of source variables as well, then

$$\rho(2\pi; 2\pi) = R(2\pi; 2\pi) \tag{2-25}$$

where $\rho(2\pi; 2\pi)$ is the hemispherical reflectance, often called the "albedo." We shall use these reflectances in subsequent sections.

The reflectances discussed here are those of greatest use in remote sensing. There is a large array of reflectance terminology in use in reflectance spectroscopy and radiometry[2] that is not generally applicable to the remote sensing problem and, therefore, will not be treated here.

PROBLEMS

2-1 Draw a set of axes which will allow you to plot relative radiation energy versus wavelength. Dimension the wavelength axis so as to cover the optical portion of the spectrum. Sketch on this set of axes the general shape of the solar extraterrestrial spectrum. Can you use the resultant curve to discuss the "reflective" and "emissive" portions of the spectrum? Include in your discussion the logic behind the use of these terms and the wavelength range of each.

2-2 Add to the graph you prepared for Prob. 2-1 lines delineating the *visible band* and the *near-*, *middle-*, and *far-infrared bands*.

2-3 Write down the key words you would use in explaining the concept of a blackbody.

2-4 (*a*) Determine the wavelengths at which the spectra of blackbodies with temperatures of 250 *K*. 1200 K, and 4000 K will peak.

(*b*) Use the results of (*a*), the Stefan-Boltzmann law, and suitable assumptions to estimate the shape of the radiation curves of blackbodies at temperatures of 250 K, 1200 K, and 4000 K.

2-5 A certain target is characterized by an emittance of 0.9 and a temperature of 350 K. The background temperature is 300 K. Compute the ratio of radiant exitance due to the target acting as a graybody to the radiant exitance due to reflection of background radiation.

2-6 Derive Eq. (2-4).

2-7 (*a*) The primary ways in which the atmosphere affects electromagnetic energy are _____ and _____.

(*b*) What happens to the energy in each case?

2-8 Match the appropriate scattering mechanism and atmospheric particle sizes:

Rayleigh	$\lambda \ll s$
Mie	$\lambda \gg s$
Nonselective	$\lambda \approx s$

2-9 When both scattering and absorption take place, which effect dominates?

2-10 What is an atmospheric window? What type of energy-matter interaction predominates in the determination of atmospheric windows?

2-11 Consider a remote sensing data-gathering instrument designed to operate at an altitude of 5000 m and having a primary lens of diameter 20 cm and an effective detector area of 25 mm². What focal length and optical speed are required to achieve a field of view of 2 mrad?

2-12 The arrangement of an energy source, target, and radiation detection instrument is shown in Fig. P2-12. Experimental procedures call for measuring the light energy coming into the instrument with the target in place. This measurement is denoted as M_T. Then the target is replaced by a barium sulfate reflectance panel and the instrument is read again to obtain M_S. From the geometry shown and the data given, determine the BRDF for the target. Comment on the amount of data given.

2-13 Calculate the radiant exitance of an ideal blackbody at a temperature of 300 K. Compare your result to the value of the peak spectral radiant exitance in Fig. 2-5. Discuss your comparison.

2-14 Determine the peak spectral radiant exitance of an ideal blackbody at a temperature of 6000 K. Compare your answer to the peak spectral irradiance of the 6000 K curve in Fig. 2-1. Explain the difference. (*Hint:* you must consider the surface area of the sun and the distance from the sun to the earth.)

2-15 A thin lens of variable focal length (a "zoom" lens) is used to form an image on a 20-mm diameter film plane. The focal length of the lens can be varied from 35 mm to 120 mm.

(*a*) Determine the field of view of the system at the focal-length extremes.

(*b*) Determine the amount of lens movement (with respect to the film plane) necessary to focus the lens system over the full range of focal lengths for objects located from 1 meter to infinity.

2-16 A perfectly diffuse surface reflects 80 percent of the energy that falls upon it. For this surface what is the value of each of the following?

(*a*) $R(\theta, \phi; \theta', \phi')$

(*b*) $\rho(\theta, \omega; \theta', \omega')$

(*c*) $f(\theta, \omega; \theta', \omega')$

(*d*) $\rho(\theta, \omega; 2\pi)$

Study objectives

After finishing Secs. 2-4, 2-5, and 2-6 you should be able to:

1. Relate a specific remote sensing instrumentation system to the general functional block diagram of Fig. 2-14.

2. Sketch the general shape of responsivity versus wavelength characteristics for photon and thermal detectors, be able to state the wavelength regions

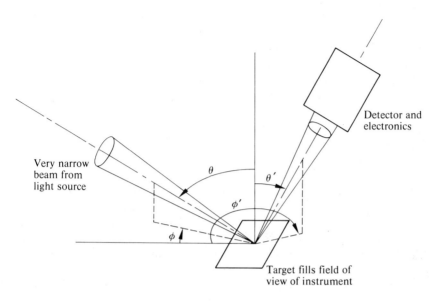

θ	θ'	ϕ	ϕ'	M_T	M_S
20°	15°	0°	0°	0.31	1.3
20°	25°	0°	0°	0.29	1.3
20°	15°	0°	45°	0.30	1.2
20°	25°	0°	45°	0.28	1.2
10°	15°	0°	0°	0.32	1.4
10°	25°	0°	0°	0.30	1.4

Figure P2-12.

over which these types of detector can be designed to operate, and select a detector for a specific application.

3. Contrast the characteristics of spectral data systems and image-forming systems.

4. Select an appropriate type of spectral data-collection system when given the data-collection environment and the quality, quantity, and calibration requirements.

5. Estimate the power flow into a scanner system when given the scanner parameters and compute the change in signal strength due to a parameter change.

6. Determine the effect changes in scanner parameters will have on the scanner signal-to-noise ratio.

7. Determine acceptable bounds on scanner-system design parameters given scanner platform stability specifications.
8. Describe the appearance of an object on color infrared film if it has a very low reflectance everywhere except in a specified portion of the spectrum.
9. Compare the relative advantages and disadvantages of multispectral scanners and electron imagers.

2-4 GENERAL RADIATION INSTRUMENTATION PRINCIPLES

In remote sensing instrumentation systems, radiation enters the aperture of the measuring instrument and there follows extensive optical and electronic processing. Although photographic remote sensing systems are important in many respects, the data they return do not lend themselves very easily to numerically oriented remote sensing data-processing systems. Therefore, those systems which are more directly suitable for use with a digital-processing system will receive the greatest emphasis here, specifically those basic components which tend to be common to all such systems. The instruments which are most frequently used in laboratory and field measurements in remote sensing research will be described on a functional basis. Once the groundwork has been laid, a detailed description of field instrumentation and airborne data-collection systems will be given in Sec. 2-6.

The *spectrometer* is the basic measuring instrument for the detection of radiation in the numerically oriented systems described in this textbook. Actually, to be more accurate, the *monochrometer*, an instrument which analyzes polychromatic radiation into its individual spectral components, is the basic measuring instrument. The spectrometer is essentially an adjustable monochrometer. In Fig. 2-14, a functional diagram of a spectrometer is shown. The polychrome radiation enters the instrument through the entrance aperture. This aperture is used to control the intensity of the radiation entering and, in some cases, is used to control the field of view of a remote sensing system in which the spectrometer might be a component. After the radiation passes through the entrance aperture, a series of collimating or focusing optics, which may be reflective, refractive, or a combination of the two (catodiatropic), is used to focus the radiation. The effective optical

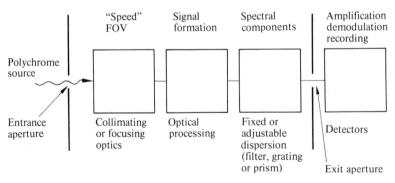

Figure 2-14 A functional diagram of a spectrometer.

speed of the system and the field of view are determined by the collimating or focusing optical system. In some instruments, the radiation passes through an optical-processing system such as an optical chopper where the optical signal is formed. This signal then passes through the dispersive portion of the instrument, such as an interference filter, a grating, a prism, or any optical component whose transmission or reflection characteristics are a function of the wavelengths of the impinging radiation. It is in this portion of the instrument that the incident radiation is resolved into its spectral components. (Optical processing can also take place after this dispersion as well as before.) Usually the radiation is then focused upon an exit aperture, which in some cases defines the field of view of the instrument or its spectral resolution. After passing through the exit aperture, the radiation falls upon one or more detectors, is converted into an electrical signal, amplified, processed in various ways, and then recorded. Actually, practical spectrometers vary widely in design, but they all contain the basic components illustrated in Fig. 2-14.

The spectrometer described above is a basic component in a number of instruments used in spectral analysis. For example, a *spectrophotometer* (Fig. 2-15) is an instrument that contains a spectrometer as well as an internal source of illumination which may generate both visible and infrared radiation. The instrument is designed so that the internally generated radiation can be focused upon a target of interest. The spectrometer is then used to analyze the radiation coming from the target. The spectrometer and the internal radiation source are coupled so that the output of the spectrometer can be normalized in terms of the incident radiation. Usually, the internal radiation source illuminates a reference

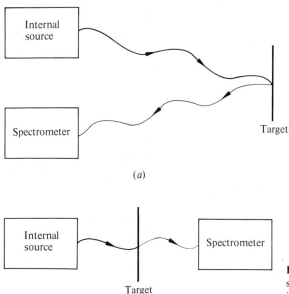

(a)

(b)

Figure 2-15 Arrangement of a basic spectrophotometer into different instruments. (*a*) A spectroreflectometer. (*b*) A spectrotransmissometer.

target as well, so that unknown target characteristics can be expressed in terms of the known reference.

If the instrument is geometrically arranged so that the spectrometer is receiving radiation *reflected from* the target, then the instrument is called a *spectroreflecto-meter*. If the instrument is arranged so that the spectrometer is viewing radiation that has passed *through* the target, the instrument is a *spectrotransmissometer*. Usually the spectroreflectometer and the spectrotransmissometer are designed so that the target is in a fixed position with respect to the impinging radiation; at most, only a limited geometric adjustability between the spectrometer and the internal source is possible. If, however, the instrument is arranged so that the angle of the internal source with respect to the target and the angle of the spectrometer with respect to the target can be varied, then the instrument is referred to as a *goniometric spectrophotometer*.

An instrument that does not contain an internal light source but depends upon an external source for its illumination is referred to as a *spectroradiometer*. This instrument is used to examine targets that are located in a natural environment so that spectral analysis can be made under natural conditions. A functional diagram of this instrument is shown in Fig. 2-16. The spectroradiometer usually has provisions for making spectral measurements of the external source so that reflectance calculations can be made and for establishing some type of internal reference calibration so that absolute radiation measurements may be made. Details of such an instrumentation system are described in Sec. 2-6.

The instruments discussed up to this point produce data which are directly interpretable in terms of spectral irradiance and wavelength. Instruments called *interferometers* produce extremely high spectral resolution data, but the data must be subjected to complex processing before they can be interpreted in terms of spectral irradiance and wavelength. Interferometers are discussed in Sec. 2-6.

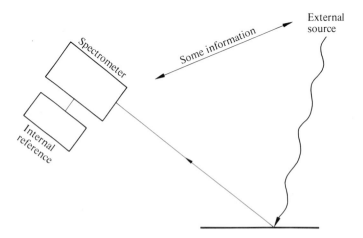

Figure 2-16 Functional diagram of a spectroradiometer.

2-5 RADIATION DETECTORS

Radiation detectors are critical components of any remote sensing system. The spectral components of the radiation, having been resolved by the optical portion of the system, fall upon the radiation detectors and are converted into a form capable of being processed into data. The radiation detectors in a remote sensing system may be photographic film in which a photochemical process is used to convert the radiation into an image that can then be used for subsequent processing. Photon detectors convert the impinging radiation into electrical signals that are especially useful in a digital remote sensing system. The nature of the radiation detectors in a remote sensing system, especially one that is numerically oriented, exerts an important influence upon the quality of the resulting data. It is essential that the analyst be aware of the effects of various radiation detectors upon data quality so that he or she does not attribute detector-related effects to anomalies in the target. This section of the chapter will describe the various kinds of radiation detector most commonly used in remote sensing systems and provide a relatively simple description of their operating principles.

Photon and Thermal Detectors

Radiation detectors commonly used in remote sensing technology fall into two broad categories: photon detectors and thermal detectors. A *thermal detector* essentially changes its temperature in response to incident radiation, and in most cases its electrical resistance is a function of its temperature. The advantage of a thermal detector is that its response is not a function of the wavelength of the impinging radiation; it responds to all wavelengths of radiation. The disadvantages are that the thermal detector is incapable, in general, of rapid response to a rapidly changing radiation input and is generally less sensitive than a photon detector.

Photon detectors, on the other hand, are capable of rapid response and for this reason are frequently employed in numerically oriented remote sensing detection systems. In this text we will consider some of the elementary quantum-mechanical characteristics of these photon detectors in order to gain a basic understanding of their characteristics and limitations.

Basically, photon detectors operate on the principle that incoming radiation excites electrical charge carriers, causing them to move from one energy level to another energy level within the crystal lattice of the detector. Charge carriers which are at lower energy levels are said to exist in the valence band of the crystalline detector. If a charge carrier is excited into an energy level such that it can move freely throughout the crystalline structure of the detector, it is said to exist in the conduction band. Due to the quantum-mechanical nature of the device, the charge carrier jumps from the valence-band energy level to the conduction-band energy level across a "forbidden" energy gap. A simple schematic representation of this energy-band structure is shown in Fig. 2-17.

When discussing radiation in terms of quantum-mechanical concepts, the

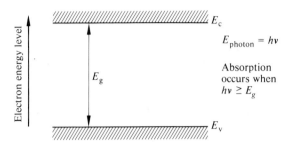

Figure 2-17 Photon detector energy band diagram.

radiation can be considered either in terms of its wave-like character or its particle-like character. When discussing radiation in terms of wave-like character, we speak of wavelength, trains of waves, etc. However, when discussing radiation in terms of its particle-like character, we speak in terms of "particles" called *photons*, each photon having an energy equal to $h\nu$, where h is Planck's constant and ν is the frequency of the radiation. This duality in the character of radiation is one of the basic tenets of modern physics.

When a photon (or quantum) of radiation falls upon a photon detector, an electrical charge carrier is kicked from the valence band to the conduction band whenever $h\nu$ is greater than or equal to E_g, where E_g is the "distance," in energy, between the valence band and the conduction band. More precisely, two charge carriers are actually produced. One of these, a positive charge carrier (a "hole"), remains behind in the valence-band energy level; the negative charge carrier (an electron) moves up to the conduction band. Both of these charge carriers contribute to the production of the electrical signal in the photon detector. Since an electron is moved into the conduction band only when the incoming photon has energy $h\nu$ that is greater than or equal to E_g, then the detector absorbs radiation and is opaque to that radiation only when $h\nu \geq E_g$.

The response of a photon detector is proportional to the number of charge carriers in the conduction band; i.e., the response is proportional to the number of electron transitions induced by incoming photons. The *responsivity* of the detector is a measure of the electrical response (i.e., the number of electron transitions) per watt of incoming radiation.

Since the photon detector responds only when $h\nu \geq E_g$ and $\nu = c/\lambda$ (c = speed of light, λ = wavelength), then detector response occurs whenever

$$\frac{hc}{\lambda} \geq E_g \qquad \text{J}$$

or

$$\lambda \leq \frac{hc}{E_g} = \lambda_c$$

This defines the *cutoff wavelength* of the detector, λ_c. The incident beam power Φ is given by

$$\Phi = N_\Phi \frac{hc}{\lambda} \qquad \text{W}$$

where N_Φ is the number of photons per second falling on the detector and, of course, hc/λ is the energy per photon. Solving for N_Φ:

$$N_\Phi = \frac{\Phi\lambda}{hc} \quad \text{s}^{-1}$$

The responsivity of the detector R_Φ (in volts per watt) is given by

$$R_\Phi = k_1\eta N_\Phi \quad \text{V/W}$$

where η = quantum efficiency
k_1 = proportionality constant

Therefore,

$$R_\Phi = \frac{k_1\eta\Phi\lambda}{hc} \quad \text{V/W}$$

$$\lambda \leq \lambda_c$$

A plot of this relation is given in Fig. 2-18. For comparison, the relative response of a thermal detector is plotted on the same axes. Note that the responsivity of the thermal detector is wavelength independent and is smaller than that of the photon detector in the wavelength region near λ_c. In general, the peak responsivity of a photon detector is superior to that of a thermal detector and the speed of response of a photon detector is greater than that of a thermal detector.

Several different photon detectors are usually needed to cover a wide band of wavelengths, since photon detectors are incapable of detecting radiation beyond their own cutoff wavelength. In addition, as the wavelength becomes shorter, the responsivity of a given photon detector declines to such a value that a photon detector with a smaller λ_c is usually employed. Detector crystal materials that are frequently used are silicon, lead sulfide, indium antimonide, mercury cadmium telluride, lead tin telluride, and mercury-doped germanium. Figure 2-19 illustrates

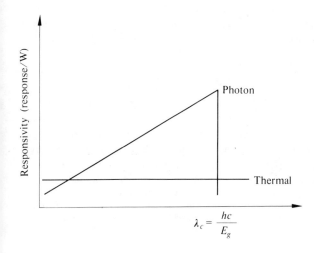

Figure 2-18 Idealized responses of thermal and photon detectors.

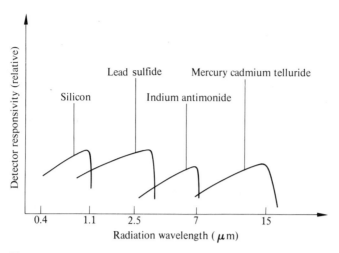

Figure 2-19 Relative responses of crystalline photon detectors.

a typical arrangement of such detectors showing how a multiplicity of detectors could be used to detect radiation over a wide band of wavelengths.

A figure of merit is needed to aid the sensor designer in the choice of a detector for a given wavelength region.† It is useful if this figure of merit is relatively easy to measure and is dependent on only the material from which the detector is fabricated and not upon the dimensions of the detector. Two basic factors determine the "merit" of a detector: the sensitivity of the detector, i.e., the magnitude of its response, and the intrinsic internal noise generated by the detector. The former factor is quantified by the previous equation. The latter factor, noise, is an unwanted signal generated within the detector due to thermally induced, randomly generated charge carriers. (Complete theories on noise generation in crystalline photon detectors are available in the literature.[11]) The internal noise factor is quantified in terms of the noise-equivalent power (NEP) of the detector, defined as the power required to produce a signal-to-noise ratio for the detector equal to 1. Since we usually like to have a figure of merit become *larger* as a detector becomes *better*, another symbol, D, is defined which is the reciprocal of NEP:

$$D = \frac{1}{\text{NEP}}$$

In many detectors the NEP is proportional to the square root of the area, but, as noted above, we want a figure of merit independent of the detector area. Therefore, we define *detectivity*, denoted D^*, by $D^* = \sqrt{A}/\text{NEP}$, where A is the area of the detector. D^* depends on the wavelength (λ), the electrical bandwidth (BW) of the recording system attached to the detector, and the chopping rate (f) of the

---

† The remote sensing data analyst needs to be well aware of the sensitivity and noise performance of the detectors employed, since it may be necessary to select data-analysis procedures that will perform well in spite of the presence of noise (a theme which will recur frequently throughout this book).

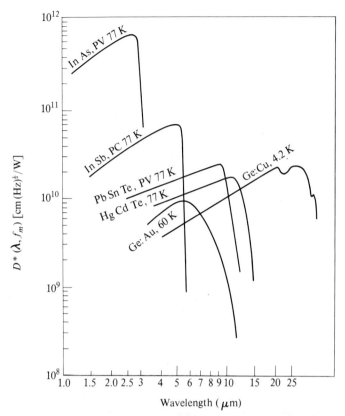

Figure 2-20 Detectivities of several photon detectors (2π steradians field-of-view, 295 K background temperature) (Source: Hughes Aircraft Company, Santa Barbara Research Center. Used with permission.)

optical-processing and electrical-amplifying systems associated with the detector. In functional form, therefore, the detectivity appears as $D^* = D^*(f, \lambda, \mathrm{BW})$. The field of view (usually 2π steradians) is sometimes stated as well.

Usually detectivity is plotted as a function of wavelength for a given chopping rate and electrical bandwidth, as shown in Fig. 2-20. Detectors are compared by noting the maximum value of their D^*($D^*_{\max}$) and the wavelength at which this maximum occurs ($\lambda_{\max}$). Some typical values are shown in Table 2-3.

Table 2-3 Characteristics of typical photon detectors

Detector material	$D^*_{\max}$, cm(Hz)$^{1/2}$/W	$\lambda_{\max} \simeq \lambda_c$
Silicon	9×10^{11}	1.0 μm
Lead sulfide	6×10^{11}	2.1 μm
Indium antimonide	5×10^{11}	5.0 μm
Mercury cadmium telluride	4.5×10^{10}	12.0 μm

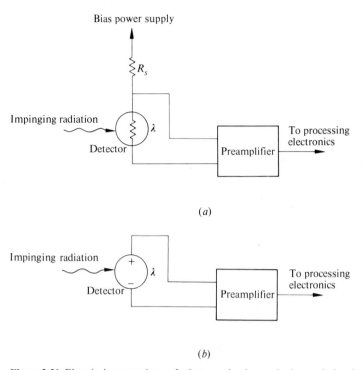

Figure 2-21 Electrical connections of photoconductive and photovoltaic photon detectors: (*a*) a photoconductive detector circuit and (*b*) a photovoltaic detector circuit.

Photon detectors frequently have to be operated at temperatures considerably below the effective radiation temperature of the object whose radiation they are sensing. This is necessary to reduce the thermally induced, internally generated noise of the detector sufficiently below the effective radiant temperature of the signal. In the near- and middle-infrared portions of the spectrum, it is usually not necessary to cool the detectors below ambient temperature, since the apparent radiation temperature of the sun, which is the source of the reflected energy dominant in this portion of the spectrum, is much higher than that of the detector itself. However, in the far-infrared portion of the spectrum the effective radiation temperature of the target is approximately equal to the ambient temperature and noise temperature of the detector; therefore, it is necessary to cool the detector to a temperature considerably below ambient. Liquid nitrogen is often used as a coolant because of its relative accessibility. Detectors cooled by liquid nitrogen operate at approximately 77 K ($-320°$F).

The previous discussion has dealt with the production of electrical-signal output from the detector. Now we consider the ways in which the electrical signal is generated.

Photon radiation detectors manifest their electrical behavior in two basic ways. The first way is a change in the electrical resistance of the detector as the radiation level and wavelength changes (photoconductive, PC). The second way is a change in the output voltage of the detector as the radiation level and wavelength changes (photovoltaic, PV).

In order to detect a change in the resistance of a detector it is necessary to furnish a bias voltage to the detector, as shown in Fig. 2-21(*a*). The voltage across the detector is held constant and the current through the detector changes as the detector resistance changes. Most detectors are photoconductive. The noise associated with the bias supply limits the performance of these detectors.

Photovoltaic detectors produce an output voltage proportional to the level of incident radiation [Fig. 2-21(*b*)]. These detectors generally offer higher signal-to-noise ratios, but only certain materials are capable of photovoltaic operation, e.g., silicon, indium antimonide, and germanium.

Photomultipliers

The photomultiplier tube (PMT) is most useful in the visible and near-infrared portions of the spectrum. It is a vacuum electronic device which operates by producing *photoelectrons* when the incoming radiation strikes a *photocathode* (Fig. 2-22). The interaction of radiation with the photocathode resembles that in

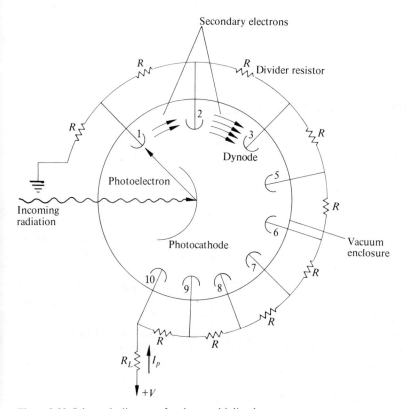

Figure 2-22 Schematic diagram of a photomultiplier detector.

a crystalline photon detector, except that the photoelectron is emitted from the photocathode into the surrounding vacuum. A series of special electrodes, called *dynodes*, are connected to a resistor network and a power supply so that the first dynode is approximately 100 V above ground level and each succeeding dynode (2 through 10) is at 100 V higher potential than its predecessor. The internal geometry of the PMT is arranged so that the photoelectrons strike only the first dynode. The dynodes emit secondary electrons when they are struck by incident electrons, the number of secondary electrons emitted by the dynode being greater than the number of incident electrons striking the dynode. The secondary electrons from the first dynode stimulate emission of more electrons from the second dynode, and this process continues around the circle of dynodes. As a result, the number of electrons passing through R_L is much greater than the number of photoelectrons released originally by the photocathode.

The PMT amounts to a photocathode merged with an electron amplifier (the dynode structure) and is used in the visible and near-infrared wavelength regions (up to 0.9μm) when sensitivity superior to that offered by silicon detectors is desired. It requires a high-voltage power supply (approximately 1200 V) and, since it is a vacuum electronic device, is not as rugged as the solid-state crystalline photon detector. Nevertheless, the PMT has been used in both satellite and aircraft sensor systems because of its high sensitivity in the visible spectrum.

In Sec. 2-6 the impact of radiation detector characteristics on the design and performance of multispectral scanners will be considered.

2-6 REMOTE SENSING INSTRUMENTATION

Instruments used in remote sensing fall into two broad categories, which we shall refer to as *spectral data systems* and *image-forming systems*. Spectral data systems generally do not form images but give detailed spectral information about the target. Image-forming systems yield information regarding the spatial structure of the target and, usually, some spectral information as well.

Spectral data systems acquire data by spectral scanning (as opposed to spatial scanning in image-forming systems). In remote sensing, spectral data systems are commonly used in field research, and that aspect will be emphasized in this section.

Image-forming systems are of two types: framing systems and scanning systems. The picture elements, or *pixels*, in a *framing system* are acquired simultaneously in the basic image unit, the *frame*. In a *scanning system*, the pixels are acquired sequentially, but may be arranged after acquisition into a frame format. Both kinds of image-forming system yield spectral information, usually by producing multispectral pixels consisting of a set of measurements from selected wavelength regions in the spectrum. The image-forming systems considered in this section are multispectral scanners, photographic systems, and electron imagers.

Spectral Data Systems

The technical strategy used in the analysis of multispectral remote sensing data is often related closely to the specific problem under study. For example, a multispectral classification of a scene involving vegetation, soil, and water is, relatively speaking, a straightforward analytical problem requiring simple implementation of the principles discussed in Chap. 3. However, if one is concerned, for example, with the problem of separating various spectral classes of wheat in a crop-yield analysis, then a far more difficult task is at hand. This latter category of problem might be referred to as a "second-order problem." Very often the strategy that a data analyst uses in approaching a second-order problem must be predicated upon some a priori knowledge of the detailed spectral characteristics of the significant elements that will be present in the scene. More important, the spectral information that the analyst uses in planning his technical strategy must be based upon spectral data from actual field situations rather than the idealized spectral data obtained from plants in a laboratory setting.

The spectral data are most valuable if acquired in field situations under both controlled and natural conditions. In the case of cultivated crop species, such data are often acquired from carefully controlled experimental plots on an experimental research farm as well as from commercial fields operated by farmers in a standard, agribusiness setting. The data from these two situations can be correlated and analyzed in order to properly assess the role of the various agronomic variables and also to properly assess the amount of variability to be expected in a real, commercial situation. The process of obtaining these spectral data under field conditions is quite specialized and requires the design of spectral field instruments and field instrumentation systems that can produce reliable, calibrated data when operated in a field environment.

A field instrumentation system first acquires the spectral data and later produces calibrated data from the original raw data. It is necessary to make provisions in the system for the orderly acquisition of supporting data, an important aspect of preparing a complete data set that will be of maximum utility to the analyst. Therefore, such a system useful for producing spectral data for remote sensing research consists of the following major components:

1. An instrument capable of measuring radiation over the wavelength region(s) of interest, for example, 0.4 to 15 μm.
2. A way to provide calibration standards under field conditions and properly account for the effects of solar radiation upon the spectral characteristics of the scene under investigation.
3. A set of subsystems and procedures for acquiring data relevant to the meteorological and other specialized variables pertinent to the description of the scene under investigation.
4. A subsystem and procedure for verifying data produced by the system and collating it into a format that can easily be used by a data analyst, usually in some type of computer processing.

The characteristics of the basic spectral measuring instrument used in a field instrumentation system differ from the usual laboratory instrument used to acquire spectral data. In the early days of remote sensing research, standard laboratory instruments were often used to acquire spectral data under field conditions and the procedures were often complicated by the incompatibility of a laboratory instrument with field conditions. The features necessary to make spectral data instruments usable in a field-operable remote sensing system are summarized below:

1. A rapid spectral scan is essential. Whereas some laboratory instruments require as much as 5 min to develop a spectral scan over the complete spectral range of the instrument, a field instrument usually has to produce a spectral scan in 1 s. This is necessary due to the rapidly changing environmental variables which will affect the spectral characteristics of the scene.
2. The sensitivity of the instrument used in the field situation must be compatible with the signal levels produced by direct and reflected solar radiation.
3. The instrument must be quite rugged so that it can withstand the physical abuse to which it may be subjected in a field environment.
4. The instrument must have an internal calibration system that is compatible with field calibration standards necessary to produce reliable, calibrated field spectral data.

Three different kinds of field spectral instrument will now be discussed. All of these instruments are spectroradiometers in that they rely upon the sun as the source of radiation rather than being equipped with internal radiation sources. The three basic kinds of instrument are the *interferometer*, the *prism* or *grating spectroradiometer*, and the *filter wheel spectroradiometer*. These instruments differ principally in the way that they disperse the radiation entering the instrument into its spectral components. The differing methods of dispersion affect the ways in which the internal references of the instruments are established.

Figure 2-23 illustrates schematically the arrangement of the detector and disperson elements in an interferometer. The interferometer is an instrument which has been used for many years in high-precision spectroscopy. The field version of the interferometer differs from the laboratory version primarily in the way in which the movable mirror is actuated. In a laboratory instrument a very finely pitched screw is used to move the mirror ; in the field version of the instrument, the movable mirror is rapidly driven by an electrodynamic coil arrangement that permits several spectral scans per second. The detector is positioned so that it can observe the beam-splitter array at the point at which the two reflected beams of radiation converge. The two beams of radiation will interfere with each other either constructively or destructively, depending upon the length of the path of each beam as it traverses its respective optical path. The beam-splitter array is designed so that the length of the optical paths of the two split beams is nearly the same, differing only by a fraction of a wavelength. Therefore, the detector will "see" a Newton interference ring formation (the interferometer is usually designed to be circularly symmetric). The detector is arranged so that it looks only at the

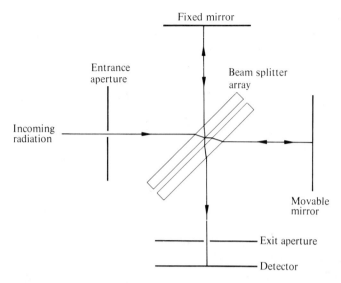

Figure 2-23 Ray pattern in an interferometer.

central spot of the Newton ring formation. The instrument does not form an image of the scene in its field of view but merely observes interference patterns due to the radiation from the scene.

Suppose for the moment that the instrument is illuminated with a beam of monochromatic light. As the position of the movable mirror is varied, a spot which is alternately bright and dark will appear at the converging point of the beam splitter and, hence, will be observed by the detector. This occurs because constructive and destructive interference occur alternately at the beam splitter as the path length involving the movable mirror, relative to that of the fixed mirror, is altered. If the movement of the movable mirror is made linear with respect to time, then the detector will be exposed to an alternately darkening and brightening interference pattern such that the electrical signal out of the detector will be time varying, with frequency dependent upon the velocity of the mirror and the wavelength of the incoming beam of light. As the wavelength of the incoming monochromatic beam is changed, the number of wavelengths traversed by the moving mirror in a given distance (or time) will change; therefore, the number of intensity changes of the interference pattern for a given mechanical movement of the mirror will change.

Now let the incoming beam be polychromatic (consisting of several wavelengths of light). By an extension of the preceding description, we see that each frequency component of signal coming out of the detector will be related to a wavelength component of the incoming beam of radiation. Thus, if one performs a frequency analysis on the output signal, the wavelength structure of the incoming beam of radiation can be reconstructed. In addition, the average brightness of the interference pattern is directly proportional to the intensity of the beam. Numerical Fourier transforms can be used to accomplish this analysis.

Interferometers are used in connection with applications requiring very high resolution spectra. However, the Fourier analysis step complicates the use of interferometer data in remote sensing applications. Furthermore, although the movable mirror makes several scans per second, the transformation process requires averaging the results of several scans in order to acquire sufficient energy to produce an output signal with an adequately high signal-to-noise ratio. Therefore, the effective spectral scan rate of the instrument more closely approximates one spectral scan every 3 or 4 s. Also, mechanical requirements in the construction of the instrument make it quite difficult to use in the visible and near-infrared portions of the spectrum. Interferometers tend to be delicate and require great care in their calibration and use in the field.

A second category of field instruments frequently used in remote sensing research is one with instruments that use prisms and gratings as the basic dispersion elements. These were the first types of instrument to be used in field work, since the original field research in remote sensing was done with laboratory instruments modified for field applications. These instruments usually use an optical chopper system for converting the optical signal into an alternating signal more suitable for processing in the instrument's electronics. In addition, the optical chopper makes it convenient to establish an internal calibration reference for the instrument.

A schematic diagram of the dispersion arrangement for the grating and prism instruments is shown in Fig. 2-24. As indicated in the figure, the grating instrument usually employs a reflection grating; the incoming radiation strikes the grating and is reflected off, dispersed into its various spectral components. A characteristic of the grating instrument is that several orders of the spectra are reflected off in a given direction. Frequencies of radiation that are integer multiples belong to the same order. It is necessary to do "order sorting" using filters over the detectors to separate competing orders of the spectra from the reflection grating. On the other hand, the prism produces a single order in a given direction and order sorting is not necessary. However, the spatial dispersion of the prism instrument is less than that of the grating instrument, and it is therefore mechanically more complicated to arrange the detectors in the prism instrument. Additionally, the grating instrument tends to be spectrally more precise since the resulting spectrum is spread out over a wider area. The grating instrument tends to be more delicate and is usually not capable of as rapid a spectral scan as the prism instrument, since the mechanism used to mount the grating must be quite massive and mechanically complex in order to properly position the grating at a particular portion of the spectrum. The prism mechanism tends to be more rugged and simpler and, therefore, is inherently capable of more rapid spectral scans. However, the spectral coverage of the prism instrument is limited by the material from which the prism is fabricated. Since a prism is basically a refracting instrument, several different types of prism material must be used to cover the optical portion of the spectrum.

Still another category of field instruments is that which utilizes interference filters as the basic dispersion element. An interference filter is a multiple-layer dielectric structure which permits radiation to flow through the structure. Due to multiple reflections and transmissions, interference phenomena result. Only one

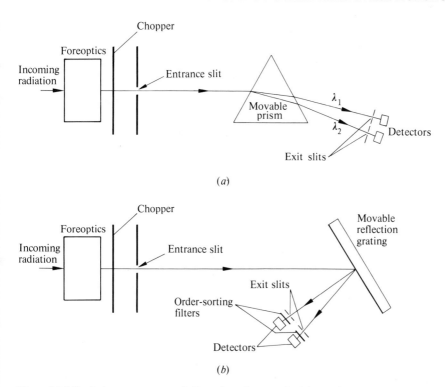

Figure 2-24 Typical arrangements of dispersion elements in (*a*) a prism spectroradiometer and (*b*) a grating spectroradiometer.

spectral band is of the proper wavelength to constructively interfere and thereby pass through the multiple-layer structure without substantial attenuation. The phenomenon is illustrated in Fig. 2-25. This particular method of dispersion requires a form of order sorting similar to that used in a grating system. Any wavelength which is a multiple of the primary wavelength passed through the interference filter will also pass through, since constructive interference will also occur for the multiple-wavelength components. It is necessary to provide a filter to eliminate all multiple orders of radiation that might be incident on the filter. A wavelength that is to be passed through the interference filter is related to the thickness of the dielectric elements. Therefore, a convenient way to make an adjustable dispersion element is to produce a tapered interference filter as illustrated in Fig. 2-26. Rather than employing multiple wavelength-defining slits as shown in the figure, such a filter can be positioned in front of the entrance slit of the detector instrumentation and the relative position of the filter changed in order to select the spectral component from the incoming beam of radiation.

An especially convenient form of the interference filter is a circularly variable filter (CVF), in which the thickness of the dielectric elements of the filter vary with angular position about the rim, as illustrated in Fig. 2-27. The order-sorting filters are coated on the surface of the filter wheel and over the detector used to sense the

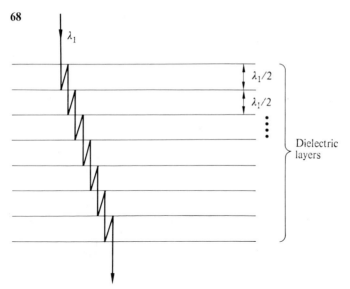

Figure 2-25 Interference in a multiple-layer dielectric interference filter.

radiation passing through the filter. The filter wheels can be rotated to produce rapid spectral scans and represent an inherently rugged method for achieving dispersion in a field instrument. Fig. 2-28 illustrates a method for integrating a filter wheel assembly with a detector and chopper to permit coverage of the spectral region from 0.4 to 2.8 μm. In this system the chopper provides both a method of optically producing an alternating signal for subsequent electronic processing as well as establishing an internal reference in the instrument. Note that a given

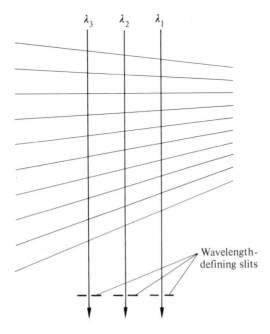

Figure 2-26 A tapered interference filter. Wavelengths are selected by positioning of the filter or by wavelength-defining slits.

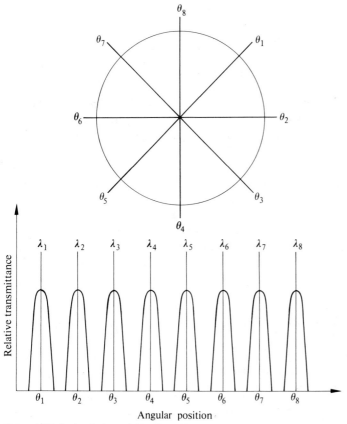

Figure 2-27 A circularly variable filter. The wavelength selected depends on the angular position of the filter wheel.

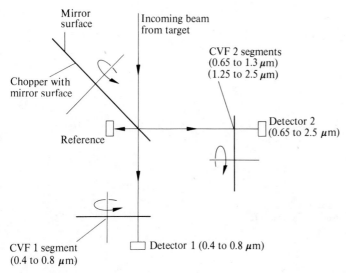

Figure 2-28 Arrangement of CVF wheels in a spectroradiometer. Detector 1 alternately sees the target through the chopper or sees the reference by reflection off the chopper teeth. The reverse is true for detector 2.

segment on a CVF can only transmit an octave of wavelengths due to order-sorting problems. Therefore, two segments are required to cover 0.62 to 2.5 μm.

Figure 2-29 (between pages 36 and 37) shows a CVF instrument capable of operating under field conditions and measuring radiation coming from a target as well as from the sun. Figure 2-30 (between pages 36 and 37) gives an overview of the entire instrumentation system utilizing this CVF instrument.

Table 2-4 contains a summary of the four types of spectroradiometer we have discussed here. Included is a summary of relative advantages and disadvantages of the four types of instrument.

Calibration A critically important aspect of field spectral measurement is calibration of the data. Over the years, calibration procedures for laboratory situations have been established and well documented.[12] Figure 2-31 illustrates some of the radiance and irradiance standards that are used in laboratory calibration procedures. A *radiance standard* is one in which the radiation produced by the standard is accurately known, usually by comparison to a primary standard or, in some

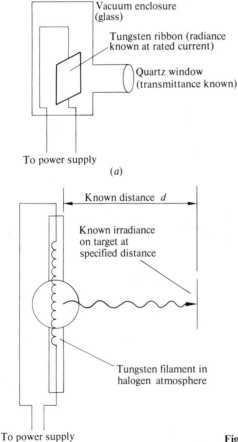

Vacuum enclosure (glass)

Tungsten ribbon (radiance known at rated current)

Quartz window (transmittance known)

To power supply

(*a*)

Known distance *d*

Known irradiance on target at specified distance

Tungsten filament in halogen atmosphere

To power supply

(*b*)

Figure 2-31 Schematic diagram of typical radiance (*a*) and irradiance (*b*) standards.

Table 2-4 Performance summary of various spectroradiometers

Instrument	Dispersion	Spectral scan speed	Spectral resolution	Mechanical strength	Sensitivity	Special problems
Interferometer	Interferometer plate structure	Moderate	Very high	Very delicate	Moderate	Fourier transform required
Prism	Prism fabricated from transmitting material	Rapid	Moderate	Moderate	Moderate	Multiple dispersion elements inconvenient
Grating	Reflecting or transmitting grating	Slow	High	Delicate	High	Order sorting required
Interference filter	Variable interference	Rapid	Low	Rugged	High	Order sorting required

cases, from calculations applied directly to the standard. In Fig. 2-31(a), the radiance standard is a tungsten ribbon accurately positioned with respect to a quartz window and inside a vacuum envelope which permits accurately known transmission of the radiation produced by the tungsten ribbon. The current through the ribbon is carefully controlled so that the temperature of the ribbon is known and the radiance from the ribbon can be calculated using blackbody radiation theory. These radiance standards are suitable for laboratory purposes because the area of the standard, although very small, is usually sufficient to fill the field of view of a laboratory spectral-measuring instrument.

Another form of laboratory standard is the *irradiance standard*, such as illustrated in Fig. 2-31(b). The standard is basically a tungsten halide lamp which has been carefully calibrated at a standards laboratory. It differs from the radiance standard in that the actual radiance produced by the lamp is not known, but the irradiance at a specified distance from the lamp has been precisely determined. If the lamp is positioned in a carefully prescribed manner with respect to the laboratory measuring instrument, then the irradiance falling upon the entrance slit of the instrument is known. In the case of the radiance standard the irradiance could be calculated from knowledge of the geometry relating the instrument and the radiance standard. Irradiance standards have the advantage that they can be extended under certain conditions to a quasi-field situation.

If the field instrument being calibrated has some provision for measuring diffuse radiation, an irradiance standard can be used in the field to fill the aperture of the instrument. For example, if the field instrument has a special aperture intended for measuring solar radiation, then the instrument can be arranged to view the irradiance standard instead of viewing the sun and sky. This situation is illustrated schematically in Fig. 2-32. Therefore, it is possible to accurately calibrate a field instrument in a situation outside of an optical standards laboratory. One need only be reasonably certain that secondary reflections from the irradiance standard are not reaching the aperture of the instrument being calibrated. A typical procedure is to flood the solar port of the field instrument with the irradiance standard at night in an enclosure where secondary reflections are

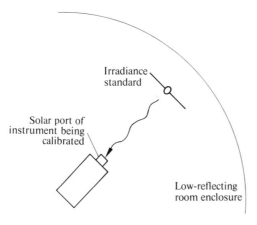

Irradiance standard

Solar port of instrument being calibrated

Low-reflecting room enclosure

Figure 2-32 Irradiance calibration of a field instrument.

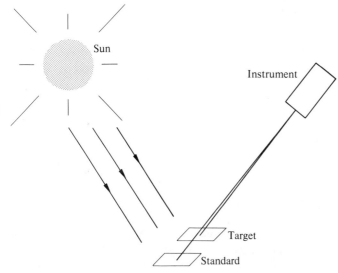

Figure 2-33 Reflectance calibration of a field instrument.

minimal. This procedure can be done to a field instrument at intervals of a few days to assure the experimenters that instrument calibration is being maintained.

Another procedure for calibrating field instruments uses a barium sulfate reflectance standard. The theory behind the procedure is contained in Sec. 2-3 in the discussion of the bidirectional reflectance factor. Operationally, the panel and target are arranged as shown in Fig. 2-33. It is important to ensure that the irradiation and viewing geometry for both the target and the reflectance standard are as nearly identical as possible.

In addition, if the solar irradiance is measured at the same time that the instrument is viewing the reflectance panel, then the absolute radiance of the scene can also be established by combining the reflectance information for the panel with the incoming solar irradiance measurement. The solar irradiance can be measured absolutely by the instrument if it has been previously calibrated with an irradiance standard.

The field calibration procedures discussed above deal with the reflective portion of the spectrum, from 0.4 to 2.5 micrometers. Field calibration is also needed in the thermal portion of the spectrum. A convenient way to establish a thermal calibration reference is through the use of reference blackbodies which flood the aperture of the instrument with a known quantity of thermal radiation. In the thermal infrared portion of the spectrum this requires that a body be produced that has very high emittance and can be accurately controlled with respect to its temperature. The radiation coming from that body is accurately known and can be used to calibrate the instrument. A convenient blackbody reference is a conical structure as illustrated in Fig. 2-34. A cone is used because an emittance multiplying factor results if a surface of high emittance is formed into such a structure.[13] A particularly high-emittance structure (approximately 0.999)

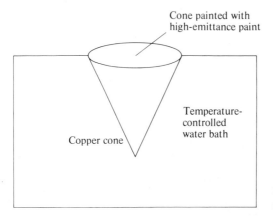

Cone painted with
high-emittance paint

Temperature-
controlled
water bath

Copper cone

Figure 2-34 Field thermal calibration reference.

occurs if copper coated with a high-emittance lacquer such as Eppley Parsons Black[14] is used. Due to its high thermal conductivity, the copper substrate provides a uniform temperature over the structure. The cone can then be immersed in a well-insulated water bath whose temperature can be controlled quite accurately to within 0.1°C. A stirring motor is used to distribute the water uniformly around the thermally insulated structure and the temperature of the water is measured with precision thermometers. The blackbody is then placed beneath the input aperture of the instrument and the temperature of the blackbody is recorded in conjunction with the output signal produced by the instrument. Two blackbodies are normally used in a field situation with temperatures set to bracket the scene temperature. That is, one of the blackbodies has its temperature set slightly below the minimum temperature expected in the scene and the other blackbody has its temperature set slightly above the highest temperature expected in the scene. If the temperature range in the scene is small enough with respect to the absolute mean temperature of the scene, then a linear relationship may be established between the instrument output and the effective radiant temperature of the scene. If the temperature range in the scene is a significant fraction of the absolute mean temperature (say one-tenth), then the Stefan-Boltzmann radiation law (Sec. 2-1) must be used in a non-linear calibration procedure to provide the radiant temperature as a function of wavelength. This assumes, of course, that the emittance of the scene is constant over the wavelength band. Conversely, one could assume a constant temperature of the scene and establish a wavelength-dependent emittance from 7 to 15 μm. The mode of interpreting the thermal data is a matter of choice and depends considerably on the particular problem under study.

Spectral information systems The acquisition of spectral information in a field situation appears at first glance to be a straightforward procedure. In fact, it is relatively easy to take a spectral measuring instrument into the field, acquire spectra, publish the resulting spectra, and wind up with a relatively simple experiment and result. However, the results of such a simple procedure could have very limited credibility since the ancillary data associated with the spectra would be very difficult to reproduce unless careful procedures were established to relate the spectra to other experimental variables. Figure 2-35 illustrates the flow of data in a field instrumentation system designed to permit the integration of ancillary data with spectral data. The data produced by the spectral measuring instrument occur in two basic forms: magnetic tape and hard copy. The hard-copy data serve as backup to the magnetic-tape data in case uncertainty should develop concerning processing of the latter. Field record sheets are prepared at the time that the spectral data are acquired so that punched cards may be prepared while the spectral data are being digitized (placed in computer format). The ancillary data are then integrated with the magnetically recorded spectral data and identification headers are attached to each portion of the spectral data. Photographic data simultaneously acquired with the spectral data are placed in a record-keeping system so that the identifying information in the magnetic-tape records can be integrated and compared to the photographic data. This enables a data analyst who is examining a particular spectrum produced by the field system to acquire intimate knowledge of the appearance of the scene and the agronomic or meteorological variables associated with the scene, often critical factors in the proper interpretation of spectral data.

Figure 2-36 is an example of some typical products from a field spectral data system operated by Purdue University's Laboratory for Applications of Remote Sensing. The header associated with the spectra contains all of the important information associated with the scene at the time the spectra were acquired. The

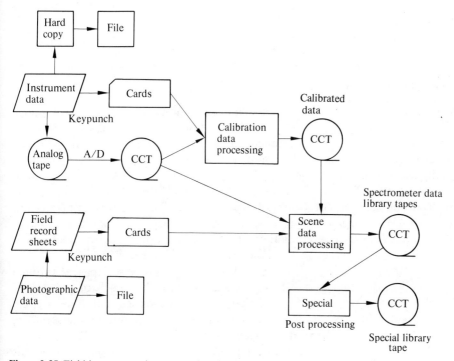

Figure 2-35 Field instrumentation system data flow (CCT = computer compatible tape).

(a)   *** LISTING OF FIELD SPECTRORADIOMETER DATA ON TAPE 4305 ***

```
RUN SEQUENCER ...................     365      RUN NUMBER ............... 76523801
EXPERIMENT NUMBER ...............76100213      OBSERVATION NUMBER ..............  602031
DATE DATA COLLECTED .....    6/ 2/76           TIME DATA COLLECTED ...........   1925
EXPERIMENT NAME ....... SPRING WHEAT NO        PRINCIPAL INVESTIGATOR          BAUER, M.E
SCENE TYPE ...................... PLCT 8       LOCATION ............... WILLISTON,ND
AIR TEMPERATURE .................   33.2       BAROMETRIC PRESSURE ............  768.4
RELATIVE HUMIDITY ...............   28.0       CLOUD COVER ........................ .25
WIND SPEED ......................   14         WIND DIRECTION ...................  180
REFORMATTING DATE ............. 1/21/77        REFORMATTING CALIBRATION CODE ........  3
SIDE ANGLE ......................    0         SWING ANGLE ........................    0
FORWARD DIRECTION ...............  230         DISTANCE TO GROUND ............... 20.00
FIELD OF VIEW ...................  15.00       LOCATION LATITUDE ........... 481000N
LOCATION LONGITUDE ........... 1034100W        NUMBER OF SAMPLE GROUPS ..............  3
LEVELS OF FACTOR 1 ..............    1         LEVELS OF FACTOR 2 ..............    1
LEVELS OF FACTOR 3 ..............    1         LEVELS OF FACTOR 4 ..............    2
REPLICATION NUMBER ..............    1         PLOT NUMBER .........................  8
SPECIES ..................... WHEAT            VARIETY .................... OLAF
MATURITY ........................  2.1         HEIGHT .....................  0.19
ROW WIDTH ....................... 0.18         PLANT COUNT 99 IN 1.00METERS OF ROW
LEAVES PER PLANT ................    4         LEAF AREA INDEX ................ 0.57
ROW DIRECTION ................... N-S          SERIES NAME ............. WILLIAMS
TEXTURE ......................... LOAM         MOISTURE (FIELD) CONTENT         DRY
SURFACE CONDITION .... ....... SMOOTH          HORIZON ....................... A
DATA QUALITY FACTOR 1    ( 0.55,   0.0993)     DATA QUALITY FACTOR 2   ( 0.65,   0.0364)
DATA QUALITY FACTOR 3    ( 1.05,   0.0397)     DATA QUALITY FACTOR 4   ( 1.65,   0.0835)
DATA QUALITY FACTOR 5    ( 2.20,   0.0746)     FACILITY NAME .......... PURDUE / LARS
INSTRUMENT NAME ....... EXOTECH MOD 20C        SCAN RATE ..................... 0.50
IRRADIANCE CALIBRATION RUN .... 76523106       HIGH SQUARE WAVE LEVEL ........... 5.000
LOW SQUARE WAVE LEVEL ............. 0.0
COMMENTS -- SCLAR    PRT 42.4
```

DETECTOR NAME	DETECTOR RANGE	DETECTOR EQUILIZATION	NUMBER OF SAMPLES	WAVE BAND COEFFICIENTS A	B	C	D	SAMPLE GROUP
SI	0.300	0.50	470	0.374	0.001	0.0	0.0	1
PB S	0.300	0.50	465	0.733	0.001	0.0	0.0	2
PB S	0.3C0	0.50	420	1.286	0.003	0.0	0.0	3

Figure 2-36 Field spectral system data products. (a) Header information. (b) (opposite) Plot of spectral data.

analyst may take the spectra, in digital form on the magnetic tape, and manipulate them using computer programs designed to extract the information in which he or she is interested.

Further aspects of the effective processing of spectral data are discussed in Sec. 4-3.

Multispectral Line Scanners

Multispectral line scanners (or, more commonly, multispectral scanners) produce imagery in a sequential fashion. The operation of a multispectral scanner can be conveniently represented by the functional block diagram shown in Fig. 2-37. The target is scanned in raster fashion (a line at a time), usually with an optical-mechanical system. The radiation passes through converging optics which establish the instantaneous field of view (IFOV). The total field of view (TFOV) is established by the scanning motion of the optical system. The radiation is then dispersed into its spectral components using prisms, gratings, dichroics, or filters. An array of detectors senses the dispersed radiation. The detectors are arranged spatially so that the appropriate detector can sense the wavelength region to which it is sensitive.

The signal from each detector is amplified and processed (passed through filters and/or digitized) and is then recorded on board the vehicle or transmitted by telemetry to a receiving station. Information regarding the calibration sources is also recorded or transmitted; the calibration sources as well as the scene are scanned by the optical-mechanical scanning system.

The motion of the vehicle carrying the multispectral scanner provides the along-track scan motion of the sensor; whereas, the line scanner itself develops the

CLASS   SYMBOL
-----   ------
ALFALFA   1
WHEAT   2
FALLOW   3

***** SAMPLE GROUP 1   RANGE   0.3500-   2.4000 MICROMETERS

RESPONSE, BI-DIRECTIONAL REFLECTANCE FACTOR

WAVELENGTH (MICROMETERS)

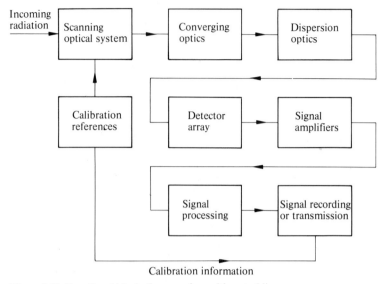

Figure 2-37 Functional block diagram of a multispectral line scanner.

across-track scan motion. Figure 2-38 illustrates several typical scanning mechanisms. Figure 2-39 illustrates how the scanning optics might be combined with the converging optics. The rotating elements in Fig. 2-38 have highly polished reflecting surfaces. The motion of the element effectively "aims" the scanner at different points on the target surface. The incoming radiation is then passed through the converging optics of the scanner which focus the radiation on the dispersion optics and detectors.

The optical system in Fig. 2-39(a) uses a primary (concave) mirror in conjunction with a secondary (convex) mirror to focus the radiation. This system resembles a *Cassegranian telescope* in its basic structure.

The optical system shown in Fig. 2-39(b) utilizes a four-sided scanning mirror, flat relay (folding) mirrors, and a paraboloidal primary mirror to focus the received radiation onto the detector. As shown in the figure, a split-beam technique is used to transfer the incoming radiation to the detector.

Most scanning systems have basically reflecting optical systems. The rotating prism type of scanner tends to have a larger optical aperture for a given rotating mass and overall size; it is also superior in terms of platform velocity/height ratio *versus* scanner shaft speed as compared to the wedge- and the oblique-type scanner systems.

Another characteristic which must be considered is the rotation of the instantaneous field of view as a function of scan angle which occurs in both the wedge and the oblique scanning systems. However, the rotating prism system is optically far more complicated, and the application of the scanner to specific problems must be considered before choosing a particular scanner optical system.

The converging optics can, of course, be either reflective, refractive, or some combination of the two (catodiatropic). Reflective systems have some distinct

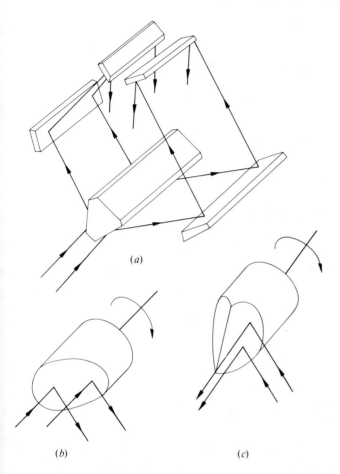

(a)

(b)　　　　　　　　　　　　　(c)

Figure 2-38 Several optical-mechanical scanning systems. (a) Rotating prism. (b) Oblique. (c) Wedge.

advantages; they are free from chromatic aberration, they have high transmission, they are relatively inexpensive, and they are somewhat more easily corrected for other aberrations. However, a refractive system can be made smaller than an equivalent reflective system for a given effective aperture; it can be made more efficient in terms of primary obscuration, and in some cases it can provide its own transmittance window and spectral filtering.†

Power flow: signal and noise A multispectral scanner can be considered an imaging sensor of radiation in a spectral fashion, the radiation coming from a remotely located target. We have discussed various methods of scanning and focusing radiation from a target of interest upon the dispersive optics (a

---

† In reflective systems, some of the secondary optical components invariably lie partially in the path of radiation which ideally should fall on the primary mirror (see Fig. 2-39). This does not block out any portion of the image, but it does dim the image somewhat. This characteristic is called "primary obscuration."

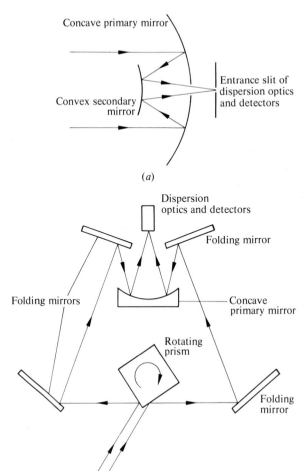

(a)

(b)

Figure 2-39 Scanner converging optical systems. (a) Converging optics, oblique or wedge. (b) Converging optics, rotating prism.

monochrometer). In Fig. 2-40, a representation of a multispectral scanner is given.[15] The power flows from the ground resolution patch through the atmosphere into the aperture of the scanner. The effective aperture size of the scanner is determined by the area of the collecting mirror. The focal length of the scanner, and hence its field of view, is determined by the focal length of the collecting mirror and the optics of the monochrometer entrance slit. The electromagnetic power into the monochrometer and in the wavelength range λ to $\lambda + \Delta\lambda$ is given by

$$\Phi = \tau_a L_\lambda A_p \beta^2 \, \Delta\lambda \qquad \text{W} \qquad (2\text{-}26)$$

where τ_a = atmospheric transmittance
L_λ = spectral radiance of ground patch, $\text{W}/(\text{m}^2 - \text{sr} - \mu\text{m})$
A_p = scanner aperture, m^2
β = IFOV of scanner, rad

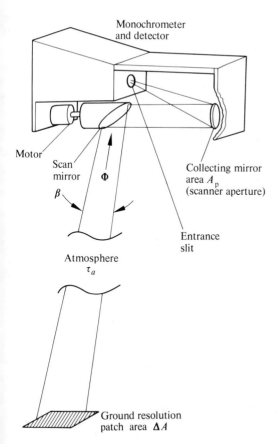

Monochrometer and detector

Motor

Scan mirror Φ

β

Collecting mirror area A_p (scanner aperture)

Entrance slit

Atmosphere τ_a

Ground resolution patch area ΔA

Figure 2-40 Aircraft system sensor layout.[15]

The spectral radiance of the ground resolution patch is either reflected solar radiation or thermal emission of the ground patch that is related to its temperature. The spectral radiance of this scene when one is considering reflected radiation from the sun is

$$L_\lambda = \frac{1}{\pi} E_\lambda R \cos \theta_S \qquad \text{W}/(\text{m}^2 - \text{sr} - \mu\text{m}) \tag{2-27}$$

where E_λ = spectral irradiance on ground patch, $\text{W}/(\text{m}^2 - \mu\text{m})$
θ_S = sun angle, rad
R' = bidirectional reflectance factor, dimensionless

Notice that R in this equation is assumed to be the BRF of the scene and is, in general, a function of the angle of view at the scanning aircraft, as in Fig. 2-41,[15] and also of the solar angle. For the sake of simplicity, we shall assume in this discussion that $R = 0.1$ and is independent of viewing angle. In the calculations that follow, we shall use the model of the solar radiation given in Fig. 2-1 to derive Fig. 2-41. When considering radiation in the far-infrared region we shall assume that the target of interest represents a blackbody with an emittance nearly equal to

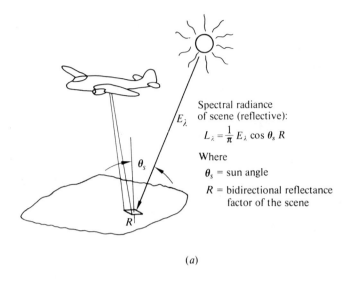

Spectral radiance
of scene (reflective):

$$L_\lambda = \frac{1}{\pi} E_\lambda \cos \theta_s R$$

Where

θ_s = sun angle

R = bidirectional reflectance
factor of the scene

(a)

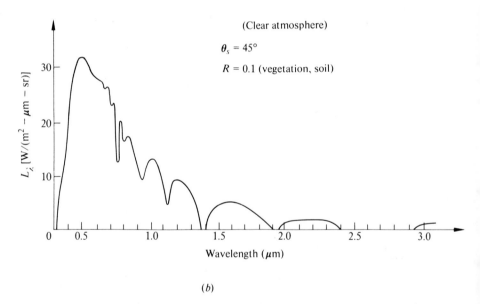

(Clear atmosphere)

$\theta_s = 45°$

$R = 0.1$ (vegetation, soil)

(b)

Figure 2-41 Reflected solar radiation measured by a scanner.[15] (a) Spectral radiance measurement. (b) Spectral radiance in the reflective region.

1 at a temperature of 300 K. Combining all this information we are then able to plot the graphs shown in Fig. 2-42.[15] To get an idea of the power flow into a scanner, let us consider a typical situation. We shall calculate Φ for the spectral band from 0.62 to 0.68 μm for a 10 percent reflectance target, assuming that the scanner has a 3-mrad instantaneous field of view and a primary mirror diameter of 0.087 m.

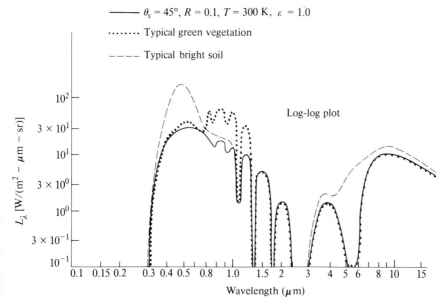

Figure 2-42 Spectral radiance in the reflective and emissive wavelength regions for three different scenes.[15]

Furthermore, we assume that

$$\tau_a = 0.5$$
$$L_\lambda = 28 \text{ W}/(\text{m}^2 - \mu\text{m} - \text{sr})$$
$$A_p\beta^2 = 5.37 \times 10^{-8} \text{ m}^2 - \text{sr}$$
$$\Delta\lambda = 0.68 - 0.62 \,\mu\text{m} = 0.06 \,\mu\text{m}$$

Then Eq. (2-26) yields

$$\Phi = 0.5 \times 28 \times 5.37 \times 10^{-8} \times 0.06 = 4.5 \times 10^{-8} \text{ W} \tag{2-28}$$

Let us consider another problem. We shall determine the power under the same conditions but for a 300 K blackbody in the spectral band from 10 to 12 μm. Here

$$L_\lambda = 9.5 \text{ W}/(\text{m}^2 - \mu\text{m} - \text{sr})$$
$$\Delta\lambda = 2.0 \,\mu\text{m}$$

and $\quad\Phi = 0.5 \times 9.5 \times 5.37 \times 10^{-8} \times 2.0 = 5.1 \times 10^{-7} \text{ W} \tag{2-29}$

These calculations give us a rough idea of the magnitude of the power flow into the multispectral scanner. Now what about the effects of noise on the scanner input? Basically, noise limits the ability of the scanner to resolve reflective changes. This is illustrated in Fig. 2-43.[15] It also limits the ability to resolve temperature or emittance changes in that portion of the spectrum where thermal effects predominate, as indicated in Fig. 2-44.[15]

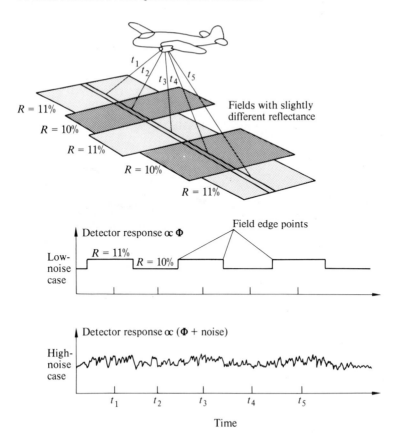

Figure 2-43 Noise limitations on reflectance change resolution.[15]

The change of power flow into the scanner due to a change in L_λ is

$$\Delta\Phi = \tau_a \Delta L_\lambda A\beta^2 \Delta\lambda = \frac{\Delta L_\lambda}{L_\lambda} \tau_a L_\lambda A\beta^2 \Delta\lambda \qquad \text{W} \qquad (2\text{-}30)$$

Therefore,

$$\Delta\Phi = \frac{\Delta L_\lambda}{L_\lambda} \Phi \qquad \text{W} \qquad (2\text{-}31)$$

But, from Eq. (2-27), we can write

$$\Delta L_\lambda = \frac{1}{\pi} E_\lambda \Delta R \cos\theta_S \qquad \text{W}/(\text{m}^2 - \text{sr} - \mu\text{m}) \qquad (2\text{-}32)$$

Therefore,

$$\frac{\Delta L_\lambda}{L_\lambda} = \frac{\Delta R}{R} \qquad (2\text{-}33)$$

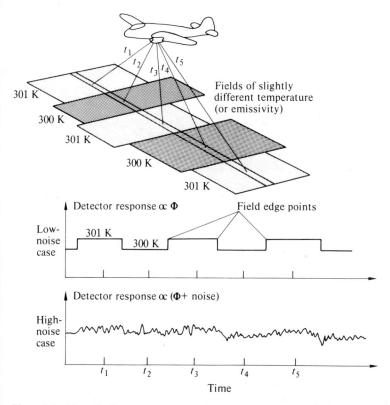

Figure 2-44 Noise limitations on emittance/temperature change resolution.[15]

and we write

$$(\Delta\Phi)_{\text{refl}} = \frac{\Delta R}{R}\Phi \qquad W \tag{2-34}$$

in the visible, near-infrared, and middle-infrared regions of the spectrum.

In the far-infrared portion of the spectrum changes in L_λ, as shown in Fig. 2-44, can arise due to changes in emittance or changes in temperature. Since

$$L_\lambda = \frac{\varepsilon 1.19 \times 10^8}{\lambda^5(e^{14388/\lambda T} - 1)} \qquad W/(m^2 - sr - \mu m) \tag{2-35}$$

we can show that

$$\frac{\Delta L_\lambda}{L_\lambda} = \frac{\Delta\varepsilon}{\varepsilon} + \frac{14{,}388}{\lambda T}\frac{e^{14388/\lambda T}}{(e^{14388/\lambda T} - 1)}\frac{\Delta T}{T} \tag{2-36}$$

For the wavelength and temperature ranges of interest this can be approximated by

$$\frac{\Delta L_\lambda}{L_\lambda} \cong \frac{\Delta\varepsilon}{\varepsilon} + \frac{14{,}388}{\lambda T}\frac{\Delta T}{T} \tag{2-37}$$

or we write

$$(\Delta\Phi)_{\varepsilon,T} = \frac{\Delta\varepsilon}{\varepsilon}\Phi + \frac{14{,}388}{\lambda T}\frac{\Delta T}{T}\Phi \qquad W \qquad (2\text{-}38)$$

We can illustrate the use of these two results, Eqs. (2-34) and (2-38), by considering the following two examples.

What is the change in power due to a 1 percent change in reflectance from 10 to 11 percent, assuming the same conditions applied in Eq. (2-28)?

$$(\Delta\Phi)_{refl} = \frac{0.01}{0.1} \times 4.5 \times 10^{-8} = 4.5 \times 10^{-9} \text{ W} \qquad (2\text{-}39)$$

What is the change in power due to a change in temperature from 300 to 301 K? At $\lambda = 11$ μm, we have

$$(\Delta\Phi)_T = \frac{14{,}388}{11 \times 300} \times \frac{1}{300} \times 5.1 \times 10^{-7} = 7.41 \times 10^{-9} \text{ W} \qquad (2\text{-}40)$$

These calculations give us an idea of the average signal power levels as well as the change in signal power levels that are involved in remote sensing work.

Now we consider the details of the scanning process and see how they affect the signals produced by the multispectral scanner. Consider Fig. 2-45. We assume a scan mirror with p sides and a detector array of q detectors. For example, in the Landsat 1 and 2 satellites, each spectral band in the multispectral scanner used six detectors. The primary focal length of the mirror is f and the effective area of the primary mirror is equal to $(\pi/4)\,D^2$. The mirror may not actually be in a circular configuration, but it is convenient to consider it so. The time interval required to move the resolution patch one unit along the scan raster line is β/ω seconds, where ω is the mirror rotation speed. The *rise time* of the detector amplifier recorder system, τ, must be some fraction of this time; that is,

$$\tau = \frac{1}{g}\frac{\beta}{\omega} \qquad s \qquad (2\text{-}41)$$

where g is a constant, usually chosen somewhere between 5 and 10. For contiguous scanning (i.e., scanning whereby the scan lines touch each other but do not overlap) the scan mirror must make one-pth of a revolution in the time it takes the scanner vehicle to advance a distance $q\beta H$, i.e., in $q\beta H/V$ seconds; thus

$$\frac{2\pi}{p} = \frac{q\beta H}{V}\omega \qquad rad \qquad (2\text{-}42)$$

The mirror must spin

$$\omega = \frac{2\pi V/H}{pq\beta} \qquad rad/s \qquad (2\text{-}43)$$

The rise time is then

$$\tau = \frac{pq\beta^2}{g2\pi V/H} \qquad s \qquad (2\text{-}44)$$

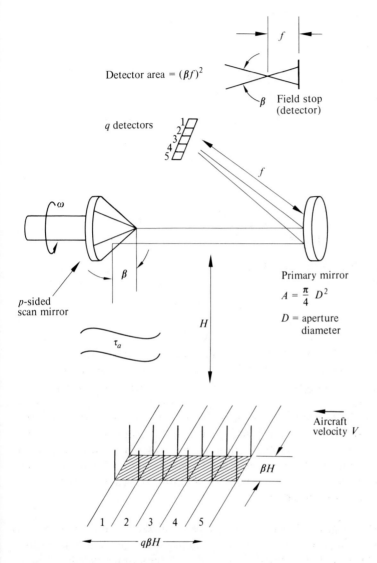

Figure 2-45 Scanner schematic diagram.

The detector-amplifier-recorder bandwidth, BW, is

$$BW = \frac{1}{a\tau} \quad Hz \tag{2-45}$$

where a is a constant, usually between 2 and 3. Therefore,

$$BW = \frac{g2\pi V/H}{apq\beta^2} \quad Hz \tag{2-46}$$

In Sec. 2-5, we discussed detectivity for photon detectors. Following that discussion the noise-equivalent power for a detector is

$$\text{NEP} = \frac{\sqrt{A \cdot \text{BW}}}{D_\lambda^*} = \frac{1}{D_\lambda^*} \sqrt{\frac{f^2 \beta^2 g 2\pi V/H}{apq\beta^2}} = \frac{1}{D_\lambda^*} \sqrt{\frac{2\pi g V/H}{apq}} f \qquad \text{W} \qquad (2\text{-}47)$$

The *optical efficiency*, η, must be taken into account. The *input-signal power* is

$$\Delta\Phi_\lambda = \tau_a \, \Delta L_\lambda \frac{\pi}{4} D^2 \beta^2 \, \Delta\lambda \, \eta \qquad \text{W} \qquad (2\text{-}48)$$

Dividing Eq. (2-48) by Eq. (2-47) yields the *signal-to-noise ratio*:

$$\frac{\Delta\Phi_\lambda}{\text{NEP}} = \frac{\sqrt{\pi} \, \tau_a \, \Delta L_\lambda DD_\lambda^* \sqrt{apq} \, \beta^2 \, \Delta\lambda \, \eta}{4\sqrt{2g} \, \sqrt{V/H} \, f/D} \qquad (2\text{-}49)$$

To simplify the subsequent discussion, we shall assume $p = 4$, $a = 3$, $g = 1.5\pi$, and $\eta = 0.4$. Then, we obtain

$$\frac{\Delta\Phi_\lambda}{\text{NEP}} = \frac{\Delta L_\lambda DD_\lambda^* \sqrt{q} \, \beta^2 \, \Delta\lambda}{20\sqrt{V/H} \, f/D} \qquad (2\text{-}50)$$

Now, let us examine this equation. Notice that the signal-to-noise ratio increases as the aperture of the scanner increases and as the speed of the optical system (D/f) increases. The signal-to-noise ratio also improves as the detectivity of the quantum detector increases and as the available signal (ΔL_λ) increases. It is dependent upon the square of the instantaneous field of view of the scanner and proportional to the spectral bandwidth. Notice that the signal-to-noise ratio is inversely proportional to the square root of the V/H ratio. The point of this is to emphasize the price in signal quality that must be paid for higher spatial or spectral resolution. However, notice that increasing the velocity/height (V/H) ratio does not place a severe demand on the signal-to-noise ratio.

Geometric considerations We shall now concern ourselves with some of the geometric characteristics of the multispectral scanner which introduce distortions in the resulting imagery.[16] First we shall consider a situation with ideal geometry and then introduce nonidealized effects.

Figures 2-46 and 2-47 illustrate the geometry of a multispectral scanner under idealized conditions. Here we are considering the recording of the jth resolution element in an arbitrary scan line, i. We shall assume that the aircraft is flying perfectly straight at a constant elevation above a reference plane called the *datum* and at a constant ground speed. The coordinates of the aircraft are (X_c, Y_c, Z_c). Also, the aircraft is subject to no exterior orientation effects; i.e., there is no roll, pitch, or yaw of the aircraft. Additionally, each scan line is assumed to be instantaneously recorded.

In Fig. 2-47, β represents the angular resolution of the scanner in the down-

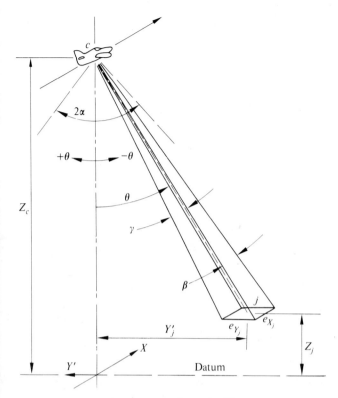

Figure 2-46 Geometric parameters of scanning.[16]

strip or X direction, and is the same as the physical resolution of the scanner. The angle γ represents the effective angular resolution in the along-scan or Y direction; it is ultimately limited by the physical resolution of the scanner, but is also affected by the sampling rate of the analog-to-digital conversion process to which the data are subjected. The angles γ and β define the instantaneous field of view (IFOV). The total angle scanned, the total field of view (TFOV), is 2α. The effective focal length of the scanner optics is f. The scanning rate is adjusted such that an overscan will occur between scans at the nominal aircraft velocity V. The average advance for each scan line will then be $(1 - S)\,dX$, where S is some overlap factor and dX is the scan width on the datum given by

$$dX = Z_c\beta \tag{2-51}$$

Here, Z_c represents the altitude of the aircraft above datum when the scanner is recording scan line i. The image element size in the downstrip direction is given by

$$dX = f\beta \tag{2-52}$$

where f is the focal length of the scanner optics. From Fig. 2-47(a),

$$V\,dt = (1 - S)\,dX \tag{2-53}$$

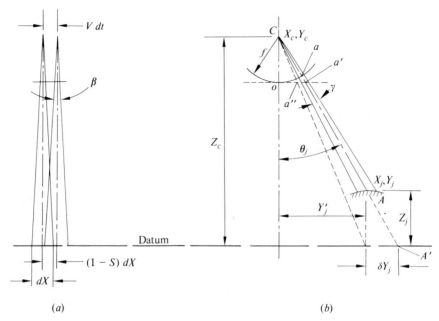

Figure 2-47 Ideal geometry of the multispectral scanner.[16] (a) Along-track geometry. (b) Across-track geometry.

Combining these results, we obtain an expression for X_j, which is also equal to X_c,

$$X_j = X_c = X_0 + (1 - S)\frac{Z_c}{f} X \tag{2-54}$$

Here, X_0 is the X coordinate of the first scan line in the flightline.

Using Fig. 2-47(b), we can calculate the ground coordinate Y_j, the distance perpendicular to the direction of flight, which is equal to

$$Y_j = Y_c + (Z_c - Z) \tan \theta \tag{2-55}$$

where Y_j is the ground coordinate of point j, Y_c is the Y coordinate of the scanner at the instant of recording the jth scan line, and Z is the elevation of the terrain (assumed level) in which Z_j in Fig. 2-47(b) has been replaced by the constant elevation Z. The angle θ_j is the scan angle at the instant of recording given by

$$\theta_j = \frac{Y_j}{f} \tag{2-56}$$

and Y_j is the image position along the circular cylindrically shaped image surface coaxial with the flightline axis. Primed quantities refer to the scanner origin.

This idealized geometry is not realized in practice for the following reasons. The effects of changing ground-resolution element size at different scan angles and effectively recording on a cylindrically shaped image surface rather than a plane surface produces scan-angle effects. The effect of neglecting differences in elevation of object points above or below the assumed datum produces errors which are called topographic effects. Each scan line is not recorded instantaneously and this produces scan-time effects. Finally, if one considers alterations from an idealized flight path, then sensor exterior-orientation effects are included in the imagery.

First, we shall consider scan-angle effects. From Fig. 2-46 the element size in the direction of flight, the X direction, is given by

$$e_{X_j} = \beta(Z_c - Z_j) \sec \theta_j \qquad (2\text{-}57)$$

In the direction perpendicular to the direction of flight, or along the scan line, the ground coordinate of the point j is given by

$$Y_j = Y_c + (Z_c - Z_j) \tan \theta_j \qquad (2\text{-}58)$$

If the sampling angle γ is small enough, it can be considered as a differential change in the scan angle θ_j so that

$$\gamma \cong d\theta_j$$

Now we can differentiate Eq. (2-58) with respect to θ_j and obtain

$$\delta Y_j = (Z_c - Z_j) \sec^2 \theta_j \, d\theta_j$$

or
$$e_{Y_j} = \gamma(Z_c - Z_j) \sec^2 \theta_j \qquad (2\text{-}59)$$

the element size along the scan line. Here the δ's refer to displacements.

Figures 2-48 and 2-49 show plots of these equations. Here, ground-resolution element size has been plotted as a percentage of flying height above the terrain in

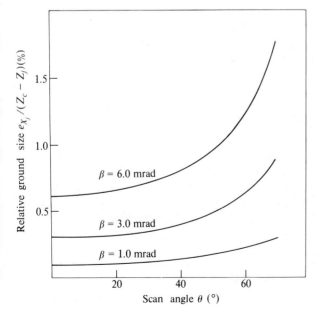

Figure 2-48 Resolution element ground size in X direction.[16]

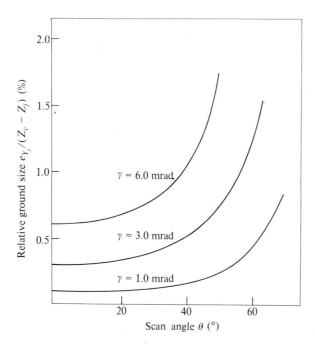

Figure 2-49 Resolution element ground size in Y direction.[16]

the X and Y directions, respectively. Serious distortions result for large resolution angles and large scan angles. This is not a problem in satellite scanners because of the relatively small scan angles used (Landsat total scan angle is approximately 11°) and small IFOV. The problem becomes considerably more serious in aircraft multispectral scanners.

An additional distortion in the data appears due to the fact that the data are effectively imaged upon a cylindrical surface rather than a plane. From Fig. 2-47(b) this may be seen to be the difference between some hypothetical image position, denoted a', and the actual image position, a. The resulting displacement is given by

$$\delta y = Oa' - Oa = f(\tan \theta_j - \theta_j) \qquad (2\text{-}60)$$

Figure 2-50 shows graphically the magnitude of this displacement relative to the focal length f for varying scan angles. Note that this displacement increases very rapidly for scan angles in excess of 45°.

Terrain variation may cause apparent image distortions at large scan angles. Referring to Fig. 2-47(b), if point A is imaged and assumed to be on the datum surface while in fact it lies at some elevation Z_j above the datum surface, the effect is to image the point as if its Y position was at A' on the datum. The effective ground displacement due to neglecting topography may be given by

$$\delta Y_j = Z_j \tan \theta_j \qquad (2\text{-}61)$$

where Z_j is the element elevation relative to datum and Y_j is the resulting Y displacement at the ground scale. The correct image position for this element would then be at a'' if the element is to retain its proper Y position after reduction to the

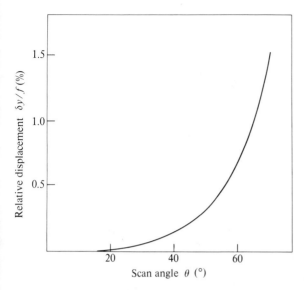

Figure 2-50 Image displacement as given by Eq. (2-60).[16]

datum surface. This error Y_j will vary from 0 at nadir ($j = 0$) to a value Z_j at 45° scan angle and would increase quite rapidly for scan angles greater than 45°. To remove this error, information regarding the elevation of the terrain beneath the scanner must be inserted in the data-analysis process.

If one considers the motion of the aircraft during the recording of the scan line, then the data would be recorded as shown in Fig. 2-51, assuming the aircraft velocity V to be constant during the very short time required to develop a single scan. The X_j coordinate at any point may then be written as

$$X_j = X_c + \frac{Y_j}{2\pi f} t_r V \tag{2-62}$$

where Y_j is the along-scan image distance to point j. The t_r term is the period of revolution of the scanning mirror and f is the focal length of the scanner optics.

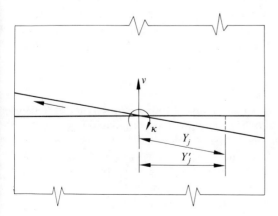

Figure 2-51 Equivalent yaw due to scan-time effect.[16]

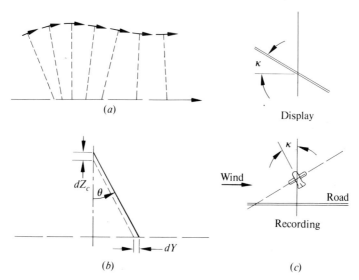

Figure 2-52 Effect of some aircraft instabilities.[16] (a) Effect of pitch. (b) Effect of flying height. (c) Effect of yaw.

Anomalies in the motion of the sensor platform are illustrated in Fig. 2-52. If one considers a theoretically grid-shaped image and superimposes these platform anomalies on the normal scanner motion, the resulting effect on the grid-shaped image is illustrated in Fig. 2-53. The mathematical formulation which describes these image perturbations is somewhat complex and will not be treated here. However, the results of the mathematical formulations are plotted in Figs. 2-54 to 2-57. Notice again that the departures from the ideal image increase monotonically as each of the perturbation variables increases. In most aircraft scanner systems the effects of roll are compensated for by an electronic adjustment of the start of each scan line in the final imagery. However, effects due to yaw and pitch are not usually compensated for. The variation due to pitch is frequently obscured by overscan in the scanner imagery. However, the effects of yaw are usually present in the imagery unless the scanner is mounted on a stabilized platform within the aircraft. In satellite imagery these parameters are very carefully controlled and are not a serious problem.

Photographic Systems

Photographic systems are often considered the grandfather of remote sensing systems in that the technology of remote sensing actually originated in the science of photo interpretation. In fact, many of the characteristics of a remote sensing system that employs a multispectral scanner are shared by the optical portion and, to some degree, the detector portion of a photographic system. In a photographic system, the film functions as the detector, and the lenses which focus the image upon the plane of the film function as the optical system. A photographic system

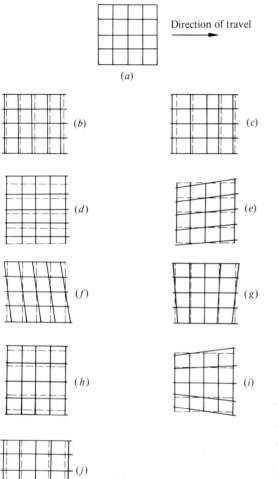

Direction of travel

Figure 2-53 Resultant earth coverage of scanner in the presence of different vehicle-orientation irregularities.[16] (*a*) Theoretical image coverage. (*b*) Pitch. (*c*) Linear velocity change. (*d*) Roll. (*e*) Linear roll change. (*f*) Yaw. (*g*) Linear yaw change. (*h*) Altitude. (*i*) Linear altitude change. (*j*) Velocity or linear pitch change.

is a framing system: all of the data in an image are acquired simultaneously. The film used as a detector in a photographic system has an additional restriction of relatively limited spectral range as compared to a multispectral scanner system. However, photographic systems feature extremely high spatial resolution compared to multispectral scanner systems. As noted in Chap. 1, photo interpretation, the science of the classification and analysis of photographic images, is a highly developed technology. Additionally, photogrammetry, which emphasizes the geometric aspects, sometimes called metric aspects, of image analysis, is also highly developed and is the primary source of maps produced by various agencies around the world. This highly developed technology, as well as the relatively low cost of photographic systems as compared to multispectral scanner systems, makes them attractive for many remote sensing applications.

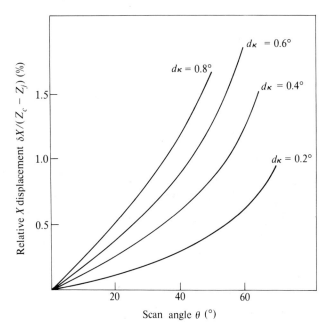

Figure 2-54 Ground displacement in X direction due to change in yaw $(d\kappa)$.[16]

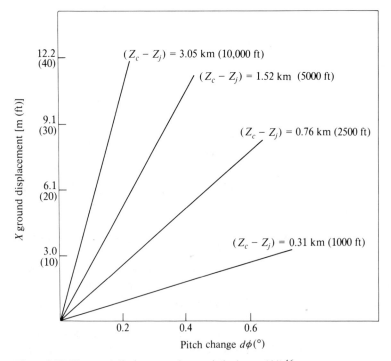

Figure 2-55 X ground displacement due to pitch change $(d\phi)$.[16]

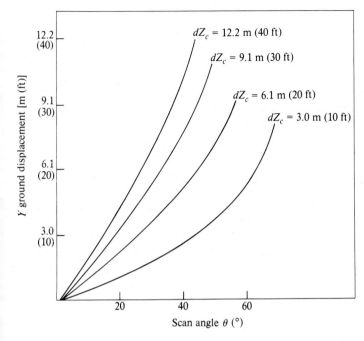

Figure 2-56 Y ground displacement due to flying height change (dZ_c).[16]

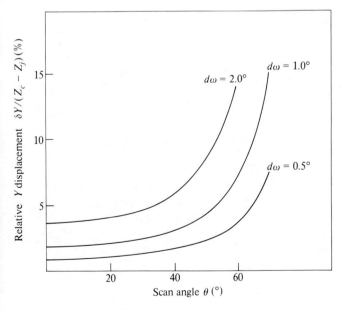

Figure 2-57 Ground displacement in Y direction due to change in roll ($d\omega$).[16]

The optical system The optics of a photographic system are designed to form a completely framed image, and therefore the field of view is relatively large compared to the instantaneous field of view of a line scanner. The field of view of some photographic systems may be as much as 80° or more. Although the formulas used for the simple-lens system described in Sec. 2-3 may be used to make approximate calculations for the optics of a photographic system, the actual lenses used consist of multiple elements, and the detailed calculations of the field of view and resolution limits are quite complex and usually are executed on a digital computer.[17] However, some of the simple geometric effects in a photographic lens arrangement can be understood by examining Fig. 2-58. For example, suppose a camera is designed to produce an image that is approximately 0.23 m on each side. If it is desired to produce a field of view of 60°, then, applying the equation in Fig. 2-10, a lens of focal length 0.20 m is required in order to fill the image format on the film. This calculation assumes a flat film plane. In practice the film plane of a high-quality aerial camera is curved in order to compensate for image distortion produced by the lens and wide field of view of the system. These image distortions are impossible to eliminate from the lens due to the laws of optics which apply to the lens design. Curving the film plane slightly produces a high metric quality of the image in the image plane.

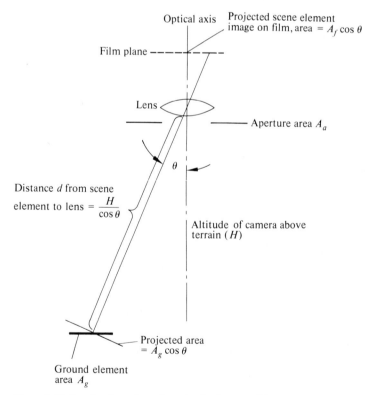

Figure 2-58 Angular dependence in a simple photographic system.

Additionally, the camera may be rocked in its mounting fixture to compensate for the forward motion of the vehicle during exposure. This is known as *image motion compensation*.

Finally, there is another significant phenomenon known as *vignetting*, which may be explained through the following simple development. Referring to Fig. 2-58, the projected area viewed on the ground by the camera is equal to $A_g \cos \theta$. Conversely, the projected area of the scene element that is imaged on the film plane is given by $A_f \cos \theta$, where A_f is the area of the image of the ground-scene element for $\theta = 0$. The distance from the camera to the ground is given by $H/\cos \theta$, where H is the height of the camera above the ground surface. The image intensity falls off according to the inverse square law (target does not fill the field of view of the lens) and in this case is inversely proportional to $H^2/\cos^2 \theta$. Combining these results one obtains an expression for the intensity of the image element upon the film plane as

$$I \propto \frac{A_g A_f}{H^2} \cos^4 \theta \qquad (2\text{-}63)$$

This $\cos^4 \theta$ law says that as one increases the angle between the image element and the optical axis of the camera, the intensity of the image on the film falls off as the fourth power of the cosine of the angle from the optical axis to the image element. This means that the edges of the image are considerably dimmer than the center of the image. This effect is compensated for either by using a special filter placed over the lens of the camera or by special steps in the design of the lens. It is important to the photointerpreter or photogrammetrist that the image intensity be consistent across the image plane if he is to use the resultant photographic product most effectively. In order to maintain metric fidelity the film is carefully positioned in the camera, and frequently a vacuum system is used to hold the film against the curved platen which supports it to insure that the geometric relationships essential for metric fidelity are maintained.

Film: the detector In a photographic system the film serves as the radiation detector in place of the photon detectors in a scanner system. Silver halide is the basic photon-detecting substance in a film. As the photons become more energetic, i.e., as the wavelength of the photon becomes shorter, the ability of the individual photons to excite the silver halides in the film become higher. The impinging photons disassociate a small amount of the silver from the silver halide structure and this process is further enhanced by subsequent chemical development. A basic black-and-white film, sometimes called orthochromatic film, is quite sensitive to blue radiation and relatively insensitive to red radiation. Films with these kinds of spectral characteristics are still in use, especially in high-resolution applications. However, most black-and-white films are produced with sensitized emulsion layers which tend to equalize the spectral response of the film over the visible spectrum. These are referred to as panchromatic films. Further, special sensitizing procedures can be applied to the emulsions of black-and-white films so that they can be made to be sensitive to infrared radiation out to wavelengths of approximately 0.9 μm.

Black-and-white panchromatic and infrared films are extensively used in remote sensing photographic systems for high-resolution metric purposes. However, color films are more widely used when the basic purpose of the operation is to classify features in the scene.

A color film is based upon the basic silver halide chemistry of black-and-white film but employs special dyes whose chemistry is linked to the silver halide disassociation process. Here the discussion will be limited to color-positive transparency films. A schematic representation of this process for natural color film is illustrated in Fig. 2-59. The film is sensitive to radiation over the visible spectrum and is designed to form colors which closely approximate the colors viewed by the human eye. The film consists of three layers, each especially designed to be sensitive to a particular band of colors (wavelengths) in the spectrum. The layer of film nearest the lens is sensitive to red radiation, the middle layer

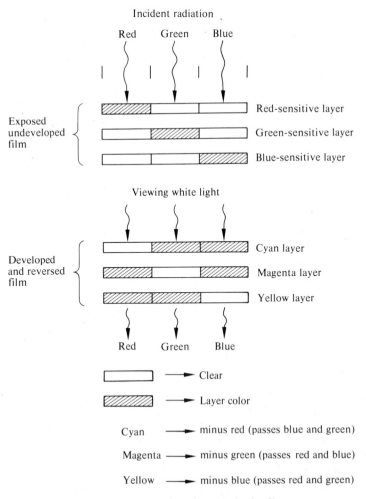

Figure 2-59 A schematic representation of a natural-color film response.

is sensitive to green radiation, and the layer furthest from the lens is sensitive to blue radiation. After the film is exposed to the incident radiation it is chemically developed and chemically "reversed." The reversal process causes the red-sensitive layer to become transparent wherever red light struck it but to turn a cyan color in those areas where it was not struck by red light but was struck by blue or green light. The green-sensitive layer becomes transparent in those areas where it was struck by green light but turns magenta in all other areas of the layer. The blue-sensitive layer becomes transparent in those areas where it was struck by blue light but turns yellow in other areas. When for viewing purposes white light is passed through the three-layer emulsion and passes through that portion of the film originally *struck by blue light*, the red component is filtered out by the cyan layer and the green component is filtered out by the magenta layer. Blue light is passed by all three layers so this portion *appears blue to the eye*. Similar processes appropriately handle red and green radiation. This color-forming process is referred to as a subtractive process, as opposed to an additive process. Additive processes use red, green, and blue filters rather than cyan, magenta, and yellow.

More information can often be obtained from a scene if the infrared radiation reflected from the scene is sensed. Color infrared film may be used for this purpose. The arrangement of such film is illustrated in Fig. 2-60. Here the layer of film closest to the lens is sensitive to infrared radiation (up to approximately 0.9 μm). The middle layer is sensitive to red radiation and the layer furthest away from the lens is sensitive to green radiation. Before the radiation is permitted to strike the film, it is passed through a yellow filter which filters out the blue component. Thus only green, red, and infrared radiation are permitted to strike the film. After the film is exposed and the original latent image is formed by the silver halides in the various film layers, the film is developed and a reversal process is applied. This processing produces dye formation in the layers of the film in a manner similar to that described above for natural color film. However, when this film is viewed with white light, in those areas of the film originally struck by green light, the cyan and magenta layers filter out the red and green components of the viewing light, permitting only blue light to pass through the film. In those areas of the film originally struck by red light, the cyan and yellow layers filter out the red and blue components of the viewing light, permitting only green light to pass through the film. And in those areas of the film originally struck by infrared radiation, the yellow and magenta layers filter out the blue and green components of the light, permitting only the red component of the viewing light to pass through the film. The resultant image is often called a *false color* image in that the apparent colors in the film do not correspond to the colors in the original scene as rendered by the human eye. This is necessary, of course, since infrared light does not produce any color perception in the human eye. It has therefore been assigned arbitrarily to the red rendition in the finished film product, and other colors must be used for the remaining colors in the visible spectrum.

A benefit derived from filtering out the blue light is that the haze-penetration characteristics of the color infrared film are considerably superior to that of conventional color film. This results from the fact that blue radiation is scattered more by the atmosphere (see Sec. 2-2) than are green, red, and infrared radiation.

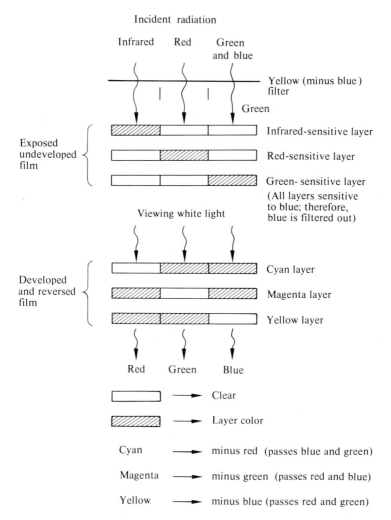

Figure 2-60 A schematic representation of a false-color infrared film response.

When photographic films are used quantitatively in remote sensing, special techniques are used to establish the quantitative response of the film to incident radiation. The relationship of the relative response of the density of a typical film to the amount of light striking the film is shown in Fig. 2-61. If the slope of the film characteristic is known, together with the value of the radiation at which the linear portion of the characteristic begins, then the relationship between the absolute value of the radiation falling upon the film and the density of the film can be established. However, normal manufacturing tolerances in the production of the film compounded by tolerances encountered in the processing of the film can produce variability in the effective slope and threshold values of the film characteristic.

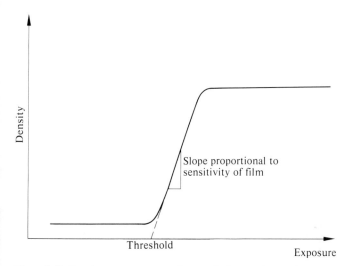

Figure 2-61 Density versus exposure for a typical monochrome film.

A convenient way to establish the processing control for a particular film is through the use of *sensitometry*. A portion of the film is exposed to a calibrated light source that can produce several intensity levels of light. After the film has been exposed to the scenes of interest, that portion of the film that was previously exposed to the calibration source, usually referred to as a *calibration wedge*, is developed before the rest of the film. If the calibration wedge does not have the desired density in the finished film, then the developing process is altered slightly and another calibration wedge is developed. This is repeated until the desired results are obtained. Then the rest of the film containing the images of the scenes of interest is processed. In this manner the relationship between absolute light intensity and film density is established and the film can be used in a quantitative fashion. Usually the optimal image quality is obtained in the process as well. The relationship between the calibration wedges and the resulting film density is shown schematically in Fig. 2-62.

No spectral dispersion is produced intentionally in the optics of a photographic system. The spectral separation that does occur is produced by the multiple-emulsion layers of the color film products discussed previously. Remote sensing photographic systems also use multiband techniques employing black-and-white films and filters to produce spectral separation. In such a system, for example, four cameras are boresighted, each with its own lens, film, and filter. Three of the cameras contain black-and-white panchromatic film and are equipped with a blue, green, and red filter, respectively. The fourth camera contains black-and-white infrared film and is equipped with an infrared filter, i.e., a filter passing infrared radiation. In this way, with four black-and-white images, a complete data set is produced for a given scene. The optical components of this photographic system can be designed to provide extremely high spatial resolution, since no compromises need be made to provide for spectral dispersion. Any aberrations caused by the

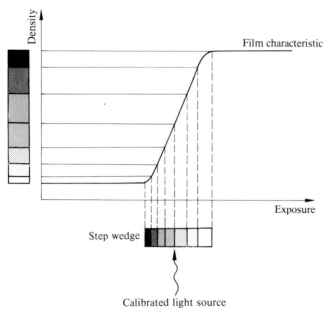

Figure 2-62 Schematic representation of the use of step wedges in the calibration of monochrome-negative film.

optical system are accurately known and can be compensated for in the geometric arrangement of the film platen and lens components of the optical system. The images produced by the system can be conveniently scaled through enlargement or reduction processes after the original film products are developed. Additionally, sterographic techniques are available to permit contouring and metrification of the photographic images. A detailed discussion of this topic is covered in texts on photogrammetry.[18]

Photographic film products can be processed by digital methods after pre-processing by a film digitizer. A photographic film digitizer resembles in many ways the equipment used in the graphics industry to produce multicolor separations for printing from color photographic film products. The operation of such a film digitizer is illustrated in Fig. 2-63. In order to digitize the areas of the film having red tones in the multiemulsion structure, a red filter is placed over the photodiode in the digitizer. The light received by the photodiode is then inversely proportional to the density of red in the film being scanned. The electrical signal produced by the photodiode is then digitized together with the coordinates of the point being scanned in the film. The process is repeated by successively placing a green filter and then a blue filter over the photodiode. In this way, three digital arrays are produced, one each corresponding to the red, green, and blue spectral bands of the image. The resulting digital arrays are in excellent registration with each other since the emulsion layers of the film are registered spatially.

Any slight registration error is due to tolerances in the operation of the digitizer. The same digitization technique can be applied to a set of black-and-white

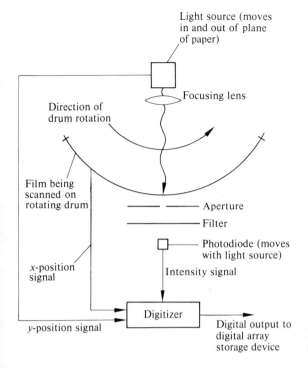

Light source (moves in and out of plane of paper)

Focusing lens

Direction of drum rotation

Film being scanned on rotating drum

Aperture

Filter

Photodiode (moves with light source)

x-position signal

Intensity signal

Digitizer

y-position signal

Digital output to digital array storage device

Figure 2-63 A polychrome film digitizer.

single-emulsion films which have been exposed through filters to form a multiband set of photographic images. However, such single-emulsion products must be carefully registered using the geometric registration processes discussed in Chap. 4 in order to form a multispectral data set suitable for the multivariate analysis methods discussed in Chap. 3. Even in that segment of the remote sensing community principally concerned with the digital processing of imagery produced by multispectral scanners, photographic products comprise a valuable part of remote sensing data systems. The high spatial-resolution characteristics of photographic films make them very useful for rendering finely detailed portions of the scene so that "training" information can be developed for processing of the digital multispectral scanner data. Further practical considerations relative to the use of photographic products are discussed in Chaps. 5 and 6.

Electron Imagers[19]

Electron-imaging systems resemble photographic systems in that they frame an image on a photoelectric surface much like a photographic system frames an image on a photochemical surface. The systems usually involve a shutter, an optical system, possibly an image motion compensation system similar to those employed in a standard photographic camera. Since an electron imager is a framing device, collecting a full frame of data practically instantaneously, it is not necessary to maintain as close an attitude control of the sensor as is required for a line scanner.

A few basic types of electron imager will be discussed in this section, together with some of the more common spectral dispersion or separation mechanisms that are employed in order to produce a multispectral system. Although electron-beam imagers usually acquire the image in a fashion similar to that of a photographic system, the images induced on the photoelectric surface are electronically rather than chemically processed and are amenable to rapid electronic transmission from the sensor platform to a receiving station. Alternatively, the images may conveniently be recorded on magnetic tape for replay at a later time when the sensor platform is located near a receiving station.

Electron imagers involve the conversion of an electronic image on a photo-electric surface into a series of electrical signals that can be transmitted or recorded and later rearranged to reproduce an image of the original scene. A basic (direct-beam readout) *vidicon* is illustrated in Fig. 2-64. The optical system of the sensor focuses the image on a photoconductive surface, referred to as the *target*, in a fashion similar to that employed in photographic systems. However, before the shutter is opened to expose the photoconductive target to the scene, the back side of the target is negatively charged by the electron-scan beam. After the shutter has opened and closed, the image is retained as a charge pattern on the photoconductive target. An electron beam is then used to produce an electrical signal by scanning over the back side of the target. The target is coated on its front side with a transparent conductor which is used to conduct the signal current from the photoconductive target to the signal-processing electronics.

After exposure the light pattern that was focused on the target increases the conductivity of the photoconductive material in those areas which were illuminated, causing those areas on the back side of the target to become charged to more positive values. When the electron beam is then scanned over the back side of the target, it deposits a negative charge on the positively charged areas, producing a capacitively coupled signal in the transparent electrode coating the

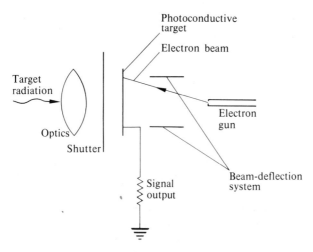

Figure 2-64 A basic (direct-beam readout) vidicon.

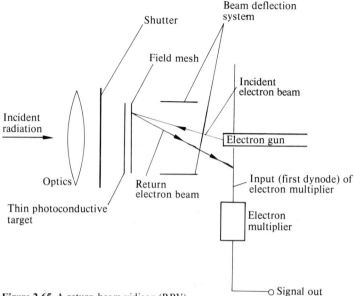

Figure 2-65 A return-beam vidicon (RBV).

front side of the target. Electrical signals encoding the location of the scan beam are synchronized with the electrical signal coming off the transparent conductor, and these signals together carry sufficient information to permit reconstruction of the image.

The *return-beam vidicon* (RBV) also uses a photoconductive target as a light detector. In this device, however, it is arranged for the electron beam *reflected from* the back side of the target to carry the image-describing signal (Fig. 2-65). The reflected beam is detectably weaker when charge is left behind in the light-sensitized areas of the photoconductor.

The *image dissector* is based on a different principle. The photosensitive surface in an image dissector (Fig. 2-66) is a photocathode which emits photoelectrons when struck by photons. An electromagnetic focusing system is used to focus the emitted photoelectrons on an aperture, behind which is located an electron multiplier to amplify the signal carried by the photoelectrons. Encoded position information from the focusing system plus the output of the electron multiplier are sufficient to permit reconstruction of the image formed on the photocathode. The image dissector is capable of extremely high spatial resolution and its sensitivity is competitive with other vidicons. It is often used in slow scan systems in which the satellite or aircraft platform itself provides part of the scanning mechanism.

Notice that the family of electron imagers discussed here electronically and/or mechanically scan an electrical image (pattern of charge). They are refered to as *image-plane scanners*. The line scanners discussed previously scan the image in the object plane and are called *object-plane scanners*.

A charge-coupled device (CCD) represents one of the more recent developments in electron imaging.[20] A photosensitive surface, usually composed of an

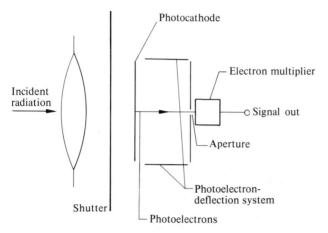

Figure 2-66 An image dissector.

array of silicon diodes, is located behind a shutter and optical system. Each silicon diode is connected to a charge storage cell in an array of integrated circuit MOS (metal oxide semiconductor) devices, as shown in Fig. 2-67. When light strikes a diode in the array it generates a quantity of charge (current) proportional to the light striking it, and the charge is transferred to the charge cell behind the diode.

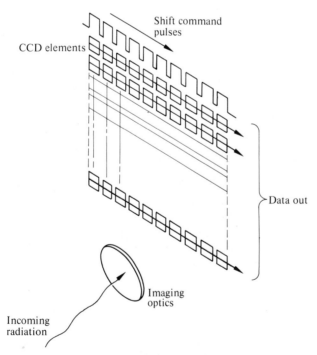

Figure 2-67 A charge-coupled device (CCD) imager.

The charge cells can all be part of an electronic shift register which can be activated to read out the charge cells in a sequential fashion. The output signals are correlated with the shift pulses to reconstitute the image.

Spectral separation in electron-beam imagers is accomplished in much the same manner as in photographic systems. In a multiband photographic system utilizing panchromatic and infrared films, spectral separation is produced by using photographic filters with multiple-lens systems. An analogous procedure is used in some multiband-vidicon systems. Since the photosensitive surface does not have a multiple-emulsion characteristic like color photographic film, it is necessary to have a separate photocathode for each spectral band together with a corresponding optical system and filter assembly. A typical arrangement of such a system is shown in Fig. 2-68. A multiple-photocathode arrangement such as is employed in commercial color television systems is usually not utilized in earth resources remote sensing systems due to resolution limitations. However, a form of color production employed in early color television systems may be used. In the single-tube vidicon system shown in Fig. 2-69, a moving color filter is positioned sequentially so that each time an image is framed, the color filter changes to a different portion of the spectrum. Thus spectrally different frames are produced in rapid sequence rather than simultaneously. Provision must be made to account for the small amount of displacement of the sensor between spectral frames, but this can be compensated

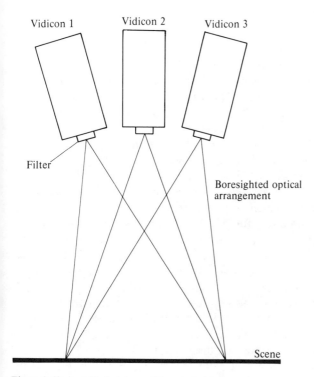

Figure 2-68 A multiple-tube multispectral vidicon system.

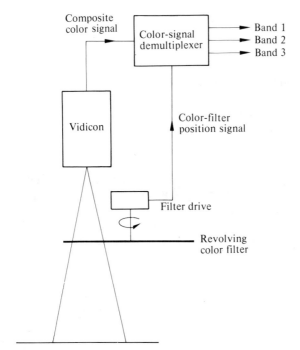

Figure 2-69 A single-tube multispectral vidicon system.

for in the filter/decoder system used to decode the sequential color-frame information as it comes from the sensor.

It is possible to prepare photosensitive surfaces to produce spectral dispersion and utilize the motion of the platform to help provide scanning information. A system that utilizes these concepts is the wide-range image spectrophotometer (WISP).[19] The WISP uses a two-dimensional photosensitive surface to record spatial information along one axis and spectral information along the other axis. This is done by taking a one-dimensional line image and spreading it spectrally along the axis perpendicular to the direction of the line image. Figure 2-70 illustrates the basic layout of such a system. The objective lens focuses the radiation from the scene on a narrow slit which determines the field of view of the system. A dispersion element disperses the beam into its individual spectral components. The beam is dispersed along the vidicon photosensitive surface at right angles to the orientation of the slit. The vidicon (any type described previously) scans the image and develops an image that has spatial information along one axis and spectral information along the other axis.

This section has dealt with only a few of the electron imagers that are available. Table 2-5 summarizes the relative advantages and disadvantages of the devices discussed here, along with a number of other electron-beam imagers. All of these systems have one thing in common: they require a photosensitive surface capable of responding to radiation in the spectral bands of interest. It is technically difficult to produce a surface of area sufficient to provide adequate spatial

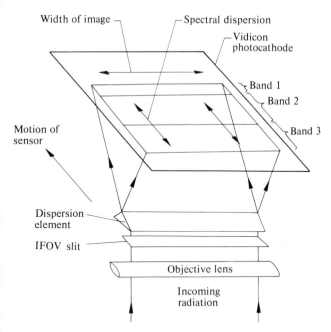

Figure 2-70 A wide-range image spectrophotometer (WISP).

resolution at wavelengths longer than 1.2 μm. This fact tends to restrict the applicability of electron-imaging systems with respect to earth resources problems due to the importance of information in the spectral regions beyond 1.2 μm.

PROBLEMS

2-17 Relate each component of the spectroradiometers shown in Fig. 2-24 to the spectrometer functional block diagram (Fig. 2-14). Explain the function and/or purpose of each component in terms of Fig. 2-14.

2-18 Repeat Prob. 2-17 for the CVF spectrometer shown in Fig. 2-28.

2-19 Consider the design of a six-channel, aircraft-borne multispectral scanner with spectral bands:

$$0.49–0.57 \ \mu m$$
$$0.57–0.66 \ \mu m$$
$$0.67–0.80 \ \mu m$$
$$0.80–0.93 \ \mu m$$
$$1.70–2.25 \ \mu m$$
$$7.00–11.50 \ \mu m$$

What type of detector would you recommend for each spectral band? In the case of photon detectors, what detector material would you recommend? Support your recommendations by stating the reasoning used to arrive at your choice.

2-20 Describe the distinguishing features of spectral-data and image-forming instrumentation systems. Include considerations such as type of data collected, format, volume of data, and how the data impacts remote sensing.

Table 2-5 Electron imager performance summary[19]

Type	Sensitivity,† J/m²	Spectral range, μm	Dynamic range‡	Resolution,§ line pairs/mm	Relative response speed
Return-beam vidicon (RBV)	3×10^{-4}	0.4–0.9	80:1	60	High
Direct-beam readout vidicon	3×10^{-5}	0.4–0.9	50:1	40	High
Focus projection and scanning (FPS) vidicon	2×10^{-4}	0.3–0.75	90:1	50	High
Image dissector	2.6	0.25–1.1	100:1	60	Low
Secondary electron-conduction (SEC) vidicon	2×10^{-7}	0.1–0.9	80:1	40	Moderate
Electron-bombarded silicon (EBS) vidicon	2×10^{-7}	0.1–0.9	100:1	30	Moderate
Dielectic tape camera	4×10^{-6}	0.4–0.9	100:1	30	High
Charge-coupled device (CCD) vidicon	1×10^{-3}	0.4–1.0	100:1	20	High
Wide-range image spectrophotometer (WISP)	Same as image dissector or CCD used				

† At signal-to-noise ratio of 10:1.
‡ Ratio of brightest measurable radiance to darkest measurable radiance.
§ At 50 percent contrast.

2-21 Consider the two spectral data-gathering missions described below. In each case make a recommendation concerning the type of instrument best suited for the mission. Discuss both the positive and negative implications of your choice.

(*a*) The spectral characteristics of forest canopies are to be determined from a helicopter-mounted instrument. Several missions are planned in order to determine the effects of seasonal changes. Changes in spectral characteristics due to seasonal variations or differences in species or forest-management techniques are considered to be more important than spectral detail.

(*b*) Data are required to support a fundamental study of a plant disease known as potato leaf blight. Well-controlled plots of potatoes have been established on an experimental farm. The farm has good access roads and electrical power distribution. When disease infestation occurs, leaf cell structure is affected so that subtle changes in spectral response need to be monitored.

Study objectives

Upon completion of Secs. 2-7 and 2-8 you should be able to:

1. Draw the data-flow block diagram for a remote sensing data-collection mission.
2. Compare the design parameters of the four scanner systems described and give plausible reasons to account for the design differences.

2-7 GENERAL CHARACTERISTICS OF DATA-ACQUISITION SYSTEMS

Earlier in this chapter the interactions of solar energy with natural materials on the surface of the earth and the detection of the energy reflected and emitted from these materials have been discussed. The emphasis has been upon the production of an electrical signal after the reflected or emitted energy has traveled through the sensor and impinged upon a detector. Usually this electrical signal is amplified through an electronic system and in some cases is filtered or shaped by specialized electronics in order to enhance the information content of the data. Through such techniques it is often possible to minimize the undesirable effects of noise by filtering the electrical signals in a portion of the spectrum where information content is not predominant but noise content is predominant. We are assuming that the data will ultimately be processed on a digital computer in a multispectral fashion. Therefore, the electrical signals produced by the sensor must ultimately be turned into a numerical format and stored for subsequent digital processing. The electrical signals are referred to as *analog functions*, and therefore the process of converting these functions into a digital representation is referred to as the *analog-to-digital conversion* process.

Analog-to-Digital Conversion

Figure 2-71 illustrates the systematic procedure used to convert an analog signal into its digital representation—sometimes called *digitization*. The signal is sampled frequently enough so that the digital representation of the signal will reproduce the information content of the signal sufficiently accurately for the application at

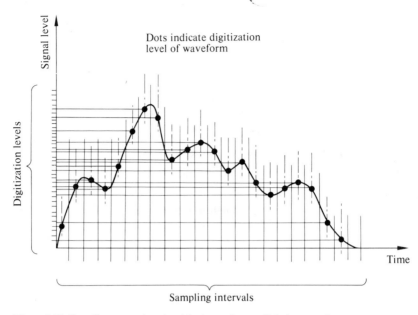

Figure 2-71 Sampling an analog signal in the analog-to-digital conversion process.

hand. The sampling rate must be greater than or equal to approximately twice the highest frequency component in the signal which is intended to be preserved by the system.[21] The highest frequency component preserved will determine the fidelity of reproduction of the signal after it has been subjected to the digitization process. The number of discrete quantization or digitization levels used to represent the signal is also related to the fidelity of reproduction of the signal and to the precision of the digital format. The number of levels is usually selected on the basis of digital-system performance specifications, assuming an array of signals from a given set of applications. Long experience in remote sensing has indicated that quantization to 256 levels (eight binary bits) is usually suitable for remote sensing data. (The discussion on measurement complexity in Chap. 7 is interesting in this respect.) The sampling rate chosen depends upon the altitude of the scanner and its instantaneous field of view, both of which affect the frequency characteristics of the electrical waveform as discussed in Sec. 2-6 (where the bandwidth characteristics and the resulting frequency performance of the scanner were discussed in the signal-to-noise calculations relating to the scanner system).

Often, in aircraft systems it is less expensive to record the signals produced by the detectors on an analog tape recorder rather than to first put the signals through an airborne analog-to-digital conversion process. The analog tape is later processed by a ground-based analog-to-digital conversion system, producing a computer-compatible digital tape. Additional reformatting of the digital tape into formats appropriate for routine processing often follows. Although this approach may lower the cost of the data-acquisition system, it does introduce an extra processing step in the input subsystem of the digital computer-processing system. Additionally, some loss of dynamic range of the signal and loss of signal-to-noise ratio inevitably result when an analog recording step must be included in the process. An alternate approach is to digitize the signals as they come from the detectors and record the resulting digital signals directly on digital magnetic tape. Because of the nature of the recording process, no loss of dynamic range or signal-to-noise ratio occurs in the digital recording step. The signals are recorded in binary form, i.e., either presence or absence of a signal, and the only loss in signal quality results from the sampling and quantization processes.

Data Flow in an Information System

The organization of an information system with respect to data flow is illustrated in Fig. 2-72. Data flow into the input subsystem and are directed to the data-processing subsystem. After the data have been processed, they are directed to the output subsystem and eventually appear in the form of output data or information. All of these various subsystems are controlled by an interactive subsystem which provides an interface with the data analyst. This model of the data-processing system is discussed in detail in Chap. 4; the input subsystem in particular is discussed in Sec. 4-3. In this section we will devote most of our attention to the interaction between the data-acquisition system and the input subsystem.

In a satellite system, the data from the satellite generally come from the telemetry link in digital form; the electrical signals from the detectors are digitized on board the satellite. In some cases, other sensors such as multispectral scanners mounted in aircraft acquire data coincidentally with the data recorded by the satellite. As we have seen in earlier discussions, in these cases the multispectral signals may be recorded in analog form. Surface observations are often made simultaneously with these aircraft and satellite overpasses, and these data may be recorded in tabular fashion on reporting forms. Thus it is possible for the data collected during any given data-acquisition procedure to occur in diverse forms, including analog and digital signals recorded on magnetic tape, photographic products, surface observations recorded in tabular form, data from special field

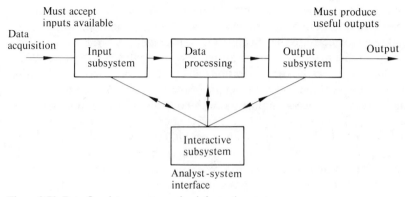

Figure 2-72 Data flow in a remote sensing information system.

sensors recorded on strip charts, and still other forms of data from various sources (e.g., meteorological data from the United States weather service). All of these data must be related to each other and must somehow be merged into a cohesive data set for subsequent analysis.

The way in which these data are handled depends to a large extent on the characteristics of the data-processing system that will be used to handle the data after they are placed in the appropriate formats. If the data-processing system is equipped with several geographically dispersed terminals, then it may be beneficial to catalog the data in a data library accessible by many users. Use of standard and well-documented formats can greatly enhance the accessibility of the data to potential users. Notice that this means that the system constrains the way in which the data are placed into the appropriate form for processing. The steps followed to rearrange or preprocess the data for storage in a data bank or a data library are discussed in Chap. 4.

Data Calibration

Absolute calibration of data is becoming increasingly important as greater attention is paid to multitemporal collection of remote sensing data. If the data taken at different times are properly related to an absolute reference, subtleties in the data that are related to the temporal effects may be preserved. In multispectral scanner (MSS) systems the usual procedures involve the inclusion of reference standards somewhere in the multispectral scanner structure. The most frequently used standards are hot and cold thermal reference plates, incandescent standard-lamp sources, and solar reference sources. The solar sources are usually obtained by directing sunlight from outside the scanner platform into an optical system in such a way that the solar radiation can be viewed during each scan of the mirror.

Figure 2-73 illustrates schematically an arrangement of calibration sources and a typical signal that would be produced by one of the detectors sensitive in the reflective portion of the spectrum and another detector sensitive in the thermal portion of the spectrum. Notice that the detector sensing the energy in the reflective portion of the spectrum views the two thermal references as black surfaces corresponding rather closely to the black background exhibited by the interior portions of the scanner. The solar reference and the calibration lamp will saturate the thermal detector because the effective radiant temperatures of both of these sources far exceed the maximum temperature range of the thermal detector. When proper account is taken of the saturation effects, the appropriate calibration signals can then be recorded and made a part of the data set supplied to the input subsystem of the data-processing portion of the overall information system. The spectral irradiance *versus* wavelength characteristics of the reference lamp are known to a relatively high precision. Thus, the digital data value corresponding to the signal generated by a reflective radiation detector can be calibrated in terms of the average spectral radiance of the reference lamp in any given spectral band. On the other hand, when the detector is exposed to the black background of the interior of the scanner as the scan mirror rotates, a "zero" reference level is obtained.

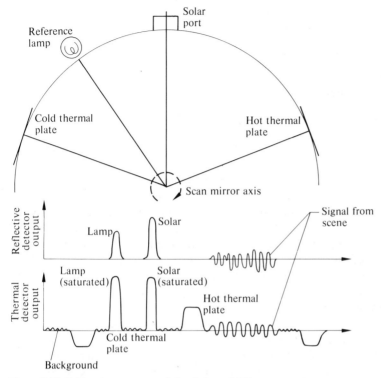

Figure 2-73 Calibration sources and signals in an MSS.

Combining these two pieces of information, one is able to establish a functional relationship between digital data values and radiance for that particular spectral channel. Subsequently, every digital data value produced by the detector and analog-to-digital conversion system can be related to the calibration data value produced by the reference lamp-detector analog-to-digital conversion system. In this manner each digital data value can be related to an absolute radiance which then makes it possible to overlay digital images taken on different days and properly account for different scene lighting effects that may be occurring. One should notice that this calibration or referencing step takes place during each scan produced by every rotation of the scan mirror. Effectively, the scanner is being calibrated almost continuously during the data-collection mission.

The thermal band is handled in a similar fashion with the exception that two references are required. Two thermal reference plates have their temperatures adjusted so that one plate has an effective radiant temperature slightly below the expected lowest radiant temperature in the scene, and the warmer thermal plate has its temperature set at a radiant temperature just above the highest radiant temperature expected in the scene. In this way, two signals corresponding to effective radiant temperatures bracketing the radiant temperatures of the scene are obtained. The calibration procedure is analogous to that described in Sec. 2-6 for thermal-measuring field instruments.

To conclude this section we shall examine three remote sensing systems from the viewpoint of tracing the flow of data through each of the systems into the data product that is ultimately to be used in the input subsystem of a data-processing system.

Data Flow in a Satellite System

A block diagram which illustrates the flow of data in the Landsat satellite system is shown in Fig. 2-74.[22] The Landsat 1 and 2 multispectral scanners have four spectral bands: two in the visible portion of the spectrum and two in the infrared. (Further details on the instrument are given in Sec. 2-8.) The outputs of the detectors are digitized on board the satellite, coded in a modulation system, and transmitted to a receiving station on the earth's surface. The telemetry signals are received by the ground station using a steerable-dish microwave-receiving antenna. Several receiving stations are available for the Landsat satellite; a line-of-sight path is necessary for the signals to be picked up by the receiving station. In those cases where it is impossible for the multispectral data to be transmitted as they are generated, a video tape recorder aboard the satellite can be activated by ground command to record the data for subsequent transmission when the satellite is in view of a receiving station. However, the video recording step is to be avoided when possible since a loss in the dynamic range of the detector signals results when they are passed through the video recording process.

The telemetry signals from the satellite are received, amplified, and demodulated into their digital format. As noted in Sec. 2-8, the data rate from the satellite is quite high so that a special tape recorder is needed, capable of high tape speeds and extremely high density recording. This tape recorder produces digital video tapes used to drive a processing system that converts the data into computer-compatible tapes (CCTs) suitable for processing on most digital computers. Since many potential users of Landsat data are not equipped to process digital multispectral image data, special electron-beam systems are used to produce black-and-white multiband imagery.

In the process of producing the CCTs the ancillary data corresponding to the particular scene being sensed and recorded are merged with the multispectral data. These ancillary variables include the limits of the particular spectral band being recorded, the coordinates of the center point of the scene, the time of day and solar elevation at which the data in the scene were acquired, and a variety of special calibration signals which indicate the dynamic range of the data in the scene. These data are also printed on the photographic products. The black-and-white multiband imagery can be combined into color-coded imagery which is often the most convenient form to use in photo interpretation of the data. The black-and-white images are also helpful in locating features in the data as part of digital analysis techniques.

Data Flow in an Aircraft System

The data flow in an aircraft system which uses ground-based analog-to-digital conversion is illustrated in Fig. 2-75. The signals from the multispectral scanner

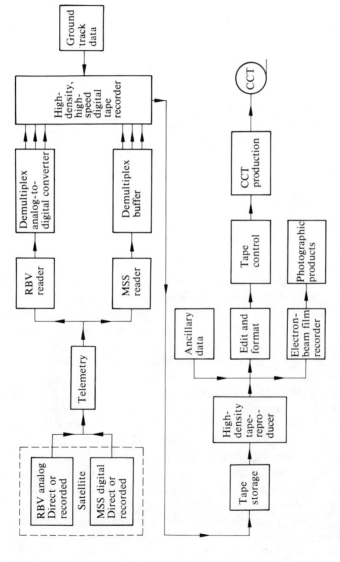

Figure 2-74 Multispectral data flow in the Landsat system.

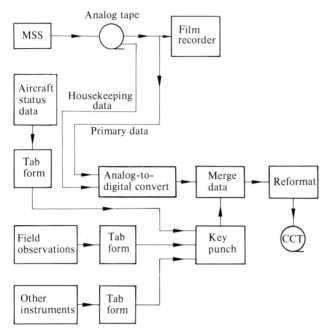

Figure 2-75 Data flow in an aircraft MSS system.

are recorded on a wideband analog tape recorder aboard the airplane. The analog-to-digital system located on the ground is usually a flexible system capable of being used in a multipurpose fashion (e.g., to digitize field data as well as the aircraft data); if the analog-to-digital conversion system were located in the aircraft, it would probably have to be a single-purpose, dedicated system.

The electrical signals produced by the airborne multispectral scanner resemble those produced by a satellite-borne scanner. The signals are recorded on a multichannel tape recorder along with ancillary information that is recorded onto a "housekeeping" channel. Information such as sensor switch and potentiometer settings, aircraft heading and altitude, and scene-descriptor information are recorded on the housekeeping channel in analog form. After the flight mission, the analog tape is transferred to the analog-to-digital conversion system located in the data-processing center. The tape is replayed on an analog-tape reproducer. Some of the ancillary data that have been collected, usually in tabular form from several sources, may be keyed into the analog-to-digital conversion system to be merged with the image data. A computer-compatible digital tape is produced having the appropriate descriptor information for the particular run of data on the digital tape. The digital data are converted into a format suitable for interfacing with the input subsystem of the data-processing system.

Usually the operation of the multispectral scanner in the aircraft system is more nearly under the control of the experimenter than is the case in a satellite system. Also there is more human intervention in the production of the digital tapes, with the net result that the ultimate data product may have many more features helpful to the ultimate user than does the data produced by a satellite scanner system. Since the satellite system has to serve a much wider spectrum of users, it must produce standardized products with fixed formats. The aircraft system, on the other hand, can have more flexibility designed into it but, of course, will not be as capable as the satellite system of producing the volume of data suitable for distribution to as wide a spectrum of users. As a result, the satellite system is potentially more cost effective and characteristically more operational in nature; the more flexible aircraft system, though less cost effective, may be more useful for supporting research.

Data Flow in a Field Spectral System

The last remote sensing system that we will consider in this section is a field spectroradiometer system. This differs from the previous two systems in that it includes a spectrally scanning, rather than spatially scanning, instrument. That is, the sensor looks at one spot on the ground at a time and provides very detailed spectral information about it.

Figure 2-76 illustrates the flow of data in a field spectroradiometer system.

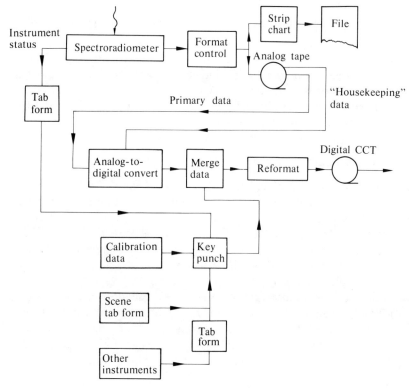

Figure 2-76 Data flow in a field spectroradiometer system.

This system is used almost exclusively for research and is therefore capable of producing a wide variety of extremely detailed data products. The system is very flexible, which is characteristic of research-oriented systems. The data from a series of four detectors are recorded on a multichannel analog tape recorder. In this system a hard copy of the analog signals is also produced on a strip chart recorder and used for verification purposes in establishing acceptable data quality. The hard copy also serves as a backup data source in case the analog tape-recording system fails. Ancillary data are recorded in tabular form by the experimenters and instrument operators. Instrument status is recorded on a "housekeeping" tape channel in a fashion similar to that of the aircraft system. The analog tape is then taken to an analog-to-digital conversion system in the data-processing center. This analog-to-digital system may be the same system used in converting aircraft scanner tapes. This is an example of a very flexible multipurpose analog-to-digital system. The ancillary data acquired during the data-taking process is read into the analog-to-digital system at the time that the analog tape is being converted. The resulting product is a computer-compatible tape having all of the scene descriptors and other ancillary data recorded in digital form together with the digitized spectral radiometric data. The computer-compatible tape has a format suitable for interface with the input subsystem of the data-processing system.

2-8 SOME REMOTE SENSING DATA-ACQUISITION SYSTEMS

In this section four multispectral scanners will be examined : the M-7 airborne scanner built and operated by the Environmental Research Institute of Michigan (ERIM), the scanner on board the Landsat satellites, the S-192 space-borne scanner associated with the Skylab earth resources experimental package (EREP), and the proposed thematic mapper (TM) scanner.

The M-7 Airborne Scanner[23]

A layout of the M-7 multispectral scanner system is shown in Fig. 2-77. Signals from the scanner are monitored and controlled in flight at the operator console and are recorded in analog form by a wideband magnetic tape recorder. The recorded signals are usually digitized and reformatted later on the ground.

Figure 2-78 shows a schematic diagram of the scanner optical system. This is an object-plane scanner with an oblique mirror-scanning system. As in all scanners of this type, the resolution element image rotates during the scan, a second-order geometric complication. As the scan mirror rotates, it sequentially looks at the ground (90° FOV), a calibration lamp for the reflective portion of the spectrum, a thermal graybody reference plate, a solar reference port, and another graybody reference plate. At all other times it is looking at the inside of the scanner casing.

After the incoming radiation is reflected off the oblique scanning mirror, a portion of it is reflected off a folding mirror into a dichroic beam splitter. The folding mirror recovers that portion of radiation that would have been lost by obscuration due to the convex mirror in the Dall-Kirkham telescope. The

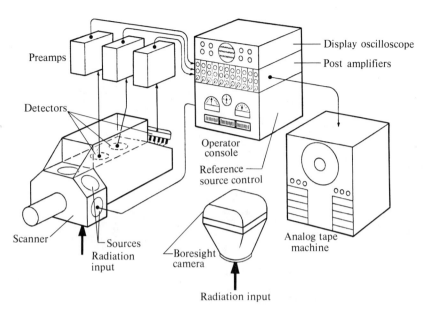

Figure 2-77 ERIM experimental multispectral scanner system.[23]

dichroic is ground into a spherical mirror so that the reflected radiation is imaged by a folding mirror into detector 1B. The field stop is an aperture located immediately in front of the silicon diode detector. The radiation passing through the dichroic mirror is imaged through a refracting system onto detector 1A, a mercury cadmium telluride (HgCdTe) element. Here the detector itself is the field stop. The detector and its foreoptics are cooled to 77 K with liquid nitrogen. It should be noted that the dichroic performs the spectral channel separation between detectors 1A and 1B. The detector at position 1A is sensitive to a portion of the wavelength region between 2 and 14 μm which is determined by the specific nature of the HgCdTe material used; the detector at position 1B is sensitive from 0.33 to 1.0 μm.

The portion of incoming radiation that was not reflected out of the main beam strikes first the primary mirror (spherical 12.3-cm diameter) of the Dall-Kirkham telescope and then the secondary mirror (hyperbolic). The aperture of the scanner is established by the diameter of the primary mirror. The effective focal lengths are established by the focal lengths of the beam splitter and lens described above and the focal length of the Dall-Kirkham telescope. The radiation reflects off the first folding mirror (the one shown with the hole), through a flat beam splitter, and then through a condenser lens detector into position 2 (0.9 to 2.6 μm) occupied by an indium arsenide detector cooled to 77 K. The size of the field stop is determined by the detector size. The portion of radiation reflected by the dichroic passes through the hole in the first folding mirror, reflects off the second folding mirror, and is imaged by the Dall-Kirkham telescope onto the entrance slit of the 12-channel spectrometer. This slit serves as the field stop for the radiation in the 0.4 to 0.9-μm wavelength region. After passing through the entrance slit, the radiation is

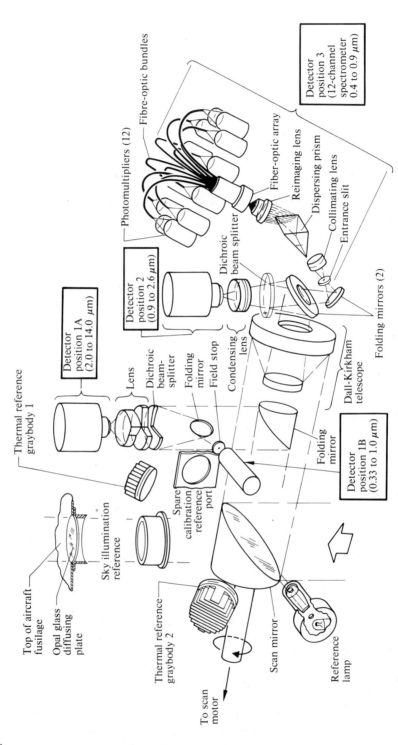

Figure 2-78 Optical schematic of ERIM experimental M-7 multispectral scanner.[23]

Top of aircraft fusilage

Opal glass diffusing plate

Sky illumination reference

Thermal reference graybody 1

Lens

Dichroic beam-splitter

Folding mirror

Field stop

Spare calibration reference port

Condensing lens

Detector position 1A (2.0 to 14.0 μm)

Detector position 2 (0.9 to 2.6 μm)

Detector position 1B (0.33 to 1.0 μm)

Folding mirror

Dall-Kirkham telescope

Thermal reference graybody 2

Scan mirror

To scan motor

Reference lamp

Dichroic beam splitter

Folding mirrors (2)

Photomultipliers (12)

Fibre-optic bundles

Fiber-optic array

Reimaging lens

Dispersing prism

Collimating lens

Entrance slit

Detector position 3 (12-channel spectrometer 0.4 to 0.9 μm)

collimated (made parallel), dispersed through a quartz prism, and reimaged onto the input of a fiber optic array. The geometric position of each fiber optic bundle selects a spectral band of radiation and directs it to one of twelve photomultiplier tubes.

Table 2-6 summarizes the specifications of the scanner. This scanner has two fixed scanning speeds which give contiguous scanning at operating limits and over-scan otherwise. This is characteristic of scanners which record in analog form. Scanners which record in digital form may operate in a fixed mode or have a continuously variable scan speed that permits contiguous scanning at any V/H within the design range of the scanner.

By using Eq. (2-43) and applying the specifications of Table 2-6, the maximum V/H may be calculated as follows:

$$\frac{V}{H} = \frac{pq\beta\omega}{2\pi} \quad \mathrm{s}^{-1}$$

$$= 0.2 \ \mathrm{s}^{-1}$$

since $p = 1$, $q = 1$, $\beta = 2 \times 10^{-3}$ rad, and $\omega = 2\pi \times 100$ rad/s.

The M-7 is mounted in an aircraft with a nominal speed of 60 m/s; therefore,

$$H = \frac{V}{0.2} = \frac{60}{0.2} = 300 \ \mathrm{m}$$

for contiguous scanning. That is, the minimum altitude for contiguous scanning at nominal aircraft velocity is 300 m. The aircraft is sometimes operated for calibration purposes at 150 m at which time 50 percent underscan is obtained. At altitudes above 300 m overscan is obtained, and at altitudes above 1500 m the scan rate is reduced to 60 scans/s in order to reduce the overscan.

A typical output signal from the scanner is shown in Fig. 2-79. The *roll-stabilized* pulse is shown that permits line-to-line synchronization in the finished imagery. Pitch effects are reduced by overscanning. The various calibration signals are also shown. Note that the two thermal reference signals are of opposite polarity, which indicates that the reference temperatures are set to bracket the scene temperature as described in Sec. 2-7.

Table 2-6 Nominal M-7 scanner-performance characteristics[23]

1. 12 spectral bands in ultraviolet, visible, infrared regions
2. 90° external FOV ($\pm$45° from nadir)
3. 2 mrad maximum spatial resolution, 3 mrad nominal
4. 0.1°C nominal thermal resolution
5. 1 percent nominal reflectance resolution
6. 5 radiation reference ports
7. 12.25-cm-diameter collector optics
8. 60 or 100 scans/s
9. Direct current to 90 kHz electronic bandwidth
10. Roll-stabilized imagery

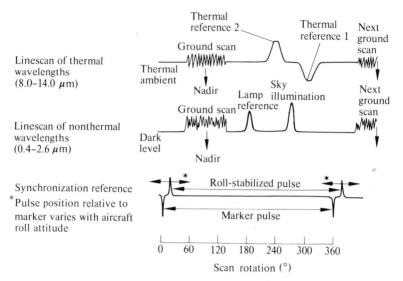

Figure 2-79 Scanner voltage output versus time for the ERIM M-7 scanner.[23]

The Landsat Multispectral Scanner[22]

Figure 2-80(a) shows a schematic diagram of the MSS on board the Landsat 1 and 2 satellites. The scanner has four spectral bands (five on Landsat 3) and forms six scan lines during each mirror sweep. The scanning mirror is of the plane oscillating type rather than rotational. The satellite is in a sun-synchronous orbit with a ground-track speed of 6.47 km/s. Table 2-7 lists some of the parameters of the scanner. The spectral bands used in the Landsat MSS are:

Band 1†   0.5–0.6 μm
Band 2   0.6–0.7 μm
Band 3   0.7–0.8 μm
Band 4   0.8–1.1 μm
Band 5   10.4–12.6 μm (Landsat 3 only)

Since the V/H of Landsat is constant and on-board analog-to-digital conversion is employed, contiguous scanning is preferred to overscanning in order to minimize the data rate which must be handled by the analog-to-digital converter.

A plane mirror, when rotated through an angle θ about an axis in its plane, causes the reflected beam to rotate through an angle of 2θ if the plane of incidence is normal to the axis of rotation, the case in Landsat. Landsat is at an altitude of 915 km with a swath width of 185 km; the FOV will be

$$2\theta = 2 \, \text{Tan}^{-1} \frac{185}{2 \times 915} = 11.56°$$

† In the literature, bands 1 to 5 are sometimes referred to as channels 4 to 8, the communications channels of the telemetry link.

Therefore, the mirror must turn through a total angle of 5.78°, or $\pm 2.89°$ about its neutral position. To calculate the oscillation frequency of the mirror, we note that the satellite moves $\beta Hq = 474$ m along its path [see Fig. 2-80(b)] during the active scan-plus-retrace cycle of the mirror. Therefore, the mirror cycle time must be $\beta Hq/V = 73.26$ ms, which is a frequency of 13.65 Hz (819 oscillations per minute).

Photomultiplier tubes are used as detectors for spectral bands 1, 2, and 3 and silicon photodiodes are used for band 4 (six detectors are used for each spectral band). In Landsat 3, band 5 uses a two-element HgCdTe array maintained at 100 K through use of a passive radiation cooler. To satisfy signal-to-noise ratio requirements, the IFOV of band 5 is 0.258 mrad.

The analog video signals from each detector are sampled, multiplexed, and digitized by a constant-rate, on-board analog-to-digital system. The mirror scan

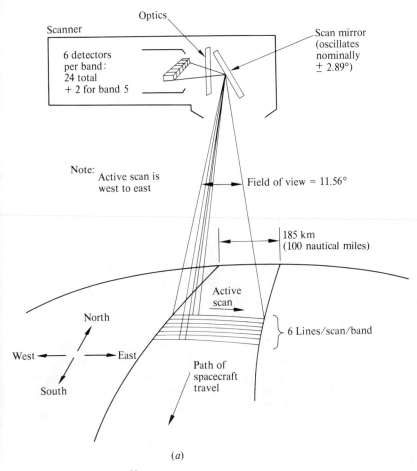

Figure 2-80 Landsat MSS.[22] (a) Schematic. (b) (page 128) Scan pattern.

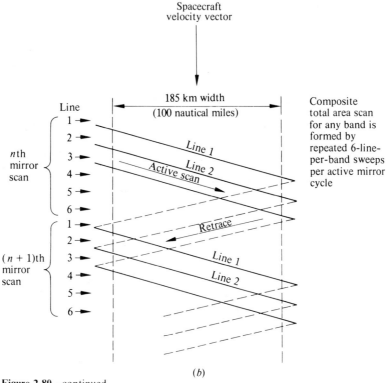

(b)

Figure 2-80—continued.

speed is not precisely constant; therefore, a slightly variable number of samples per scan line results. This effect is accounted for by monitoring the mirror position electronically and later removing the error by processing on the ground. Since (1) the satellite is at an orbit inclination of about 86 degrees, (2) the scan lines are slightly canted with respect to the path of the satellite [Fig. 2-80(b)], and (3) the earth rotates beneath the satellite during its passage, a distorted image results which also can be rectified by ground processing. Various signal compression and expansion techniques are used to enhance the signal quality of the data.

Comparison of the M-7 and Landsat scanners is interesting.

1. The entrance aperture of the M-7 scanner is approximately 0.13 m in diameter as opposed to approximately 0.23 m for the Landsat scanner. Assuming equal detector areas in the two scanners, the ratio of the focal lengths $f_{Landsat}/f_{M-7} = \beta_{M-7}/\beta_{Landsat} = 23$. Therefore, the ratio of the f numbers of the two scanners is $(f/D)_{M-7}/(f/D)_{Landsat} \simeq 1/13$. That is, the optical "speed" of the M-7 scanner is 13 times that of the Landsat scanner.

2. The $\sqrt{V/H}$ of the M-7 scanner is $0.2 \text{ s}^{-\frac{1}{2}}$, whereas the Landsat scanner has a $\sqrt{V/H}$ of $0.08 \text{ s}^{-\frac{1}{2}}$. The M-7 scanner has $q = 1$ and Landsat has $q = 6$. Also, β^2 for Landsat is 0.0074 $(\text{mrad})^2$ and for the M-7 is 9 $(\text{mrad})^2$. By applying

Table 2-7 Landsat MSS performance parameters[22]

Parameter, units	Band	Nominal
Instantaneous field of view, mrad	1–4	0.086
	5	0.258
Mean photocathode sensitivity over spectral band, mA/W	1	34.5
	2	25.0
	3	12.0
PMT sensitivity enhancement factor	1–3	2.40
Electrical bandwidth, kHz	1–4	42.3
	5	14.1
Optical efficiency including obscuration	1–4	0.26
	5	0.34
Electron-multiplier noise factor	1–3	1.40
Preamplifier noise factor	4	1.35
	5	1.30
Noise-equivalent power, 10^{-14} W/Hz$^{1/2}$	4	11.0
	5	1.0
Radiance into sensor to produce full-scale output, mW/(cm^2 − sr)	1	2.48
	2	2.00
	3	1.76
	4	4.60
Entrance aperture, cm	1–5	22.82
Ratio of filter effective noise bandwidth to information bandwidth	1–5	1.05
f number	5	2.0
Blackbody radiance change per unit temperature change, 10^{-4} W/(cm^2 − K − sr − μm)	5	0.131

these facts to Eq. (2-49) and assuming the other parameters to be equal, one can see why the spectral channel bandwidth of the Landsat has to be wider than that of the M-7 to achieve comparable signal-to-noise performance.

3. Both scanners use reflecting optics and fiber optic bundles for conveyance of visible and near-infrared signals. Both scanners have calibration lamp references.
4. The M-7 scanner has resolution element image rotation whereas the Landsat scanner does not. The M-7 scanner suffers from greater scanner-induced image distortions (discussed in Sec. 2-6) than Landsat due to its wider total FOV.

The Skylab Multispectral Scanner[24]

The S-192 13-channel multispectral scanner on board the Skylab space station is shown pictorially and schematically in Figs. 2-81 and 2-82, respectively. The S-192 is an object-plane scanner with a 110° conical scan (see the reference for a

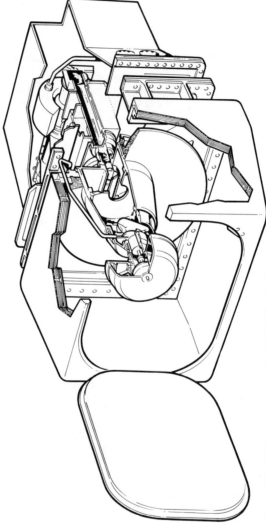

Figure 2-81 Multispectral scanner for S-192 (Skylab).[24]

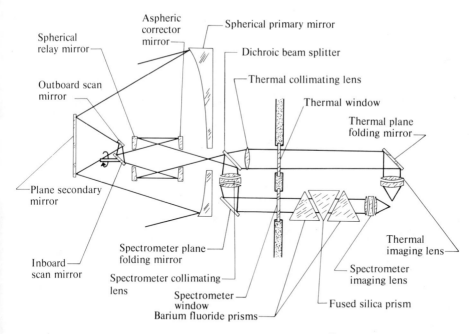

Figure 2-82 Multispectral scanner optical schematic for S-192 (Skylab).[24]

description of conical scanning). The ground swath traced out by the scanner is shown in Fig. 2-83. The scanner uses a mercury cadmium telluride array of detectors cooled by a refrigeration system to develop the 13 spectral bands listed in Table 2-8. The primary mirror has a diameter of 0.43 m, which is the aperture stop of the scanner; the field stops are the detector areas.

Table 2-8 S-192 Skylab MSS spectral bands[24]

Band number	Coverage, μm
1	0.41–0.46
2	0.46–0.51
3	0.52–0.56
4	0.56–0.61
5	0.62–0.67
6	0.68–0.76
7	0.78–0.88
8	0.98–1.08
9	1.09–1.19
10	1.20–1.30
11	1.55–1.75
12	2.10–2.35
13	10.20–12.50

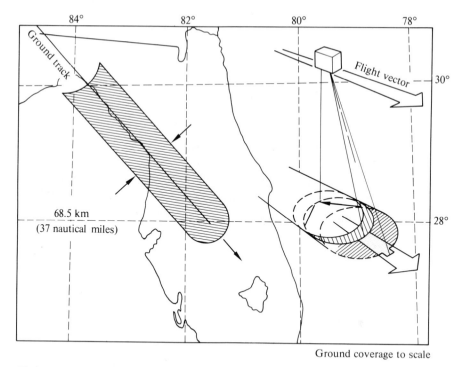

Ground coverage to scale

Figure 2-83 Ground coverage of the S-192 multispectral scanner.[24]

The ground resolution of the S-192 is approximately equal to that of Landsat. The altitude of Skylab is about 435 km; therefore, the IFOV is 0.18 mrad.

The scanning action of the S-192 is rotational with the forward 110° of rotation being active. The other portion of the scan encounters calibration lamps, thermal plates, and zero references. The rotation of the outboard scan mirror about the inboard scan mirror provides the scanning action. The rest of the imaging and dispersion system is conventional.

Comparing the Skylab S-192 multispectral scanner to the Landsat instrument yields the following:

1. The optical speed of S-192 is four times that of the Landsat scanner.
2. β is larger by a factor of 2.
3. V/H is larger by a factor of 2.5.
4. The spectral bands are about one-third as wide.
5. $q = 1$ for S-192; $q = 6$ for the Landsat scanner.
6. All other parameters are about equal.

In view of these comparisons, examination of Eq. (2-49) shows that the signal-to-noise ratios of the two scanners are approximately the same.

The Thematic Mapper Scanner[25]

Finally, a proposed multispectral scanner still in the design and construction stages is referred to as the thematic mapper (TM). The scanner is to be orbitted on board the Landsat D satellite which will be launched in 1980 or 1981. It is to be retrievable and repairable by means of the Space Shuttle.

The proposed spectral bands for the TM are:

$$0.45–0.52 \; \mu m$$
$$0.52–0.60 \; \mu m$$
$$0.63–0.69 \; \mu m$$
$$0.76–0.90 \; \mu m$$
$$1.55–1.75 \; \mu m$$
$$2.08–2.35 \; \mu m$$
$$10.40–12.50 \; \mu m$$

The visible and near-IR bands will use silicon detectors, the middle-IR bands indium antimonide, and the far-IR mercury-cadmium-telluride. The latter three detectors will be cooled to 100 K with radiation coolers. The projected signal-to-noise ratio of the TM (derived from noise-equivalent reflectance requirements†) is approximately 30 in the visible and near-IR bands and 13 in the middle-IR bands. The projected noise-equivalent temperature† of the far-IR band is 0.5 K (at 300 K).

The scanner aperture will be approximately 40 cm. The spatial resolution of the scanner is to be 30 m in the visible, near-IR, and middle-IR bands and 120 m in the far-IR band. A four-detector array will be used for the far-IR band; the other bands will use 16 detectors each.

The TM will be placed in a near-polar sun-synchronous orbit at an altitude of 705 km. The scan mirror will sweep a ground swath width of 185 km. The scan is bidirectional; i.e., data are gathered in both back-and-forth sweeps of the mirror (recall that the Landsat MSS scan is unidirectional; see Fig. 2-80). The orbit provides earth coverage in 16 days, compared to 18 days for Landsat 1, 2, and 3.

The radiometric data from the TM will be digitized to 8-bit precision (256 levels), compared to 6-bit precision for the Landsat MSS.

The TM system was originally intended primarily for vegetation mapping applications, which substantially influenced the choice of spectral bands.

The reader will have a chance to study the design of the TM in Prob. 2-26 on the following page.

---

† *The noise-equivalent reflectance* is that reflectance change in the scene required to produce a signal-to-noise ratio of 1.0 at the output of the scanner. *The noise-equivalent temperature* is that temperature change in the scene required to produce a signal-to-noise ratio of 1.0 at the scanner output.

PROBLEMS

2-22 Propose a data-flow diagram for a series of data-collection missions which are to be conducted in support of a large-scale global remote sensing experiment. Each mission is scheduled to be coincident with the passage of a satellite-borne multispectral scanner. The satellite data are augmented with data from a helicopter-mounted multispectral scanner and a truck-mounted spectroradiometer, local meteorological variables, visual ground observations, and photography.

Include in your design provision for converting analog electrical signals to digital form, keypunching of tabulated data, assignment of identification numbers, etc.

2-23 The M-7 scanner predated the Landsat scanner systems. Why, then, do you suppose that only four channels were used on the Landsat 1 and 2 scanners and that there was no channel in the thermal infrared region of the spectrum? It would seem that engineers could have improved upon or at least duplicated the M-7 specifications.

2-24 Table 2-7 indicates that the Landsat 3 thermal band has an instantaneous field of view of 0.258 mrad whereas the visible and near-infrared bands have a 0.086 mrad IFOV. Explain how this is accomplished if all of the incident energy passes through the same primary optical system.

2-25 When flown at its nominal altitude of 1500 m the ground resolution of the M-7 scanner is about 4.5 m. The Landsat 1 and 2 scanners have a ground resolution of approximately 80 m. Give some possible reasons that went into the 80-m Landsat resolution design decision.

2-26 Investigate the design of the thematic mapper (TM) scanner described in Sec. 2-8. Determine or estimate all of the parameters in Eq. (2-49) and use them to compare the performance of the Landsat MSS and the TM. Determine the angular deflection and the vibration frequency of the TM mirror.

2-27 Suppose an analog-to-digital system were to be installed in the aircraft carrying the ERIM M-7 scanner. The aircraft is operating at an altitude of 330 m (approximately 1000 ft). The analog-to-digital converter produces 10 bits (1024 levels).

 (*a*) Determine the required sampling rate of the analog-to-digital converter (remember there are 12 channels of data).

 (*b*) What is the bit rate of the converter?

REFERENCES

1. Air Research Development Command, U.S. Air Force: "Handbook of Geophysics," The Macmillan Company, New York, 1961.
2. Richmond, J. C., I. W. Ginsberg, T. Limperis, and F. W. Nicodemus: "Geometrical Considerations for Reflectance Nomenclature," National Bureau of Standards, Draft Copy, 1973.
3. Richtmyer, F. K., E. H. Kennard, and T. Lauritsen: "Introduction to Modern Physics," 5th ed., McGraw-Hill Book Company, New York, 1955.
4. Abramowitz, M., and I. A. Stegun (Eds.): "Handbook of Mathematical Functions," 4th ed., U.S. Department of Commerce, National Bureau of Standards, Washington, D.C., 1965.
5. Parsons, C. L., Jr., and G. M. Jurica: Correction of Earth Resources Technology Satellite Multispectral Scanner Data for the Effect of the Atmosphere, *LARS Information Note 061875*, Laboratory for Applications of Remote Sensing, Purdue University, West Lafayette, Ind., 1975.
6. Coulson, K. L., J. V. Dave, and Z. Sekera: "Tables Related to Radiation Emerging from a Planetary Atmosphere with Rayleigh Scattering," University of California Press, Berkeley, Calif., 1960.
7. Kerker, M.: "The Scattering of Light and Other Electromagnetic Radiation," Academic Press, New York, 1969.

8. DeWitt, D. P., and B. F. Robinson: Description and Evaluation of a Bidirectional Factor Reflectometer, *LARS Information Note 091576*, Laboratory for Applications of Remote Sensing, Purdue University, West Lafayette, Ind., 1976.
9. DeHoop, A. T.: A Reciprocity Theorem for the Electromagnetic Field Scattered by an Obstacle, *Appl. Sci. Res.*, sec. B, vol. 8, no. 2, pp. 135–140, 1960.
10. Grum, F., and G. W. Luckey: Optical Sphere Paint and a Working Standard of Reflectance, *Appl. Opt.*, vol. 7, no. 11, pp. 2289–2294, 1968.
11. King, R. A.: "Electrical Noise," Chapman and Hall, London, 1966.
12. Hammond, H. K., III, and H. L. Mason: "Precision Measurement and Calibration: Radiometry, Photometry," National Bureau of Standards, Special Publication 300, vol. 7, U.S. Government Printing Office, Washington, D.C., 1971.
13. Sparrow, E. M., and V. K. Jonsson: Radiant Emission Characteristics of Diffuse Conical Cavities, *J. Opt. Soc. Am.*, vol. 53, pp. 816–821, 1963.
14. Radiation Instruments and Measurements: "IGY Instruction Manual," part IV, Pergamon Press, New York, 1958, p. 436.
15. Laboratory for Agricultural Remote Sensing: "Remote Multispectral Sensing in Agriculture," vol. 3, Research Bulletin no. 844, Agricultural Experiment Station, Purdue University, West Lafayette, Ind., 1968.
16. Baker, J. R.: "Geometric Analysis and Restitution of Digital Multispectral Scanner Data Arrays," Ph.D. Dissertation, School of Civil Engineering, Purdue University, West Lafayette, Ind., 1975.
17. Born, M., and E. Wolf: "Principles of Optics," Pergamon Press, Oxford, England, 1964.
18. American Society of Photogrammetry: "Manual of Photogrammetry," 3d ed., Falls Church, Va., 1966.
19. Scientific and Technical Information Office: "Advance Scanners and Imaging Systems for Earth Observations," SP-335, NASA Publication, Washington, D.C., 1973.
20. Steckl, A. J., R. D. Nelson, B. T. French, et al.: Application of Charge-Coupled Devices to Infrared Detection and Imaging, *Proc. IEEE*, vol. 63, no. 1, pp. 67–74, 1975.
21. Cooper, G. R., and C. D. McGillem: "Methods of Signal and System Analysis," Holt, Rinehart and Winston, Inc., 1967.
22. National Aeronautics and Space Administration: "Earth Resources Technology Satellite: Data Users Handbook," Goddard Space Flight Center, Greenbelt, Md., 1972.
23. Hasel, P. G., Jr., et al.: "Michigan Experimental Multispectral Mapping System: A Description of the M7 Airborne Sensor and Its Performance," Environmental Research Institude of Michigan, Ann Arbor, Mich., 1974.
24. National Aeronautics and Space Administration: "EREP Investigators' Data Book," Science and Applications Directorate, Manned Spacecraft Center, Houston, Tex., 1972.
25. National Aeronautics and Space Administration: "Landsat-D Thematic Mapper," Technical Working Group Final Report JSC-09797, Johnson Spaceflight Center, Houston, Tex., 1975.

THREE

FUNDAMENTALS OF PATTERN RECOGNITION IN REMOTE SENSING
Philip H. Swain

The simple classification decision rule mentioned in Chap. 1 for assigning data to various classes of ground cover is an example of a data-analysis methodology called *pattern recognition*. The essence of this methodology is a two-step process in which:

1. The classes of interest are characterized through the analysis of data which are representative of the classes.
2. All remaining data are classified by means of numerical rules which utilize the class characterizations.

There are many reasons why pattern recognition provides an ideal approach for most applications of quantitative remote sensing: pattern recognition is basically a computer-oriented methodology, permitting rapid and repeatable analysis; it allows for statistical treatment of multivariate data; it is easily tailored to a wide range of problems; and it produces quantitative results. Pattern recognition is by no means the only quantitative analysis technique which could be applied to remote sensing data; statistical methods such as correlation analysis and regression are possible alternatives. Pattern recognition is most applicable when the goal is to categorize or classify each elementary observation into one of a limited number of discrete classes. Thus, classification of agricultural data into crop species is an example of an appropriate use of pattern recognition.

Many remote sensing applications involve classification problems of this nature and, as a result, pattern recognition has been widely adopted as the basis for remote sensing data-processing systems.

In this chapter we shall describe in relatively simple terms the fundamental principles and some of the basic mathematics of pattern recognition as it is commonly used in quantitative remote sensing. The concluding section of the chapter describes a typical remote sensing data-analysis procedure. Chapter 4 discusses many of the considerations essential for evaluating computer systems used to implement the data-analysis procedure.

Study objectives

After studying Secs. 3-1, 3-2, and 3-3, you should be able to:

1. State a definition of *pattern* which takes into account the means by which a machine can be designed to recognize patterns.
2. Sketch a block diagram or model of a pattern-recognition system and briefly describe, in a remote sensing context, the function of each block.
3. Given a graph showing a group of two-dimensional measurements from a set of discriminable patterns, sketch in the boundaries for a suitable set of decision regions.
4. Given a vector X and a set of discriminant functions $g_i(X)$, decide to which class X belongs.
5. Define *training pattern*, describing the role training patterns play in classifier design.

3-1 WHAT IS A PATTERN?

A *pattern* is most commonly thought of as something having spatial or geometrical character—two or three dimensional. Familiar examples include letters and numbers on a printed page or tanks and artillery pieces in an aerial photograph. However, when we take into account the means by which a machine can be made to recognize patterns, a more abstract concept of a pattern is possible. In particular, a pattern-recognition system makes a series of measurements on the pattern to be classified and compares these measurements against a set of "typical" measurements in a "pattern dictionary." A match or the closest match to an entry in the dictionary produces the desired classification. The most crucial aspects of this process are the pattern dictionary and the method by which the measurements from the pattern to be classified are compared to the entries in the dictionary.† As a matter of fact, once the measurements are made, it does not

† Although the idea of using a "pattern dictionary" is a convenient way to picture the workings of a pattern-recognition system, the reader is cautioned against trying to carry the analogy too far in the remote sensing context. For example, due to the randomness of nature, there would be too many "alternative spellings" (similar but distinct patterns) for any given class of ground cover to be able to compile an exhaustive list of them.

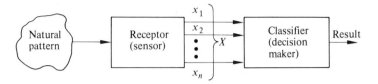

Figure 3-1 A model for a pattern-recognition system.

matter what physical form the pattern has; as far as the machine is concerned, the set of measurements *is* the pattern. In general, then, it is possible to think of a pattern simply as being any well-defined set of measurements, whether it is geometrical in character or not.

Figure 3-1 shows a simple model for a pattern-recognition system. For our purposes, the *receptor* may be an airborne or satellite-borne multispectral scanner as described in Chap. 2. The output of the receptor is a set of n measurements, each corresponding to one channel of the scanner. Of course, the n measurements for one ground resolution element are assumed to be made simultaneously. As suggested in Chap. 1, it is convenient to think of the n measurements as defining a point in n-dimensional space, which we refer to as the *measurement space*. Any point in the measurement space can be represented by the n-component *measurement vector* X,

$$
X = \begin{bmatrix} x_1 \\ x_2 \\ \vdots \\ x_n \end{bmatrix}
$$

where x_i corresponds to the ith measurement (the measurement from the ith wavelength band or scanner channel) on a given ground resolution element.†

The decision maker or *classifier* in Fig. 3-1 assigns the measurement vector to one of a set of prespecified classes according to an appropriate *classification rule*. One of the goals of this chapter is to describe a classification rule commonly used in remote sensing applications of pattern recognition.

3-2 A GEOMETRICAL INTERPRETATION OF PATTERN RECOGNITION; DISCRIMINANT FUNCTIONS

As we have previously noted, the vector of n measurements produced by the multispectral scanner for each ground resolution element can be plotted in n-dimensional space. Some examples are shown in Fig. 3-2 for the case $n = 2$.

---

† Throughout this chapter, capital letters will denote vectors while lower case letters will denote single measurements. Often these single measurements will be components of a vector.

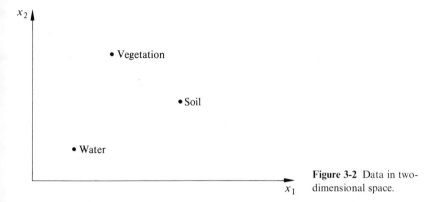

Figure 3-2 Data in two-dimensional space.

(Many of the concepts introduced in this chapter are illustrated with two-dimensional examples. All of the concepts can be generalized to measurement spaces of arbitrary dimensionality.)

When several such measurement vectors are plotted, all nominally of the same ground-cover type but from different locations on the ground, it is very likely that they will plot as a localized cluster or cloud of points (Fig. 3-3) rather than as a single point. This phenomenon, which is characteristic of remote sensing data, results from the inherent randomness of nature, e.g., random variations in leaf orientation, in atmospheric conditions, and even "noise" in the remote sensing equipment itself. Still, the clusters corresponding to specific ground covers are usually observed to be more or less distinct, and it is possible in such cases to associate localized *regions* of the measurement space with specific ground covers. These ground covers are said to be *discriminable*; all others are *indiscriminable*.

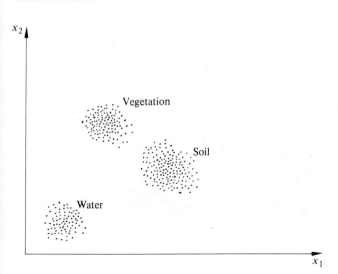

Figure 3-3 Clusters of data points.

The job of designing the pattern classifier consists of first dividing the measurement space appropriately into *decision regions*, each region corresponding to a specific discriminable class (Fig. 3-4), and then constructing the classifier so that it will identify any measurement vector as belonging to the class corresponding to the decision region in which it falls. The first part of this task is often much more difficult than might be suggested by the relatively simple situation shown in Fig. 3-4, since in many cases there is some overlap of the patterns belonging to the classes of interest. How we can deal effectively with such situations is the major thrust of this chapter. But the process of actually constructing the classifier is conceptually rather simple, and we shall take this opportunity to outline it briefly.

Assume there are m classes and that the decision regions corresponding to these classes have been determined. Suppose we can find a set of m functions of X, which we call *discriminant functions* and which we denote by $g_1(X)$, $g_2(X), \ldots, g_m(X)$, having the property that $g_i(X)$ has a larger value than any other of these functions whenever X is a point in the ith decision region (we will show later how such functions can be found). Then if we wish to classify *any* point X_u, that is, determine which decision region X_u belongs to, all we need to do is to calculate the values $g_1(X_u)$, $g_2(X_u), \ldots, g_m(X_u)$. The point X_u belongs to the class having the largest g value.

Once found, the discriminant functions can be used to construct a classifier as shown schematically in Fig. 3-5. Formally, we have the following:

AG

Classification rule Let ω_i denote the ith class. Decide $X \in \omega_i$ if and only if $g_i(X) \geq g_j(X)$ for all $j = 1, 2, \ldots, m$.

(*Read*: decide X belongs to class i if and only if $g_i(X)$ is greater than or equal to $g_j(X)$ for all classes $j = 1, 2, \ldots, m$.) Actually, two discriminant functions can

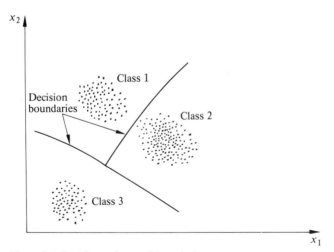

Figure 3-4 Decision regions and boundaries.

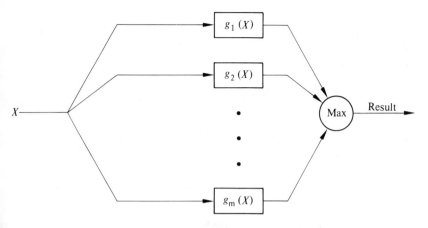

Figure 3-5 A pattern classifier defined in terms of discriminant functions.

have equal values only at points on the boundaries between the decision regions. When this occurs, a rule for tie-breaking must be defined.

As a simple example, consider the situation shown in Fig. 3-6. Assume we are given that the region below the line $x_1 + x_2 - 2 = 0$ corresponds to class 1 and the region above the line corresponds to class 2. If we define $g_1(X)$ and $g_2(X)$ according to Eq. (3-1),

$$g_1(X) = x_1 + x_2 + 2$$
$$g_2(X) = 2x_1 + 2x_2$$

(3-1)

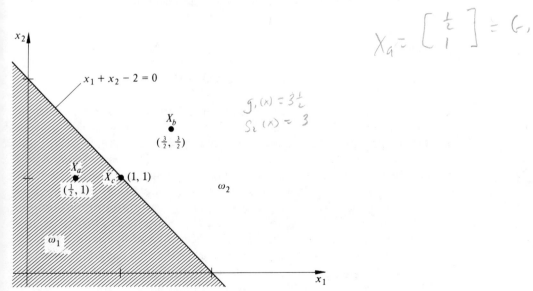

Figure 3-6 A simple example illustrating discriminant functions.

we can show that these functions will serve as discriminant functions, since $g_1(X)$ is larger than $g_2(X)$ everywhere below the line, whereas $g_2(X)$ is larger than $g_1(X)$ everywhere above the line. If, for instance, we wish to classify the point

$$X_a = \begin{bmatrix} \frac{1}{2} \\ 1 \end{bmatrix}$$

i.e., the point for which $x_1 = \frac{1}{2}$ and $x_2 = 1$, straightforward calculation yields $g_1(X_a) = 3\frac{1}{2}$ and $g_2(X_a) = 3$. Thus, $g_1(X_a)$ is greater than $g_2(X_a)$, indicating that this point is in the region corresponding to class 1—which is indeed the case (Fig. 3-6). Similarly, the point

$$X_b = \begin{bmatrix} \frac{3}{2} \\ \frac{3}{2} \end{bmatrix}$$

yields $g_1(X_b) = 5$ and $g_2(X_b) = 6$, indicating that X_b is from class 2. The point

$$X_c = \begin{bmatrix} 1 \\ 1 \end{bmatrix}$$

is on the boundary, and we note that $g_1(X_c) = g_2(X_c) = 4$. We must arbitrarily specify that the boundary points [all points for which $g_1(X) = g_2(X)$] shall belong to either class 1 or class 2. Usually this assignment can be made arbitrarily because the frequency with which data points will actually fall on the boundary is exceedingly small.

Taking this approach, the classifier design problem becomes the problem of selecting an optimal set of discriminant functions.

3-3 "TRAINING" A PATTERN CLASSIFIER

Although it is sometimes possible to determine discriminant functions solely on the basis of either theoretical considerations, knowledge about the physical problem, or perhaps even intuition, most commonly the discriminant functions are derived using information distilled from a set of *training patterns, i.e.,* measurement vectors of known identity which are assumed to be representative of the classes of interest. In this case, the process of classifier design is called *training* the classifier.

The term "training" arose from the fact that many of the earliest pattern-recognition systems were "trainable"; i.e., they "learned" the discriminant functions by reacting adaptively (adjusting themselves) when shown a pattern together with its true identity. This is illustrated by the following. Suppose that for the simple example of the previous section involving two classes in two-dimensional measurement space, the discriminant functions are unknown but are assumed to be linear in form, that is,

$$g_1(X) = a_{11}x_1 + a_{12}x_2 + b_1$$

$$g_2(X) = a_{21}x_1 + a_{22}x_2 + b_2$$

(3-2)

where the a and b values are unknown. Assume further that it is not known where the linear boundary separating the classes is actually located, but that the class membership is known for several points on each side of the boundary (these points are the "training patterns"). It can be shown (Nilsson,[1] Chap. 5) that the following procedure will "converge to a solution," i.e., produce a set of a and b values such that the resulting discriminant functions can be used to correctly classify all of the training patterns.

Procedure Initially select a and b values arbitrarily. For example, let

$$a_{11} = a_{12} = b_1 = 1$$

$$a_{21} = a_{22} = b_2 = -1$$

Then take the first training pattern (say it is from ω_1, that is, from class 1) and calculate $g_1(X)$ and $g_2(X)$. If $g_1(X) > g_2(X)$, the decision is correct; go on to the next training sample. If $g_1(X) < g_2(X)$, a wrong decision would be made. In this case alter the coefficients so as to increase the discriminant function associated with the correct class and decrease the discriminant function associated with the incorrect class: if X is from ω_1 but ω_2 was decided, let

$$a'_{11} = a_{11} + \alpha x_1 \qquad a'_{21} = a_{21} - \alpha x_1$$

$$a'_{12} = a_{12} + \alpha x_2 \qquad a'_{22} = a_{22} - \alpha x_2$$

$$b'_1 = b_1 + \alpha \qquad b'_2 = b_2 - \alpha$$

(3-3)

where α is a convenient positive constant. If X is from ω_2 but ω_1 was decided, change the signs in Eq. (3-3) so as to increase g_2 and decrease g_1. Then go on to the next training pattern. Cycle through the training patterns until all are correctly classified.

This process will always stop, with all training patterns correctly classified, after a finite number of cycles. The process is easily implemented on a computer or with special-purpose hardware.

This training procedure is simple and automatic, but convergence to a solution is guaranteed only when the training patterns are separable by a linear boundary (a straight line in the two-dimensional example). The procedure can be generalized to allow for boundaries of somewhat greater complexity, but when the pattern classes actually overlap in the measurement space, i.e., when different pattern classes can produce identical measurement vectors, this approach is generally not suitable. Since in many remote sensing applications the spectral responses of the classes of interest do in fact overlap, this simple classifier training method is inappropriate. The statistical techniques described subsequently will provide an alternative method for deriving discriminant functions from training patterns in such complex situations.

PROBLEMS

3-1 Give a definition of a *pattern* which makes sense in the context of a pattern-recognition machine.

3-2 In earth resources applications of remote sensing, the components of the measurement vector are most commonly values of relative reflectance in various wavelength bands. List four or five other types of remote sensing measurements which might be useful as components of the measurement vector.

3-3 Complete the block diagram of the pattern-recognition system started in Fig. P3-3 and briefly describe in a remote sensing context the function of each block.

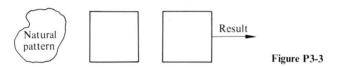

Figure P3-3

3-4 Measurements from four types of ground cover are plotted in two-dimensional space in Fig. P3-4. Based on these data, sketch in a set of decision boundaries which may be used to classify each measurement into one of the four classes.

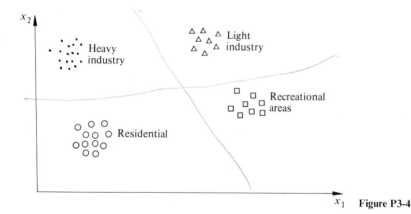

x_1 **Figure P3-4**

3-5 To separate the four classes in Prob. 3-4, you could have drawn straight-line boundaries between the decision regions. What is an advantage of using straight lines rather than more complex boundaries? What is a potential advantage of more complex boundaries?

3-6 A three-dimensional measurement,

$$X = \begin{bmatrix} 2 \\ 3 \\ 1 \end{bmatrix}$$

is to be classified by means of the following set of discriminant functions:

$$g_1(X) = x_1 + 0.5x_2 + 2x_3 + 1$$

$$g_2(X) = 2x_1 + x_2 + 0.5x_3$$

$$g_3(X) = 3x_1 + 2x_2 - 4x_3 - 3$$

To which class does X belong?

3-7 What is a *training pattern*?

3-8 Why would it generally be desirable to have several training patterns for each of the classes to be identified?

Study objectives

After studying Secs. 3-4, 3-5, and 3-6, you should be able to:

1. List three reasons why a statistical approach is appropriate in applying pattern recognition for remote sensing data analysis.
2. Distinguish between *parametric* and *nonparametric* statistical pattern-recognition techniques and give at least one advantage and disadvantage of each.
3. Sketch a one-dimensional (univariate) normal density function and give the two parameters which are necessary to specify the density function. Describe how the generalization is made to the multivariate case.
4. Cite two precautions which must be observed if the probability functions associated with the pattern classes are to be approximated by normal density functions.
5. Discuss the terms used in the maximum-likelihood decision rule, relating the statistical expressions to physical phenomena which can actually be observed in a remote sensing context.
6. Explain how the pattern-recognition system can be made to detect many of the points in the area to be classified which actually are not represented by any of the training classes.

3-4 THE STATISTICAL APPROACH: RATIONALE AND AN EXAMPLE

Statistical pattern-recognition methods are particularly appropriate for remote sensing applications because:

1. Due to the inherent randomness of nature, remote sensing data exhibit many incidental variations which tend to obscure the characteristic differences among the classes of interest. Statistical analysis helps to account for these variations and to reduce their potentially adverse effects on classification accuracy.
2. In practice there is often uncertainty, however small, concerning the true identity of the training patterns used to determine the discriminant functions. For example, some of the training patterns from a "corn field" might be more

appropriately identified as "weeds" or "bare soil." Statistical methods are tolerant of such errors as long as their frequency of occurrence is relatively low.

3. The pattern classes of interest may actually overlap in the measurement space, i.e., some of the measurements from one class may be indistinguishable from some of the measurements from other classes. For example, at certain times during the growing season, *some* corn fields are spectrally indistinguishable from soybean fields. In such situations, statistical pattern-recognition methods allow for classifications which are "most often" or "most probably" correct.

Statistical pattern-recognition techniques generally make use of probability functions associated with the pattern classes. However, the probability functions are usually unknown and must be estimated from a set of training patterns. In some cases the form of the probability functions is assumed to be known (e.g., normal, Rayleigh) and only certain parameters associated with the functions need to be estimated from the training patterns (e.g., means, variances). The method is then referred to as *parametric*. If the form of the probability functions is not assumed to be known in advance, the method is *nonparametric*. Parametric methods are usually easier to implement but require more prior knowledge or basic assumptions concerning the nature of the patterns. Nonparametric methods are potentially more powerful in terms of their ability to accurately estimate probability functions, but this advantage is usually very expensive to achieve, requiring complicated recognition systems and very large numbers of training patterns.

A simple example will illustrate many of the concepts discussed so far in this chapter, including the use of probability functions in the classification process. This example will be referred to several times later.

> **Example: Funny dice** Two players are engaged in a game involving two pairs of dice, one of which is a "standard" pair (with one to six spots on each face), the other an "augmented" pair with two extra spots on each face (three to eight spots per face). Player 1 selects a pair of dice at random and rolls the dice, concealing them from player 2. Player 1 announces (honestly) the *sum* resulting from the roll. Player 2 then guesses which pair was rolled, betting a dollar that he or she can guess correctly.
>
> What decision rule should player 2 follow to maximize winnings (or minimize losses)?

In this example, the sum produced by each roll of the dice constitutes a measurement. Player 1 acts as the sensor, reading the dice. The measurement space is one dimensional [Fig. 3-7(a)].

Assuming that player 2 is somewhat of a statistician and is fully aware of the nature of the dice (both sets), he or she has enough information to know the probability functions associated with the sums produced by the dice. Figure 3-7(b) shows these probability functions. In this figure, the notation $p(x \mid \text{standard dice})$

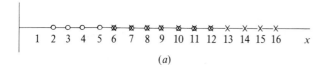

o   A possible outcome for the
    standard pair

X   A possible outcome for the
    augmented pair

⊠   A possible outcome for
    either pair

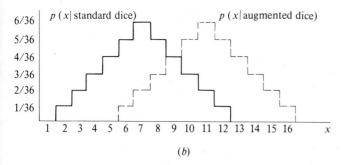

(a)

(b)

Figure 3-7 Funny dice example. (a) Measurement space: $x =$ the sum rolled. (b) Probability functions for the possible sums of two pairs of dice.

means "the probability of occurrence of sum x given that the standard dice were rolled;" and similarly for $p(x \mid$ augmented dice). The height of the histograms gives the relative frequency with which each sum may be expected to occur for each pair of dice. For instance, when the standard dice are rolled, the sum 5 occurs about 4/36 of the time, but the sum 14 can never occur. On the other hand, when the augmented dice are rolled, the sum 14 occurs about 3/36 of the time. If player 2 did not know these probability functions, they could be estimated from "training patterns," provided player 1 would allow player 2 to roll the dice many times and tabulate the number of times each sum appeared for each pair.

To get an intuitive feeling for how player 2 can use the probability functions to improve the chances of winning on each roll, consider Fig. 3-7(b) and decide for yourself which pair was probably rolled if the sum is 4; if the sum is 15; if the sum is 7; if the sum is 9. In each case, how certain do you feel of your decision?

When the sum is 4, player 2 can state with certainty that the standard pair was rolled. A 15 indicates, again with absolute certainty, that the augmented pair was rolled. Player 2 should guess "standard" if 7 is the sum and can expect to win about 75 percent of the time. When the sum is 9, player 2 can guess either "standard" or "augmented" and expect to be right about 50 percent of the time.

In other words, for a given sum, if the value of the probability function for the standard pair is greater than the value of the probability function for the augmented pair, guessing "standard" is more likely to be correct; conversely, if the value of the probability function for the augmented pair is greater, guessing "augmented" is more likely to be correct. This seems intuitively reasonable. The statistical soundness of this strategy will be shown in Sec. 3-6.

3-5 STATISTICAL CHARACTERIZATION OF REMOTE SENSING DATA

It was pointed out in the preceding example that the probability function associated with each set of dice could have been estimated simply by rolling a given set many times and tabulating the relative frequency with which each sum occurred. Suppose now that the data source were a single channel of a multispectral scanner. Instead of guessing which of two sets of dice was rolled, the objective might be to "guess" which of a set of ground-cover classes was being observed. Again the associated probability functions would be useful, and these could be estimated from training patterns. Given a set of measurements for a particular class, a tabulation would be made of the frequency with which each data value occurred for that class. The results could be displayed in the form of a histogram as shown in Fig. 3-8(a). Similar histograms would be produced to estimate the probability functions for each class, and the histograms would then be used as in

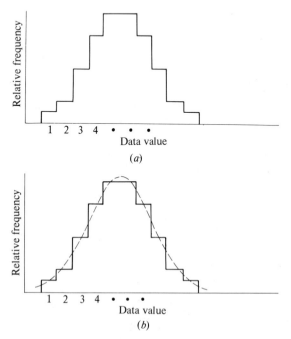

Figure 3-8 A one-dimensional data distribution and its parametric approximation. (a) Data distribution (histogram) for a single channel (b) Data distribution approximated by a normal probability density function.

the "funny dice" example to assist in classifying data of unknown identity based on spectral measurement values (although we have not yet specified in detail how this is actually accomplished).

If the number of possible data values is large, storage in the computer of the histogram representation of the class probability functions may require a considerable amount of memory. Furthermore, if we try to generalize this approach to handle additional wavelength bands, the memory requirements will rapidly get out of hand: the number of memory locations needed to save an n-dimensional histogram in which each dimension can take on p values is p^n. One way to alleviate this problem is to assume that each histogram or probability function can be adequately approximated by a smooth curve having a simple functional form. In particular, we will assume that the probability function for any class of interest can be approximated by a normal (or Gaussian) probability density function [Fig. 3-8(b)], an assumption for which we shall later provide justification. For the one-dimensional or *univariate* case, the normal density function for class i is given by

$$p(x|\omega_i) = \frac{1}{(2\pi)^{1/2}\sigma_i} \exp\left[-\frac{1}{2}\frac{(x - \mu_i)^2}{\sigma_i^2} \right] \qquad (3\text{-}4)$$

where $\exp[\ ] = e$ (the base of the natural logarithms) raised to the indicated power

$\mu_i = E[x|\omega_i]$ is the mean or average value of the measurements in class i

$\sigma_i^2 = E[(x - \mu_i)^2|\omega_i]$ is the variance of the measurements in class i

In practice μ_i and σ_i^2 are unknown and must be estimated from training samples. From statistical theory, unbiased estimators for μ_i and σ_i^2 are given by[2]

$$\hat{\mu}_i = \frac{1}{q_i} \sum_{j=1}^{q_i} x_j \qquad (3\text{-}5)$$

$$\hat{\sigma}_i^2 = \frac{1}{q_i - 1} \sum_{j=1}^{q_i} (x_j - \hat{\mu}_i)^2 \qquad (3\text{-}6)$$

where q_i is the number of training patterns available for class i and x_j is the jth training pattern for class i. Then the *estimated* probability function for class i is

$$\hat{p}(x|\omega_i) = \frac{1}{(2\pi)^{1/2}\hat{\sigma}_i} \exp\left[-\frac{1}{2}\frac{(x - \hat{\mu}_i)^2}{\hat{\sigma}_i^2} \right] \qquad (3\text{-}7)$$

Having made this parametric assumption that the probability function for each class may be approximated by a normal density function, we need only store the mean and variance for each class in the computer rather than the entire histogram. When we need the value of the probability function associated with a data value, we can compute it using Eq. (3-7).

But what if there are *two* channels of multispectral data? In this case, the *bivariate* probability function for each class could be estimated by tabulating the frequencies of occurrence of all possible *pairs* of data values, each pair consisting of a value x_1 from channel 1 and a value x_2 from channel 2. The result would be a two-dimensional generalization of Fig. 3-8(a). But as in the one-dimensional case, considerations based largely on computer-storage efficiency lead us again to make the parametric assumption that the probability function can be approximated by a normal probability density function. The two-dimensional or bivariate normal density function is given by

$$p(x_1, x_2 \mid \omega_i) = \frac{1}{2\pi(\sigma_{i11}\sigma_{i22} - \sigma_{i12}^2)^{1/2}}$$

$$\times \exp \left[-\frac{1}{2} \frac{\dfrac{(x_1 - \mu_{i1})^2}{\sigma_{i11}} - \dfrac{2\sigma_{i12}(x_1 - \mu_{i1})(x_2 - \mu_{i2})}{\sigma_{i11}\sigma_{i22}} + \dfrac{(x_2 - \mu_{i2})^2}{\sigma_{i22}}}{1 - \dfrac{\sigma_{i12}^2}{\sigma_{i11}\sigma_{i22}}} \right] \quad (3\text{-}8)$$

where $\mu_{ij} = E[x_j \mid \omega_i]$ is the mean value of the data in channel j (for class i)
$\sigma_{ijk} = E[(x_j - \mu_{ij})(x_k - \mu_{ik}) \mid \omega_i]$ is the covariance between channels j and k (for class i)

If the parameters μ_{ij} and σ_{ijk} are stored in the computer for each class (a total of five parameters for each class), the probability functions for the data can be computed as needed from Eq. (3-8). Conceptually, we can continue to generalize this approach to multispectral data involving still more channels. But given the complexity of Eq. (3-8), imagine what the formulas look like when we add a third channel, a fourth, and so on! Fortunately, the use of vector/matrix notation[3] provides for a very compact means of expressing formulas such as Eq. (3-8). For the general case of n-dimensional data, if we let

$$X = \begin{bmatrix} x_1 \\ x_2 \\ \vdots \\ x_n \end{bmatrix} \quad U_i = \begin{bmatrix} \mu_{i1} \\ \mu_{i2} \\ \vdots \\ \mu_{in} \end{bmatrix} \quad \Sigma_i = \begin{bmatrix} \sigma_{i11} & \sigma_{i12} & \cdots & \sigma_{i1n} \\ \sigma_{i21} & \sigma_{i22} & \cdots & \sigma_{i2n} \\ \vdots & \vdots & & \vdots \\ \sigma_{in1} & \sigma_{in2} & \cdots & \sigma_{inn} \end{bmatrix} \quad (3\text{-}9)$$

be the data vector, mean vector for class i, and covariance matrix for class i, respectively, then the n-dimensional multivariate normal density function can be written as

$$p(X \mid \omega_i) = \frac{1}{(2\pi)^{n/2} |\Sigma_i|^{1/2}} \exp \left[-\tfrac{1}{2}(X - U_i)^T \Sigma_i^{-1} (X - U_i) \right] \quad (3\text{-}10)$$

where $|\Sigma_i|$ is the determinant of the covariance matrix Σ_i, Σ_i^{-1} is the inverse of Σ_i, and $(X - U_i)^T$ is the transpose of the vector $(X - U_i)$. When $n = 2$, it is straightforward to show that carrying out the operations implied by Eq. (3-10) results in the expression given by Eq. (3-8).

For notational convenience later in this chapter, we will indicate the fact that $p(X \mid \omega_i)$ is the normal density function with mean vector U_i and covariance matrix Σ_i by writing

$$p(X \mid \omega_i) \sim N(U_i, \Sigma_i) \tag{3-11}$$

In practice, the mean vector and covariance matrix for each class are unknown and must be estimated from training patterns. Let $\hat{U}_i$ and $\hat{\Sigma}_i$ be unbiased estimates of U_i and Σ_i. Then $\hat{U}_i$ and $\hat{\Sigma}_i$ are given by

$$\hat{\mu}_{ij} = \frac{1}{q_i} \sum_{l=1}^{q_i} x_{jl} \qquad j = 1, 2, \ldots, n \tag{3-12}$$

$$\hat{\sigma}_{ijk} = \frac{1}{q_i - 1} \sum_{l=1}^{q_i} (x_{jl} - \hat{\mu}_{ij})(x_{kl} - \hat{\mu}_{ik}) \qquad j = 1, 2, \ldots, n; k = 1, 2, \ldots, n \tag{3-13}$$

where q_i is the number of training patterns in class i.

Summarizing, our goal is to characterize each pattern class in terms of its associated probability function, which is to be estimated from training patterns. Assuming that the n-dimensional histogram of the frequency distribution for each class can be approximated by the multivariate normal density function, it is possible to characterize the pattern classes in terms of their mean vectors and covariance matrices. This is an extremely compact and convenient representation as compared to the histogram representation. (See Prob. 3-13 at the end of Sec. 3-6.)

We have chosen repeatedly to use the normal density function to approximate the frequency distribution associated with each of the classes. Is this a reasonable thing to do in practice? To begin with, the "multivariate normal assumption" has been observed to model rather adequately the probabilistic processes observed in a large number of remote sensing applications. Also, classifiers designed on this basis are found to be "robust" in the sense that classification accuracy is not very sensitive even to moderately severe violations of this assumption. Finally, from a practical standpoint, experiments with classifiers of both greater and lesser complexity have shown that the normal assumption usually provides a good trade-off between classification performance (accuracy) and cost (speed and complexity of the classifier).

However, two points of caution must be observed in using this assumption. First, adequate training samples must be available to estimate the mean vector and covariance matrix for each class. If n wavelength bands are to be used for classification, the theoretical minimum number of training patterns per class is $n + 1$. For any fewer, the covariance matrix will be singular (its determinant will be 0 and its inverse cannot be calculated), making the computations in Eq. (3-10) impossible. In practice, however, the number of training patterns needs to be at least $10n$ or, even better, $100n$ in order to provide accurate estimates of the class parameters.

The second note of caution has to do with cases in which the normal assumption is clearly violated. In particular, classes which have distinctly multimodal probability functions (functions with more than a single maximum point) generally cannot be adequately approximated by the normal density function,

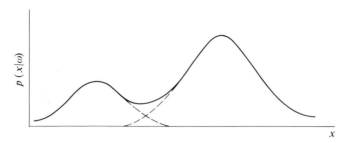

Figure 3-9 A multimodal probability function and its subdivision.

which is a unimodal (single maximum) function. A commonly practiced solution to this problem is to subdivide the offending class into a number of subclasses, one for each mode of the actual distribution, such that the probability function for each subclass can be represented by a normal density function (Fig. 3-9). A method for detecting situations involving multimodal classes and accomplishing the class subdivision will be discussed in a later section on clustering.

3-6 DISCRIMINANT FUNCTIONS BASED ON STATISTICAL THEORY

In Sec. 3-2 it was pointed out that the job of classifier design amounts to the specification of a set of discriminant functions which divide the measurement space into appropriate decision regions. Recall that the discriminant functions are defined in such a way that the discriminant function corresponding to the ith pattern class has a value larger than any other in the set at every point in that part of the feature space corresponding to the ith pattern class. Now that some essential statistical concepts have been introduced, we will show how statistical decision theory can be used to derive an optimal set of discriminant functions.

First we shall state the result we are after and discuss it briefly. Assume there are m pattern classes. Let $p(X \mid \omega_i)$ be the probability density function associated with the measurement vector X, given that X is from a pattern in class i. Let $p(\omega_i)$ be the a priori probability of class i; that is, the probability of observing a pattern from class i, independent of any other information. The following is a commonly used classification strategy:

The maximum-likelihood decision rule Decide $X \in \omega_i$ if and only if

$$p(X \mid \omega_i)p(\omega_i) \geq p(X \mid \omega_j)p(\omega_j)$$

for all $j = 1, 2, \ldots, m$.

That is, to classify a pattern X, a machine using the maximum-likelihood decision rule computes the product $p(X \mid \omega_i)p(\omega_i)$ for each class and assigns the pattern to the class having the largest product. Notice that the set of products

$p(X \mid \omega_i)p(\omega_i)$, $i = 1, 2, \ldots, m$, used in this rule constitutes a set of discriminant functions.

In the funny dice game, player 2 could use the maximum-likelihood decision rule to maximize his winnings. The quantity $p(X \mid \omega_i)$ corresponds to the probability function for a pair of dice, and $p(\omega_i)$ is the probability that player 1 will choose to roll the ith pair of dice (0.5 for each pair unless player 1 has a preference for one pair over the other).

As another example, consider a remote sensing problem in which it is desired to obtain a point-by-point classification of an area known to consist entirely of vegetation, soil, and water. Historical survey data show that the area is 60 percent vegetation, 20 percent soil, 20 percent water; that is,

$$p(\omega_1) = p(\text{vegetation}) = 0.60$$

$$p(\omega_2) = p(\text{soil}) = 0.20$$

$$p(\omega_3) = p(\text{water}) = 0.20$$

The probability functions $p(X \mid \text{vegetation})$, $p(X \mid \text{soil})$, and $p(X \mid \text{water})$ can be estimated by assuming each class to have a multivariate normal probability density and computing the mean vectors and covariance matrices from training data.

In an agricultural setting, the a priori probabilities $p(\omega_i)$ might be estimated from previous years' acreage records, seed sales records, or other Department of Agriculture information. The probability functions $p(X \mid \omega_i)$ associated with the various crops generally must be estimated from training samples.

And now to derive the discriminant functions used in the maximum-likelihood decision rule! The purpose of the following material is to lead the reader through the reasoning behind the statistical decision theory in as intuitive a manner as possible. Still, the discussion is necessarily rather mathematical. The less mathematically inclined reader may wish to skim this material or simply skip to the paragraph immediately following Eq. (3-22).

First we define a set of *loss functions* for m classes:

$$\lambda(i \mid j) \qquad i = 1, 2, \ldots, m; \ j = 1, 2, \ldots, m \tag{3-14}$$

where $\lambda(i \mid j)$ is the loss or cost incurred when a pattern is classified into class i but is actually from class j. A particular set of loss functions will be provided shortly.

The basic strategy we shall follow is to *minimize the average loss* over the entire set of classifications to be performed. Such a classification strategy is said to be *Bayes optimal*. For a given pattern X, the average loss resulting from deciding that X is from class i is given by

$$L_X(i) = \sum_{j=1}^{m} \lambda(i \mid j)p(\omega_j \mid X) \tag{3-15}$$

where $p(\omega_j \mid X)$ is the conditional probability that a pattern X is from class j.

A formula defining the relationship between joint probabilities and conditional probabilities is

$$p(X, \omega_j) = p(X \mid \omega_j)p(\omega_j) = p(\omega_j \mid X)p(X)$$

or, solving this expression for $p(\omega_j \mid X)$, we have

$$p(\omega_j \mid X) = p(X \mid \omega_j)p(\omega_j)/p(X) \tag{3-16}$$

Substituting Eq. (3-16) into Eq. (3-15) yields

$$L_X(i) = \sum_{j=1}^{m} \lambda(i \mid j)p(X \mid \omega_j)p(\omega_j)/p(X) \tag{3-17}$$

Now, Eq. (3-17) leads directly to a set of discriminant functions if the following three rules are applied:

1. *Minimizing* a set of functions is the same operation as *maximizing the negative* of the same set of functions.
2. A reasonable set of loss functions is

$$\begin{aligned} \lambda(i \mid j) &= 0 \qquad i = j \\ \lambda(i \mid j) &= 1 \qquad i \ne j \end{aligned} \tag{3-18}$$

   i.e., no cost for a correct classification, unit cost for an error.
3. If $\{g_i(X), \ i = 1, 2, \ldots, m\}$ is a set of discriminant functions, then application of any monotonic function to this set yields an equivalent set of discriminant functions $\{g_i'(X), \ i = 1, 2, \ldots, m\}$; that is, use of either set produces identical classification results.

Some examples of rule 3 which we shall make use of include the following:

$$g_i'(X) = g_i(X) \pm k \qquad \text{where } k \text{ is a constant} \tag{3-19a}$$

$$g_i'(X) = g_i(X) \cdot k \qquad \text{where } k \text{ is a constant} \tag{3-19b}$$

$$g_i'(X) = \log\left[g_i(X)\right] \tag{3-19c}$$

The Bayes optimal strategy requires making classification decisions so as to minimize Eq. (3-17). Applying rule 1 above, an equivalent strategy would be to maximize the negative of Eq. (3-17). So let

$$g_i(X) = -L_X(i) \qquad i = 1, 2, \ldots, m \tag{3-20}$$

and let the classification strategy be to classify X as belonging to that class i for which $g_i(X) = -L_X(i)$ is maximum. Referring to the definition of discriminant functions and the related general rule for classification, both in Sec. 3-2, it is apparent that the set of functions given by Eq. (3-20) qualifies as a set of discriminant functions.

But matters can be further simplified by using rules 2 and 3. First, applying the "zero-one loss function" [Eq. (3-18)] gives

$$g_i(X) = -\sum_{j=1}^{m} \lambda(i|j)p(X|\omega_j)\frac{p(\omega_j)}{p(X)}$$

$$= -\sum_{\substack{j=1 \\ j\neq i}}^{m} p(X|\omega_j)\frac{p(\omega_j)}{p(X)}$$

$$= \frac{-1}{p(X)}\sum_{\substack{j=1 \\ j\neq i}}^{m} p(X|\omega_j)p(\omega_j)$$

For any *given* X, the quantity $p(X)$ is a constant, so an equivalent set of discriminant functions is

$$g_i'(X) = g_i(X)p(X)$$

$$= -\sum_{\substack{j=1 \\ j\neq i}}^{m} p(X|\omega_j)p(\omega_j)$$

(3-21)

A simple law of probability says that

$$p(X) = \sum_{j=1}^{m} p(X|\omega_j)p(\omega_j)$$

which can be written as

$$p(X) = p(X|\omega_i)p(\omega_i) + \sum_{\substack{j=1 \\ j\neq i}}^{m} p(X|\omega_j)p(\omega_j)$$

or, solving for the rightmost term,

$$\sum_{\substack{j=1 \\ j\neq i}}^{m} p(X|\omega_j)p(\omega_j) = p(X) - p(X|\omega_i)p(\omega_i)$$

Substituting this result in Eq. (3-21) gives

$$g_i'(X) = -p(X) + p(X|\omega_i)p(\omega_i)$$

Finally, noting again that $p(X)$ is considered a constant for a given X, application of Eq. (3-19a) leads to the desired set of discriminant functions:

$$g_i''(X) = g_i'(X) + p(X)$$

$$= p(X|\omega_i)p(\omega_i)$$

(3-22)

precisely the set of discriminant functions appearing in our definition of the maximum-likelihood decision rule near the beginning of this section.

Summarizing, we began with a fundamental strategy for making decisions, namely, minimization of average loss (the Bayes optimal strategy), expressed the average loss in terms of probability functions associated with the classification problem, and converted the minimization problem to a maximization problem. Several transformations of the resulting expression then produced the discriminant functions which are implemented in the maximum-likelihood decision rule.

But we can be still more specific. If the probability functions associated with the pattern classes can be assumed to be multivariate normal density functions, as expressed by Eq. (3-10) in Sec. 3-5, then the discriminant functions become

$$g_i(X) = \frac{p(\omega_i)}{(2\pi)^{n/2}|\Sigma_i|^{1/2}} \exp\left[-\tfrac{1}{2}(X - U_i)^T \Sigma_i^{-1}(X - U_i)\right] \tag{3-23}$$

An equivalent set of discriminant functions can be derived which is somewhat simpler to implement. By repeatedly applying to Eq. (3-23) the transformations given in Eqs. (3-19), the following can be obtained:

$$g_i(X) = \log_e p(\omega_i) - \tfrac{1}{2}\log_e|\Sigma_i| - \tfrac{1}{2}(X - U_i)^T \Sigma_i^{-1}(X - U_i) \tag{3-24}$$

Once a problem has been specified and the statistical parameters estimated from training data, only the quadratic term (the rightmost term) of Eq. (3-24) varies from point to point during the actual classification process and must be calculated for each classification.

Notice that the pattern recognition system classifies *every* pattern presented to it into one of the classes it has been designed to recognize, i.e., one of the classes for which a discriminant function has been specified. In remote sensing problems, there are almost inevitably a number of points in the area to be classified which in fact do not belong to *any* of these classes. Such points may belong to classes for which there are insufficient training patterns for estimating the parameters or classes which have been completely overlooked. Although such points cannot be correctly classified since there is no discriminant function corresponding to the correct class, we can at least design the classifier to detect them, provided they are spectrally very much different from the points in the "valid" classes. As suggested by the one-dimensional, two-class example shown in Fig. 3-10, the points to be detected correspond to those which have a very low probability of belonging to any of the training classes. At the cost of "rejecting"

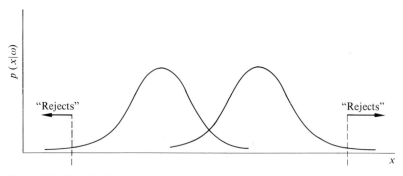

Figure 3-10 Thresholding.

a very small percentage of the points actually belonging to the training classes, it is possible to reject a comparatively large number of points not belonging to any of the training classes. This is accomplished by a technique known as *thresholding*, in which the probability value $p(X \mid \omega_i)$ associated with the data vector and the class into which it would ordinarily be classified is compared with a user-specified threshold. If the probability value is below that threshold, the data point is assigned instead to a *reject* class.

In particular, when the discriminant function has the form of Eq. (3-23), the threshold can be applied directly to the value of the multivariate normal probability density function. More commonly, however, the discriminant function is used in the form given in Eq. (3-24), in which case the normal probability density function is never actually computed. Then it is more convenient to use the fact that for a normally distributed variable X, the quadratic form in X [the rightmost term in Eq. (3-24)] has a chi-square distribution with n degrees of freedom, where n is the dimensionality of X. By using the chi-square distribution, a threshold level on X can be converted to a threshold level on the quadratic form. For further details, the reader may consult Anderson[4] (p. 112).

We have discussed only one decision strategy, the Bayes optimal strategy, giving particular attention to the special case of the zero-one loss function. This special case, called the maximum-likelihood decision rule, is widely used where pattern recognition is applied to remote sensing data analysis. It produces results which have the *minimum probability of error over the entire set of data classified*. Putting it another way, the classifier is the most accurate classifier on the average. For many typical applications, this is the appropriate strategy. But not always. To appreciate the fact that there are exceptions, look carefully again at the definition of the maximum-likelihood decision rule. The perceptive reader may conclude quite correctly that this rule tends to discriminate *against* classes which have a low probability of occurrence [low value for $p(\omega_i)$]. For some applications this can be a detriment because it may be the identification of these "rare" classes which is the main interest. This difficulty can be overcome by using a loss function more complicated than the zero-one loss function, tailoring it to reflect the characteristics of the specific application at hand.

As another practical matter, the a priori probabilities required by the maximum-likelihood decision rule may not be known. Frequent practice is to assume simply that the a priori probabilities of all classes are equal. It is not difficult to show that if the zero-one loss function is again invoked this assumption is equivalent to using the following set of discriminant functions:

$$g_i(X) = p(X \mid \omega_i) \qquad i = 1, 2, \ldots, m$$

Interestingly, this classification strategy is equivalent to a Bayes optimal strategy in which the loss function is

$$\lambda(i \mid j) = 0 \qquad i = j$$

$$\lambda(i \mid j) = \frac{1}{p(\omega_j)} \qquad i \neq j$$

i.e., the loss due to an incorrect classification is inversely proportional to the a priori probability of the correct class. Thus, on the average this strategy discriminates *in favor of* the most rarely occurring classes!

PROBLEMS

3-9 List three reasons why a statistical approach is appropriate in applying pattern recognition for remote sensing data analysis.

3-10 Here is a simple problem to develop your understanding of how decisions can be made using probabilities.

A standard deck of playing cards is altered as follows:
(a) All the red J, Q, and K cards are discarded.
(b) All the black A, 2, 3, and 4 cards are discarded.

The remaining cards are thoroughly mixed. Player 1 selects a card at random and announces its value—4, 8, 9, K, etc. Player 2 tries to guess whether the card is a black card or a red card. What strategy should player 2 use to maximize the probability of guessing the correct color of the card?

3-11 Statistical pattern-recognition techniques may be parametric or nonparametric. What are the distinguishing features of these two approaches? Why might you choose a parametric representation of a probability function over a nonparametric representation?

3-12 Sketch a one-dimensional (univariate) normal density function and give the two parameters which are necessary to specify the density function. Describe how the generalization is made to the multivariate case.

3-13 Assume a three-dimensional problem in which each dimension (measurement) can have any one of 100 possible values. Compare the number of memory locations required to store (a) a three-dimensional histogram of the data and (b) the parameters of a three-dimensional normal probability density function.

3-14 Cite two precautions which must be observed if the probability functions associated with the pattern classes are to be approximated by normal density functions.

3-15 The Landsat 1 multispectral scanner collected data in four wavelength bands. To perform a crop survey using these data, what would have been the minimum number of data points needed to estimate the mean vector and covariance matrix for each class? How many would you have recommended?

3-16 The maximum-likelihood decision rule is as follows. Decide X belongs to class i if and only if

$$p(X|\omega_i)p(\omega_i) \geq p(X|\omega_j)p(\omega_j) \qquad \text{for all } j = 1, 2, \ldots, m$$

Give an interpretation of the terms $p(X|\omega_i)$ and $p(\omega_i)$ in a remote sensing context.

3-17 Which terms of Eq. (3-24) are data dependent, and how is Eq. (3-24) used to classify an unknown pattern?

3-18 Explain how the pattern-recognition system can be made to detect many of the points in the area to be classified which actually are not represented by any of the training classes.

Study objectives

After studying Secs. 3-7, 3-8, and 3-9, you should be able to:

1. Give at least two reasons why it is desirable to be able to estimate the probability of error of a classifier.

2. Using a one-dimensional example, explain how the probability of error in classification can be related to the overlap of the class probability functions.

Give an interpretation of the terms in the equation which expresses this relationship.

3. Describe three ways of estimating the probability of error of a classifier, giving advantages and disadvantages of each.

4. Describe in general terms the concept of statistical separability and explain why it is resorted to as an indicator for probability of error.

5. Explain why it may be desirable to perform feature selection and why for this purpose J-M distance is a more useful measure of statistical separability than is divergence.

6. Explain the purpose of feature extraction in pattern recognition and describe three feature extraction transformations commonly encountered in remote sensing applications of pattern recognition.

3-7 EVALUATING THE CLASSIFIER : PROBABILITY OF ERROR

It was pointed out in Sec. 3-4 that one reason for using statistical methods in applying pattern recognition to remote sensing is that the classes we wish to discriminate may actually overlap in the measurement space. The practical implication of this is that, using the available remote sensing measurements, the classifier is bound to make some errors because some of the patterns in class 1 are indistinguishable from patterns in class 2, some in class 2 are indistinguishable from patterns in class 3, and so on. The primary purpose for taking the statistical approach is to minimize the frequency with which such errors occur, or, in other words, to minimize the probability of error. The previous section showed how statistical decision theory could be used to develop a strategy, in terms of appropriate discriminant functions, to accomplish exactly that.

Given that we have a strategy for deriving minimum-probability-of-error classifiers, there are several reasons why we need to be able to calculate, for any given problem, the corresponding probability of error. For one thing, the probability of error helps us decide how much confidence we can have in the classification results. If the probability of error is excessive, it may be necessary to seek some other means or measurements to discriminate among the classes of interest. And, as we shall see in subsequent sections, probability of error can be used as a criterion to guide in the design of an effective classifier.

To begin to get an intuitive notion of how probability of error arises and how it can be computed, let us go back to the funny dice game of Sec. 3-4. You will recall that the game involves a standard pair of dice and an augmented pair having two extra spots on each face. The objective of the game is to "classify" the dice selected for each roll based on the sum rolled. As Fig. 3-7 shows, the pattern classes do indeed overlap. Any time a sum between 6 and 12 is announced, there is at least some uncertainty as to whether the pair of dice rolled was the standard pair or the augmented pair. This uncertainty is greatest when a 9 is rolled and least when a 6 or a 12 is rolled.

Suppose that our statistician friend, player 2, has adopted the maximum-likelihood strategy. Assuming the a priori probabilities are equal [p(standard dice) = p(augmented dice) = 0.5], player 2 will follow the rule:

Rule Decide "standard" dice any time a sum x is rolled such that

$$p(x|\text{standard dice}) \geq p(x|\text{augmented dice})$$

and decide "augmented" dice any time a sum x is rolled such that

$$p(x|\text{augmented dice}) > p(x|\text{standard dice})$$

Note that all 9's are arbitrarily classified as "standard"; the rule could just as well be written to classify them as "augmented." In either case, because the probability functions are equal and the a priori probabilities are equal, the 9's will be correctly classified half of the time.

On any given roll, if player 1 has already selected the standard pair, what is the probability that player 2 will make an error by following this rule? It is equal to the probability that player 1 will roll a 10, 11, or 12, since the rule says to decide "augmented" if any of these sums is rolled. We can write this as

$$p(\text{error}|\text{standard}) = p(10|\text{standard}) + p(11|\text{standard}) + p(12|\text{standard})$$

$$= \tfrac{3}{36} + \tfrac{2}{36} + \tfrac{1}{36}$$

$$= \tfrac{1}{6}$$

In other words, about one-sixth of the times that player 1 selects the standard pair, he will roll a 10, 11, or 12 and player 2 will therefore make an erroneous classification. Similarly,

$$p(\text{error}|\text{augmented}) = \sum_{i=6}^{9} p(i|\text{augmented})$$

$$= \tfrac{4}{36} + \tfrac{3}{36} + \tfrac{2}{36} + \tfrac{1}{36}$$

$$= \tfrac{5}{18}$$

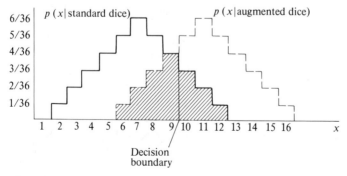

Figure 3-11 Crosshatched areas reflect probability of error.

Most important, however, is the fact that for the funny dice game, assuming that the maximum-likelihood strategy is to be followed, we can compute the probability that player 2 will make an error on any given roll:

$$p(\text{error}) = p(\text{standard})p(\text{error} \,|\, \text{standard})$$
$$+ p(\text{augmented})p(\text{error} \,|\, \text{augmented}) \tag{3-25}$$

$$= (\tfrac{1}{2} \times \tfrac{1}{6}) + (\tfrac{1}{2} \times \tfrac{5}{18})$$

$$= \tfrac{2}{9}$$

In other words,

(the probability of player 2 making an error) =

(the probability that player 1 will select the standard set) × (the probability of player 2 making an error, given that the standard set is selected)

+

(the probability that player 1 will select the augmented set) × (the probability of player 2 making an error, given that the augmented set is selected)

Alternatively, we can interpret this as the average rate at which player 2 will make errors over many successive plays of the game. Since the strategy followed is consistent with the development of the previous section, we know that the probability of error is the minimum achievable probability of error.

Figure 3-11, which is much like Fig. 3-7(*b*), illustrates a very useful concept, namely, that the probability of error is closely related to the areas under the probability functions in the region of overlap. These areas are shown crosshatched, separated by the maximum-likelihood decision boundary. Looking back at Eq. (3-25), it can be seen that the rightmost crosshatched area corresponds to the term $p(\text{error} \,|\, \text{standard})$ and the leftmost crosshatched area to the term $p(\text{error} \,|\, \text{augmented})$. Computation of the overall error simply requires weighting these areas by the respective a priori probabilities and then summing. Thus, the larger these areas are, the larger the probability of error will be, and vice versa. Since we wish the classifier to have minimum probability of error, we shall always make choices in designing the classifier so as to minimize these areas. We are going to make extensive use of this idea as we now proceed to a more general discussion of probability of error.

Let us assume that we have a single remote sensing measurement available and that we wish to discriminate between two ground-cover classes having probability density functions $p(x \,|\, \omega_1)$ and $p(x \,|\, \omega_2)$ as shown in Fig. 3-12(*a*). Multiplying every value along each curve by the corresponding a priori probability and plotting the results on a common graph, we get Fig. 3-12(*b*). The maximum-likelihood decision boundary is at point B, where the curves cross. The probability of error is given by the expression

$$p_E = E_{12} + E_{21} \tag{3-26}$$

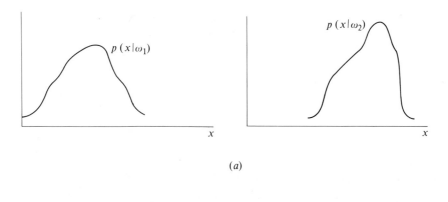

(a)

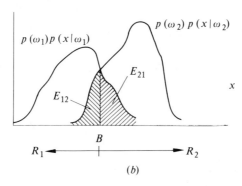

(b)

Figure 3-12 The relationship between probability of error and overlap of the class density functions. (a) Probability density functions for hypothetical ground-cover classes. (b) Overlap area (crosshatched) reflects probability of error.

where

$$E_{12} = \int_{R_1} p(\omega_2)p(x|\omega_2)\, dx$$

$$E_{21} = \int_{R_2} p(\omega_1)p(x|\omega_1)\, dx$$

The error component E_{12} is the probability associated with classifying observations as class 1 when they are really from class 2; E_{21} is the probability associated with classifying observations as class 2 when they are really from class 1. Furthermore, E_{12} and E_{21} correspond to the respective "overlap areas," shown crosshatched in Fig. 3-12(b).

Equation (3-26) applies to any two-class problem involving the maximum-likelihood decision rule. If the problem is multivariate, the only change required is to replace x by the vector X. The integrals become volume integrals.

As a practical matter, we need to be able to use Eq. (3-26) to actually compute the value for the probability of error just as we did using Eq. (3-25) for the funny dice case. Unfortunately, this is almost never possible. For most forms of density functions encountered, including the normal density function, the integrals in Eq. (3-26) cannot be evaluated analytically. If the dimensionality of the measurement space is small, numerical methods can be used on a computer to evaluate the integrals. But generally this approach is not feasible when the measurement space has dimensionality much greater than 1. It might appear, then, that although we have a method available for designing an optimal maximum-likelihood classifier, we cannot determine in practice just how good "optimal" is, and, as has been pointed out, it is important that we be able to do so. But we shall see that the situation is not really that bad.

Error Estimation from Training Data

To begin with, once the training samples have been used in the design of the classifier, the classifier may be used to classify the training samples themselves. An estimate of the probability of error associated with the classifier is obtained by dividing the number of training data points erroneously classified by the total number of points classified.

Generally speaking, this estimate is accurate only if the training samples are truly representative of the overall data set to be classified. In practice, the training-sample classification results are usually optimistic; the error rate estimated from the training samples is lower than the actual error rate for the entire data set. Data-gathering constraints in the field often prohibit the collection of a truly representative sample. And unfortunately the results of classifying the training samples do not provide any indication of just how representative of the overall data set the training samples actually are. At best we can think of the training-sample results as providing a lower bound or optimistic lower limit on the true probability of error.

Error Estimation from Test Data

Commonly, the set of data points for which the class is known with at least reasonable reliability is divided into two subsets. One subset, referred to as the training sample, is used for designing the classifier; the other subset is set aside for testing the classifier. The performance of the classifier on the latter data points provides another estimate of the probability of error associated with the classifier.

As in the case of the training samples, it is very important that the test data be representative of the overall data set to which the classifier is to be applied. Ideally, the data to be included in the test set should be drawn at random from the total area to be classified in order to avoid various biases which can impair its representativeness. For example, data analysts seem to be inclined to select homogeneous areas in the data for use in testing, whether or not homogeneity is typical of the area. Of course, the larger the test set, the more likely it is to be adequately representative. However, there are usually practical constraints on how much test data can be made available. One is the cost of collecting it, whether by ground checking or other appropriate means. Another related constraint is the need, in the face of limited quantities of data of known classification, to use adequate data for training the classifier.

The performance of the classifier on representative test data will not have inherent in it the optimistic bias which the results for the training data are likely to have. Furthermore, if there is good agreement between the performance of the classifier on the training data and the performance on the test data, then it is likely that a representative training set was indeed used. On the other hand, serious disagreement indicates that either the training set or the test set (or both) very likely was not adequately representative of the overall data set.

Indirect Estimates of Classifier Error

Although it is usually difficult or even impossible to compute the probability of error directly using Eq. (3-26), there are several ways to get an indirect indication of the probability of error. In the next section we will introduce the concept of statistical separability and show how statistical separability can be related to classifier performance. A number of "indicator functions" which are measures of statistical separability will be defined, and we will show how these can be used to compare the performances of candidate classifier designs and, in addition, to provide estimates of bounds on the probability of error.

3-8 STATISTICAL SEPARABILITY AND FEATURE SELECTION

As we pursue our search for a measure of the probability of error of a classifier, it will be enlightening to focus on a step in the data-analysis process for which such a measure is particularly useful. Specifically, suppose that the cost of using all of the available multispectral scanner channels in a given remote sensing application were prohibitively high (required too much computation or hardware) and, furthermore, that it were believed that a judicious choice of a subset of those channels would result in adequate classifier accuracy.† On what basis should the choice of channels be made? This is the problem in pattern recognition known as *feature selection*. It is a special case of the more general problem of feature extraction, which we shall look at in greater detail in the next section.

More formally, the problem of feature selection may be stated as follows.

Problem Given an *m*-class pattern-recognition problem with N available measurements, find a subset of the measurements which will provide an optimal trade-off between probability of error and cost of classification.

---

† In Chap. 7, we will discover that in practical situations involving limited availability of training data, one does not always get better classifier performance by adding channels. In fact, performance may even degrade. The reason for this is related to our earlier observation that the minimum number of points needed for training is a function of the number of channels (Sec. 3-5).

Solution of this problem requires comparison of candidate sets of measurements on the basis of the classification performance they can provide. Ideally, we would like to make this comparison on the basis of probability of error, taking into account, of course, the cost of classification as well. But we have already observed that computing probability of error is often not feasible, so we shall now set about defining an alternative predictor of classifier performance which will be referred to as *statistical separability*. We shall show how statistical separability is related to the probability of error.

To begin with, it was shown in the preceding section that probability of error is intimately related to the area under the probability functions in the region of overlap. Let us consider some factors which affect the size of this area. Figure 3-13 shows pairs of normal probability density functions having various degrees of overlap. For the purpose of this illustration (and several to follow) it is assumed that the pattern classes in question have equal a priori probabilities, in which case the decision boundary is located at the central crossover of the probability functions.

Comparing the pairs of normal densities in Fig. 3-13(*a*) and (*b*) we see that, if the variances are held constant, the overlap area decreases as the separation of the means increases. On the other hand, comparison of Fig. 3-13(*b*) and (*c*) shows that if the separation of the means is constant, the overlap area increases as the variances of the densities increase, i.e., as the data become more dispersed in the measurement space. These effects can be combined by defining a "normalized distance between the means," d_{norm}, which is the absolute value of the difference between the means divided by the sum of the standard deviations:

$$d_{norm} = \frac{|\mu_1 - \mu_2|}{\sigma_1 + \sigma_2} \tag{3-27}$$

where the subscripts refer to the respective density functions. This normalized distance can be thought of as describing quantitatively how well the density functions are "separated." It is a measure of the *statistical separability* of the pattern classes, i.e., the separability of their probability functions. From this discussion and the illustration of Fig. 3-13, it should be clear that there is an inverse relationship between statistical separability and probability of error: the greater the statistical separability of the classes, the smaller the probability of error.

We need to look further for an effective measure of statistical separability, however. For one thing, d_{norm} does not behave in quite the fashion we would like; the value of d_{norm} is 0 for *any* two density functions having equal means, even though the probability of error, as reflected by the overlap areas, varies depending on the variances associated with the densities (Fig. 3-14). Furthermore, we must have a separability measure which is multivariate in order to deal with measurement spaces of dimensionality greater than 1. Two such separability measures, divergence and J-M distance, are discussed below. For each, an attempt is made to provide both an intuitive and a mathematical justification. The special cases are treated in which multivariate normal density functions are assumed, and the relative advantages of the two measures are discussed.

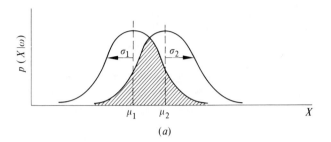

(a)

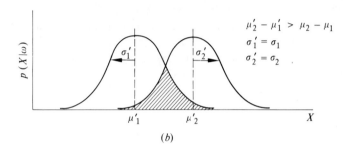

$$\mu_2' - \mu_1' > \mu_2 - \mu_1$$
$$\sigma_1' = \sigma_1$$
$$\sigma_2' = \sigma_2$$

(b)

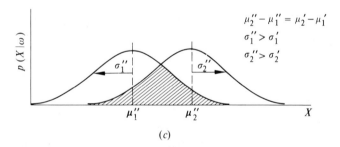

$$\mu_2'' - \mu_1'' = \mu_2' - \mu_1'$$
$$\sigma_1'' > \sigma_1'$$
$$\sigma_2'' > \sigma_2'$$

(c)

Figure 3-13 Error varies with "normalized distance." (a) Two overlapping normal densities. (b) Increasing the separation of the means decreases the probability of error. (c) Increasing the variances increases the probability of error.

Divergence

Divergence was one of the first measures of statistical separability used in pattern recognition.[5]† It is related to the *likelihood ratio* L_{ij} for two classes i and j:

$$L_{ij}(X) = \frac{p(X \mid \omega_i)}{p(X \mid \omega_j)} \qquad (3\text{-}28)$$

† The following mathematical treatment of divergence may be read lightly or the reader may skip to the first full paragraph following Eq. (3-33).

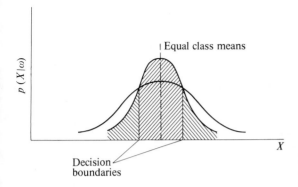

Figure 3-14 Overlap areas in the case of equal means.

Consider Fig. 3-15. Thinking back to the funny dice example (refer to Fig. 3-11), readers should have no trouble convincing themselves that the larger the interval a in Fig. 3-15, the more likely it is that the classifier will be correct in assigning measurement X_0 to class i.

But the size of a is nicely characterized by the value of the likelihood ratio at X_0: the larger a is, the larger the value of the likelihood ratio. Notice that at the crossover of the probability density functions, a is 0 and the likelihood ratio is 1.0.

To make the mathematics somewhat more tractable, it is convenient to define the *logarithmic-likelihood ratio*, which is similarly related to the size of a:

$$L'_{ij}(X) = \log_e L_{ij}(X)$$
$$= \log_e p(X \mid \omega_i) - \log_e p(X \mid \omega_j) \tag{3-29}$$

The divergence D_{ij} of classes i and j is defined in terms of this logarithmic-likelihood ratio:

$$D_{ij} = E[L'_{ij}(X) \mid \omega_i] + E[L'_{ji}(X) \mid \omega_j] \tag{3-30}$$

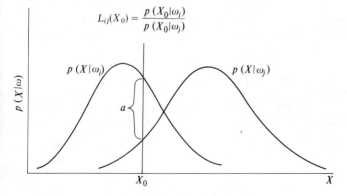

Figure 3-15 Definition of the likelihood ratio at a point.

where $E[\ ]$ denotes the expectation or average; that is,

$$E[L'_{ij}(X)\,|\,\omega_i] = \int_X L_{ij}(X)p(X\,|\,\omega_i)\,dX \tag{3-31}$$

and

$$E[L'_{ji}(X)\,|\,\omega_j] = \int_X L_{ji}(X)p(X\,|\,\omega_j)\,dX \tag{3-32}$$

In these expressions, the integrals are taken over the entire measurement space. They are volume integrals when the measurement space is multidimensional. Thus the divergence can be interpreted as the average value of the likelihood ratio with respect to the patterns in class i plus the average value of the likelihood ratio with respect to the patterns in class j.

We find that the mathematical properties of divergence are consistent with those which we might expect an indicator of probability of error to have:

1. $D_{ij} > 0$ For two different probability functions, the divergence is always greater than 0.
2. $D_{ii} = 0$ The divergence of a probability function relative to itself (or the divergence of two identical probability functions) is exactly 0.
3. $D_{ij} = D_{ji}$ Divergence is a symmetrical distance measure.
4. If the components of the measurement vector are statistically independent, then

$$D_{ij}(x_1, x_2, \ldots, x_n) = \sum_{k=1}^{n} D_{ij}(x_k)$$

5. $D_{ij}(x_1, x_2, \ldots, x_n, x_{n+1}) \geq D_{ij}(x_1, x_2, \ldots, x_n)$

The last two properties reflect the fact that adding measurements never decreases the statistical separability. In fact, if the measurements are independent, the joint separability measure is simply the sum of the separabilities for the individual measurements.

Now, as a practical matter, it might appear that we are no better off than we were when we tried to compute the probability of error directly, since the defining expression for divergence contains volume integrals, as did the expression for probability of error. But when the classes are assumed to have normal probability density functions,† that is,

$$p(X\,|\,\omega_i) = N(U_i, \Sigma_i)$$

$$p(X\,|\,\omega_j) = N(U_j, \Sigma_j)$$

the divergence can be written as an expression involving the means and covariance matrices—and no integrals. Specifically,

$$D_{ij} = \tfrac{1}{2}\,\mathrm{tr}\left[(\Sigma_i - \Sigma_j)(\Sigma_j^{-1} - \Sigma_i^{-1})\right] + \tfrac{1}{2}\,\mathrm{tr}\left[(\Sigma_i^{-1} + \Sigma_j^{-1})(U_i - U_j)(U_i - U_j)^T\right] \tag{3-33}$$

† $N(U_i, \Sigma_i)$ is the multivariate normal density function with mean vector U_i and covariance matrix Σ_i.

where tr $[A]$ denotes the *trace* of matrix A, that is, the sum of the diagonal elements of matrix A. (The other vector/matrix operations were introduced in Sec. 3-5.) Although this expression looks complicated, its evaluation is straightforward and efficiently carried out on a computer.

Notice that the expression consists of two terms which are summed. The first term represents a contribution resulting only from differences in the respective covariance matrices. The second term is a normalized distance between means. D_{ij} is never 0 unless both the means and the covariance matrices for the two classes are identical.

So now we have a measure of the statistical separability of two classes, an indicator of the probability of error in discriminating between them. Since this provides us with a measure of the relative effectiveness of candidate sets of features for discriminating between the classes, we have essentially solved the two-class feature-selection problem. That is, for the two-class case, given any pair of feature subsets, we can determine the superior subset by computing the divergence for both subsets and selecting the subset with the largest divergence.

But what about the m-class problem when m is greater than 2? Divergence is a pairwise distance measure, and an m-wise generalization has not been formulated. A fairly common strategy for the multiclass problem is to use the *average* divergence, i.e., compute the average over all pairs of classes. The average pairwise divergence D_{ave} is defined as follows:

$$D_{ave} = \sum_{i=1}^{m} \sum_{j=1}^{m} p(\omega_i)p(\omega_j)D_{ij} \qquad (3\text{-}34)$$

This is a weighted average, with the class a priori probabilities used as the weights.

Selection of the feature subset having maximum average divergence is certainly a reasonable strategy. After all, maximizing the average pairwise probability of correct recognition (probability of correct recognition is simply 1 minus the probability of error) is consistent with the average minimum-risk (Bayes) strategy we adopted earlier. As it happens, however, a problem arises with this approach because the behavior of divergence as a function of the normalized distance between classes is considerably different from the behavior of the probability of correct recognition. This is illustrated in Fig. 3-16. The general shape of the divergence curve can be explained by examination of Eq. (3-33), which gives the divergence for a pair of normally distributed classes. The second term in the summation on the right-hand side of this equation continues to increase rapidly as long as the normalized distance between the means increases. The reader may recognize this term as a multivariate generalization of the normalized distance introduced earlier [Eq. (3-27)]. On the other hand, the probability of correct recognition can increase only to 100 percent, and once the normalized distance corresponding to virtually perfect recognition is achieved, no further improvement of any significant degree is possible.

The impact of this difference in behavior can be appreciated by considering the simple example shown in Fig. 3-17. This figure suggests schematically a

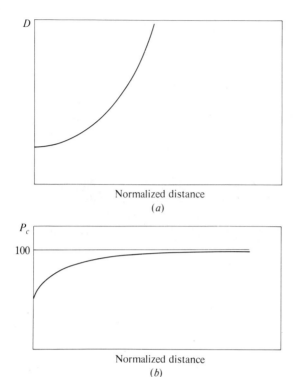

Figure 3-16 Divergence and probability of correct recognition as functions of normalized distance. (*a*) Divergence *D*. (*b*) Probability of correct recognition P_c (percent).

feature-selection problem involving two classés which are relatively hard to discriminate accurately (perhaps corn and soybeans) and a third which is easily discriminated from the first two (say, bare soil). Typical values of divergence are tabulated which correspond to the given values of probability of error.[6] The feature set (x_c, x_d) is preferable to the set (x_a, x_b) because it does a better job of discriminating between class 1 and class 2 at little or no sacrifice in discriminability of class 3. Yet if the feature selection is based on maximum average divergence, feature set (x_a, x_b), giving inferior overall classification performance, will be selected. Again, this is simply due to the fact that the behavior of divergence as a function of normalized distance of the class pairs differs very markedly from the behavior of the probability of error as a function of normalized distance.

We shall now consider another indicator function which improves on this situation.

J-M Distance

The Jeffries-Matusita (J-M) distance, like the divergence, is a measure of statistical separability of pairs of classes.† Formally, it is defined as follows:[7]

$$J_{ij} = \left\{ \int_X \left[\sqrt{p(X \mid \omega_i)} - \sqrt{p(X \mid \omega_j)} \right]^2 dX \right\}^{1/2} \qquad (3\text{-}35)$$

† The reader who is skimming the mathematical material may wish to skip to the paragraph following Eq. (3-37).

Assume $p(\omega_1) = p(\omega_2) = p(\omega_3)$

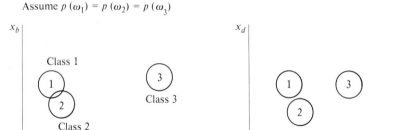

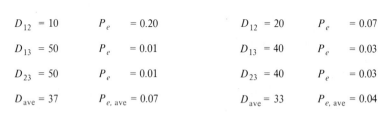

$D_{12} = 10$	$P_e = 0.20$		$D_{12} = 20$	$P_e = 0.07$
$D_{13} = 50$	$P_e = 0.01$		$D_{13} = 40$	$P_e = 0.03$
$D_{23} = 50$	$P_e = 0.01$		$D_{23} = 40$	$P_e = 0.03$
$D_{\text{ave}} = 37$	$P_{e,\text{ave}} = 0.07$		$D_{\text{ave}} = 33$	$P_{e,\text{ave}} = 0.04$

Figure 3-17 An example in which D_{ave} is not optimal. The circles suggest relative locations of the classes and not necessarily class boundaries.

Roughly speaking, the J-M distance is a measure of the average difference between the two-class density functions, which again seems intuitively a reasonable way to measure the class separability.

At this point, it is not yet obvious that the J-M distance offers any advantage over divergence. That it does, however, becomes apparent when we assume that the classes have normal density functions. In this case, Eq. (3-35) reduces to

$$J_{ij} = [2(1 - e^{-\alpha})]^{1/2} \tag{3-36}$$

where $\quad \alpha = \frac{1}{8}(U_i - U_j)^T \left(\frac{\Sigma_i + \Sigma_j}{2}\right)^{-1}(U_i - U_j) + \frac{1}{2}\log_e\left[\frac{|(\Sigma_i + \Sigma_j)/2|}{(|\Sigma_i| \cdot |\Sigma_j|)^{1/2}}\right] \tag{3-37}$

Again, as in the case of divergence, the assumption that the classes are normally distributed results in an expression involving means and covariances but no integrals; and again we recognize a term which can be interpreted as a multivariate form of the normalized distance between class means [compare with Eq. (3-27)]. There is a crucial difference, however, which arises from the negative exponential term in Eq. (3-36). The effect of this term is to give an exponentially decreasing weight to increasing differences between the class density functions. As a result, J-M distance has a "saturating" behavior, as illustrated in Fig. 3-18, and behaves much more like the percentage of correct recognition than does divergence. Now when the average is computed over class pairs, that is,

$$J_{\text{ave}} = \sum_{i=1}^{m} \sum_{j=1}^{m} p(\omega_i)p(\omega_j)J_{ij} \tag{3-38}$$

widely separated classes no longer make a disproportionately large contribution to the average. Figure 3-19 illustrates this, using an extension of the example of Fig. 3-17. In this case, typical values of J-M distance are tabulated which correspond to the given values of probability of error.[6] We see that J_{13} and J_{23} do not unduly inflate the average value, so that (x_c, x_d) is properly the selected feature subset.

The benefit of the saturating behavior of J-M distance has been verified experimentally. It has also been shown through experimentation that a saturating function of divergence can provide similar results and is somewhat easier to calculate (i.e., requires less computer time).[6] This saturating function, called the *transformed divergence* and denoted by D_{ij}^T, is given by

$$D_{ij}^T = 2[1 - \exp(-D_{ij}/8)] \tag{3-39}$$

where D_{ij} is the ordinary divergence defined previously [Eq. (3-30)]. The reader will find it interesting to compare the form of this expression with Eq. (3-36). It is the saturating effect of the negative exponential in both of these expressions which makes these distance measures useful for multiclass problems.

As we have seen, computing the average pairwise separability (D_{ave} or J_{ave}) of all pairs of classes is one possible strategy for selecting the best features in a multiclass problem. Another strategy which might be followed is:

Strategy Select the set of features for which the minimum separability between any pair of classes is largest.

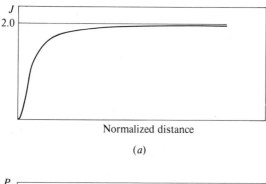

Normalized distance

(*a*)

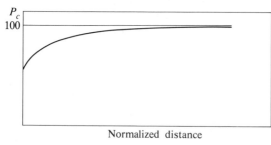

Normalized distance

(*b*)

Figure 3-18 J-M distance and probability of correct recognition as functions of normalized distance. (*a*) J-M distance *J*. (*b*) Probability of correct recognition P_c (percent).

Assume $p(\omega_1) = p(\omega_2) = p(\omega_3)$

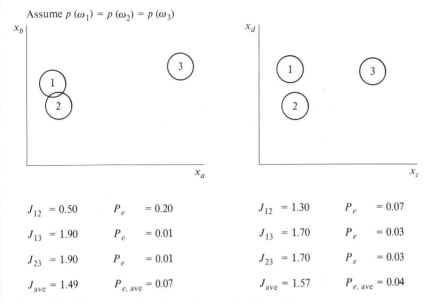

$J_{12} = 0.50$	$P_e = 0.20$	$J_{12} = 1.30$	$P_e = 0.07$
$J_{13} = 1.90$	$P_e = 0.01$	$J_{13} = 1.70$	$P_e = 0.03$
$J_{23} = 1.90$	$P_e = 0.01$	$J_{23} = 1.70$	$P_e = 0.03$
$J_{ave} = 1.49$	$P_{e,\,ave} = 0.07$	$J_{ave} = 1.57$	$P_{e,\,ave} = 0.04$

Figure 3-19 J-M distance selects the best features in this multiclass case.

In other words, select the feature set which does the *best* job of separating the pair of classes which is *hardest* to separate.

In the simple example described previously, Figs. 3-17 and 3-19, it happens that this strategy would select the desirable feature set (x_c, x_d) regardless of the separability measure used. For many practical problems, this is a reasonable strategy to follow for feature selection. Although it is not strictly compatible with the minimum average risk (Bayes) strategy on which we based the derivation of the discriminant functions, it is often the strategy followed when the application requires obtaining accurate discrimination among a few hard-to-separate classes.

Before leaving the topic of statistical separability and error probability, we can say something about the explicit relationships between the separability measures we have considered and the probability of error. It would be nice, of course, if to a given value of divergence or J-M distance between two classes we could associate a specific probability of error. This is not possible, but we can determine *bounds* on the probability of error. Specifically, it can be shown[7,8] that if D is the divergence and J is the Jeffries-Matusita distance between two classes with equal a priori probabilities, then the probability of error is bounded according to the following expressions:

$$\tfrac{1}{4}\exp\left(-\frac{D}{2}\right) \le p_E$$

and

$$\tfrac{1}{16}(2 - J^2)^2 \le p_E \le 1 - \tfrac{1}{2}(1 + \tfrac{1}{2}J^2)$$

Again we find that the properties of J-M distance better suit our needs, in this case because we can bound the error probability above and below. Divergence provides only a lower bound.

This concludes our treatment of the problem of estimating classifier error. Having dealt at some length with feature selection as an application of error estimation, we shall now consider briefly a somewhat more general approach to the determination of features best suited for classification.

3-9 FEATURE EXTRACTION

The term *feature* was used in the previous section to refer to one of the pattern measurements. More generally, a feature can be thought of as a distillation of that information contained in the measurements which is useful for deciding on the class to which the pattern belongs. The original measurements may contain a great deal of "information" which does not aid in the classification process and may even hinder it. This is "noninformation" as far as we are concerned, and it shall henceforth be referred to simply as *noise*. Roughly speaking, feature extraction accomplishes two functions: (1) separation of the useful information from the noise and (2) reduction of the dimensionality of the data in order to simplify the calculations performed by the classifier. The latter is the same function accomplished by feature selection, which, as we shall see, is a special case of feature extraction.

To gain a better appreciation of why dimensionality reduction is important, consider its impact on the amount of computation required by the maximum-likelihood classifier using multivariate normal statistics. Assuming that the time required to perform multiplications is the dominant factor in the total time required for each classification, the classification time is roughly proportional to $n(n + 1)$, where n is the dimensionality of the data (the number of features used). Historically, much work in remote sensing research has utilized the "best" four features from a 12-channel multispectral scanner system. Since

$$\frac{4 \times 5}{12 \times 13} = 0.13$$

the four-feature classifications require only 13 percent of the time that is required when all 12 multispectral measurements from the scanner are used. Thus, with respect to the cost of classification, there may be considerable savings possible through the use of feature extraction. Of course, the cost of the feature-extraction process must also be accounted for.

Up to this point in the chapter we have assumed that the discriminant functions are computed by the decision maker directly from the measurement vector X. In terms of the system model, Fig. 3-1, this means that the output of the receptor goes directly to the classifier. Now we shall modify the model by

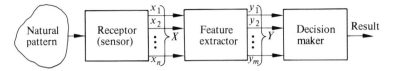

Figure 3-20 Pattern-recognition system with feature extractor.

adding a *feature extractor* between the receptor and the classifier (Fig. 3-20). We denote the *j*th feature by y_j. If there is a total of m features, we define the *feature vector* by

$$Y = \begin{bmatrix} y_1 \\ y_2 \\ \vdots \\ y_m \end{bmatrix}$$

A feature or a feature vector can be any mathematical transformation of the pattern measurements. This is a very broad definition. We shall limit our attention in the following discussion to three transformations which are often used in remote sensing applications, namely, *subset selection, ratios,* and *linear combinations.*

Subsets

Subset selection is simply feature selection as described at length in the previous section. The feature extractor transmits m of the measurements and inhibits the rest. Of course, m cannot exceed n, the dimensionality of the original measurement vector.

Ratios

In this case, each of the m components of the feature vector (again $m < n$) is a quotient of two of the pattern measurements; that is,

$$y_j = \frac{x_i}{x_k} \qquad \begin{aligned} j &= 1, 2, \ldots, m \\ 1 &\leq i, k \leq n \end{aligned}$$

In this case, situations must be avoided in which x_k might be 0. Suitable ratio features have been determined for specific remote sensing applications by reasoning based on the physical phenomena involved and by experimentation. Sometimes ratios are used in which the numerator and denominator are sums or differences of the pattern measurements, the objective usually being to "subtract out" or "divide out" the undesirable effects of the atmosphere or variable scene illumination.[9]

Linear Combinations

Each of the m components of the feature vector ($m < n$) is a linear combination of the pattern measurements; that is,

$$y_j = b_{j1}x_1 + b_{j2}x_2 + \cdots + b_{jn}x_n \qquad j = 1, 2, \ldots, m \qquad (3\text{-}40)$$

where each b_{ji} is a constant. The vector-matrix notation introduced in Sec. 3-5 can be invoked to write Eq. (3-40) in the form

$$Y = BX$$

where
$$B = \begin{bmatrix} b_{11} & b_{12} & \cdots & b_{1m} \\ b_{21} & b_{22} & \cdots & b_{2m} \\ \vdots & \vdots & & \vdots \\ b_{n1} & b_{n2} & \cdots & b_{nm} \end{bmatrix} \qquad (3\text{-}41)$$

In this case, we need to know how to choose m and also how to determine the $m \cdot n$ constants in the B matrix; as always, the goal is to maximize classification accuracy while minimizing the computation time required in performing the classification. Methods for achieving this are beyond the scope of the present discussion. The interested reader may refer to the literature.[10,11]

These three approaches to feature extraction for remote sensing are relatively straightforward. Much more complex features are conceivable, including features which account for spatial and temporal variations in the data, but the level of complexity involved precludes our dealing with these here. Effective use of such features is the subject of ongoing research in image processing as well as remote sensing.

It is important to recognize that in choosing the features we cannot ignore the assumptions underlying the classifier. For example, we have discussed at length the maximum-likelihood classifier which assumes that the data for each class are normally distributed. If the original pattern *measurements* are normally distributed, then this classifier can be used with features comprised of any subset or linear combination of the measurements, since such features are also normally distributed. However, features which are ratios of the normally distributed measurements are not normally distributed and should not be used with such a classifier.

PROBLEMS

3-19 Give at least two reasons why it is desirable to be able to estimate the probability of error of a classifier.

3-20 Using a figure similar to Fig. P3-20, explain how the probability of error in classification can be related to the overlap of the class probability functions. Your explanation should include an interpretation of the terms in the expression for the probability of error, Eq. (3-26).

Figure P3-20

3-21 Describe three ways of estimating the probability of error of a classifier, giving advantages and disadvantages of each.

3-22 Describe in general terms the concept of statistical separability and explain why it is resorted to as an indicator for probability of error.

3-23 Give two reasons why J-M distance is a more useful measure of statistical separability than is divergence.

3-24 The multispectral scanners aboard the Landsat 1 and 2 earth resources satellites were 4-channel scanners. Skylab carried a 13-channel scanner. Assuming that a maximum-likelihood classifier using normal statistics is to be used for analyzing the data, compare the computation time required for classifying the Landsat data (all 4 wavelength bands) to the time required for classifying the Skylab data (all 13 wavelength bands).

3-25 Referring to Fig. 3-20, explain the role of feature extraction in pattern recognition and describe three feature-extraction operations commonly encountered in remote sensing applications.

Study objectives

After studying the remainder of this chapter, Secs. 3-10 and 3-11, you should be able to:

1. Give a nonmathematical, intuitively oriented definition of *clustering*.
2. State two ways in which clustering can be used to aid in the analysis of multivariate data.
3. List three essential elements of the mathematical definition of a cluster and illustrate by giving two or more examples of one of these elements that clustering is an operation which can be defined in many ways.
4. Describe the differences between supervised and unsupervised classification, properly using such terms as *spectral classes*, *informational classes*, *training samples*, *decision regions*, and *cluster*.
5. State conditions under which unsupervised classification methods may be expected to work well.
6. State at least three practical limitations of supervised classification.
7. Explain the purpose or function of each step in a typical procedure for numerical analysis of remote sensing data (for example, Table 3-1).

3-10 CLUSTERING; UNSUPERVISED ANALYSIS

In Sec. 3-5 it was noted that in order to apply in practice the assumption that the classes have multivariate normal probability functions, a method is needed for detecting those cases in which the probability functions are multimodal (have more than a single maximum), since in such cases the assumption is seriously violated. When such a situation arises, we must also be able to subdivide the offending class or classes into subclasses which each have unimodal and approximately normal probability functions. We shall take up this problem now, applying a form of analysis called *clustering*, and we shall show further that this provides a method of *unsupervised classification* which is particularly useful when the supply of training samples is severely limited, perhaps even virtually nonexistent.

Clustering

The fact that a probability density function, such as the one shown in Fig. 3-8, has a mode or maximum in a particular region of the measurement space means that more of the measurement vectors tend to occur in that region than in adjoining regions. We could say that the measurement vectors tend to "cluster" near the mode. The analysis of a set of measurement vectors to detect this tendency is called *cluster analysis*, or simply *clustering*.

Clustering methods were developed to detect inherent or natural structure in data, and this is precisely what we are after. To illustrate what is meant by "natural structure," suppose we were to plot in a two-dimensional measurement space the height and weight of many randomly selected adults. The outcome might be as shown in Fig. 3-21(*a*). Without any further information about the individuals measured, we would be inclined to conclude that the resulting clusters correspond respectively to men (taller and heavier on the average) and women (shorter and lighter on the average). In fact, had we a reason to do so, we might even draw a decision boundary between the clusters and subsequently classify all individuals whose measurements fell on one side of the boundary as men and all with measurements on the other side as women. To be sure, such a classification would sometimes be in error, but we would justifiably have some confidence that our classification would more often be correct than in error. Notice that in arriving at this classification strategy we did not use "training samples" in the sense described earlier. That is, in establishing the decision boundary, we did not use prior knowledge of the actual classification of each individual set of measurements. We did, however, use other knowledge and assumptions about the population we measured: we knew they were adults and we assumed they were men and women.

Now suppose a set of multispectral training patterns were available, known from ground observations to be from fields of wheat. Suppose, further, that the results of plotting a visible wavelength band versus a near-infrared wavelength band for this set of data appeared as in Fig. 3-21(*b*). Applying our knowledge of (1) the locations of the wheat fields, (2) the time of year, and (3) the physiology of wheat, we might guess that the "natural structure" of this data set is related to the presence of widely different stages of crop maturity, perhaps further related to differences in variety. In any case the class "wheat" has a multimodal probability density function, and this is one of those instances in which the single multimodal class must be decomposed into unimodal subclasses.

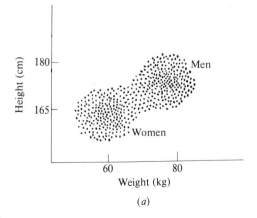

(a)

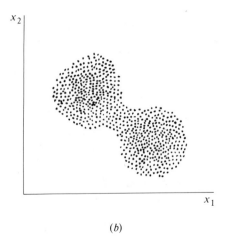

(b)

Figure 3-21 Examples of data which tend to cluster. (a) A distribution of adult heights and weights. (b) Multimodal data from wheat fields (hypothetical illustration).

In the two examples just presented, we *visually* detected the data structure once we had constructed a convenient display of the data. Clustering provides a means for the computer to do the same thing. Furthermore, the computer implementation extends the capability beyond the two- or three-dimensional limitation which is inherent in visually oriented processes.

We have developed an intuitive notion of what a cluster is. To implement the clustering process on the computer, we need to define the concept of a cluster in mathematical terms. A myriad of such definitions can be found in the literature, each fairly well specialized to a particular application or class of applications.[12,13] We shall discuss one of these which has been widely used for multispectral remote sensing applications.

Actually, we shall define what we mean by a cluster in terms of three auxiliary definitions: (1) the distance between points in the feature space, (2) the distance between collections of points (candidate clusters), and (3) a clustering criterion.

To begin with, there are several ways to measure *distance between points,* two of which are shown in Fig. 3-22. The most familiar point-to-point distance measure is *Euclidean distance,* computation of which in two-dimensional space is like computing the length of the hypotenuse of a triangle by using the Pythagorean theorem. The *n*-dimensional generalization is shown in the figure.

Figure 3-22 also defines L_1 *distance,* the sum of the distance components. On some computers, this interpoint distance is somewhat easier to calculate than Euclidean distance, and its use can therefore result in a faster algorithm if properly implemented.

Still other interpoint distance measures are available, some of which apply different weights to the various components (e.g., Mahalanobis distance; see Duda and Hart,[13] pp. 24, 213–217). But the Euclidean distance and L_1 distance are the most straightforward to calculate and have been widely used in clustering algorithms adopted for remote sensing data analysis.

There are also many ways to define and measure the *distance between collections of points.* Suppose it is desired to determine the distance between a collection labeled *A* and another labeled *B* (groups *A* and *B* might be candidate clusters). Conceptually, the simplest way to characterize the distance between *A* and *B* is to compute the average distance between all pairs of points for which one element of each pair is from *A* and one is from *B*. In Sec. 3-8 several other intergroup distance measures, including divergence, J-M distance, and transformed divergence, were defined in terms of the probability functions associated with the groups. These statistical distance measures have the advantage of accounting for within-group variability in the process of computing between-group distance; i.e., they are *normalized* distance measures, as we have already seen.

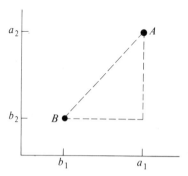

Euclidean distance

$$D_{AB} = \left[\sum_{i=1}^{n} (a_i - b_i)^2 \right]^{1/2}$$

In two-dimensional space, $n = 2$.

L_1 distance

$$D_{AB} = \sum_{i=1}^{n} |a_i - b_i|$$

Figure 3-22 Examples of interpoint distance measures.

A *clustering criterion* associates a measure of quality with each assignment of data points to clusters. Without such a measure, there is no way to determine whether one assignment is better than another or whether one method of clustering has more merit than another.

Many clustering criteria have been studied.[13] Generally, however, they reflect the assignment of points to clusters in such a way that the distances between points within a cluster are minimum and the distances between clusters are maximum. Notice how this draws on our previous discussion of interpoint and intercluster distances. We shall look at two clustering criteria very briefly.

A widely used clustering criterion, to which many other clustering criteria have been shown equivalent, is the "sum-of-squared-error" criterion. Suppose there are c clusters. Let M_i be the mean vector for the ith cluster [computed as in Eq. (3-12)] and let C_i denote the set of data points belonging to the ith cluster. Then the sum of squared errors (SSE) is defined by

$$\text{SSE} = \sum_{i=1}^{c} \sum_{X \in C_i} \| X - M_i \|^2 \tag{3-42}$$

where $\| X - M_i \|$ is the Euclidean distance between X and M_i. In other words, SSE is the cumulative distance between each point in the data set and the mean of the cluster to which that data point is assigned. Thus, when we use minimization of SSE as a clustering criterion, our goal is to assign points to clusters so that the clusters are as "tight" as possible.

Another approach to specifying a clustering criterion begins by defining the "scatter" of points within a cluster, S_W, and the "scatter" of clusters, S_B (between-cluster scatter). The object is to minimize S_W while simultaneously maximizing S_B. A very interesting result is that this clustering criterion is identical to minimization of the sum of squared error criterion.[13]

With this background, we are now prepared to consider the clustering algorithm described by Fig. 3-23. The algorithm proceeds as follows:

1. Select c vectors to serve as initial "cluster centers" $\hat{M}_i$, $i = 1, 2, \ldots, c$. The selection is arbitrary, except that no two of the initial cluster centers may be identical. The number of cluster centers must be specified by the data analyst, the consequences of which we shall have more to say about later.
2. Assign each vector in the data set to the nearest cluster center. This step requires the specification of a point-to-point distance measure (typically Euclidean distance).
3. Compute the mean vectors for the data assigned to each cluster. Denote the means by M_i, $i = 1, 2, \ldots, c$.
4. If the new cluster means are identical with the cluster centers, go to step 5. Otherwise, set $\hat{M}_i = M_i$, $i = 1, 2, \ldots, c$, and return to step 2.
5. Clustering complete. Analyze the separability of the clusters produced. This requires the use of an intercluster distance measure (perhaps transformed divergence).

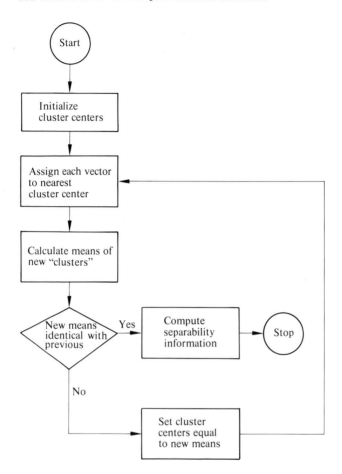

Figure 3-23 A basic clustering algorithm.

On every loop through steps 2, 3, and 4, the cluster centers move in such a way as to reduce the SSE. Since there is a lower limit on how small the SSE can be made, namely zero, there is no danger of looping infinitely; the algorithm is guaranteed to terminate.

Figure 3-24 shows an example of how the algorithm might perform when applied to a two-dimensional bimodal data set. In Fig. 3-24(a), the arbitrary selection of cluster centers results in partitioning of the data into the two "candidate" clusters shown. The means of the data in the candidate clusters become the cluster centers shown in (b), and reassignment of the data results in a new partitioning. The algorithm proceeds until in (d) the means of the clusters resulting from the partition are identical to the cluster centers which produced that partition. The process terminates when this occurs. Although this example suggests that only a few iterations are required (the number of times around the loop in Fig. 3-23), in practice tens or even hundreds of iterations may be required before the process arrives at the final cluster assignment.

The purpose of step 5 of the algorithm is to determine whether the clusters produced are actually distinct or whether some of them lie so close together in the measurement space as to constitute an unnecessary subdivision of the data set. This separability analysis is necessary because in step 1, as we pointed out earlier, the data analyst must specify how many clusters are to be determined, and usually can only make an educated guess at the appropriate number; the iterative part of the algorithm as described contains no provision for adjusting that number. The strategy, then, is to guess high and, in step 5, to consolidate any superfluous clusters after the iterative part of the algorithm terminates. This may be done by computing the intercluster distances and merging those clusters

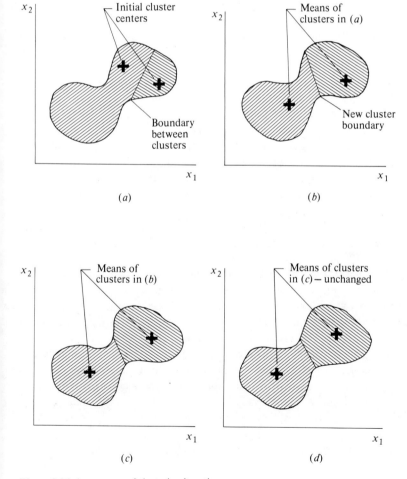

Figure 3-24 A sequence of clustering iterations.

which are separated by less than a prespecified amount. The threshold for merging must be set small enough so that resulting composite clusters will not have multimodal probability functions.

Unsupervised Analysis

When we categorize a remote sensing data set by dividing up the measurement space (or feature space) into nonoverlapping decision regions, we are actually defining *spectral* classes, i.e., classes which are discriminable because the multi-spectral properties of the corresponding ground covers are different. Remote sensing is successful because in many instances these spectral classes coincide with *informational* classes, i.e., classes of ground covers which are meaningful, such as crop species, major land uses, soil types, etc. Two different methods for establishing spectral classes have now been discussed in this chapter. The first method, which uses training samples to learn the spectral characteristics of the informational classes, is referred to as *supervised classification* because the data analyst in a sense "supervises" the establishment of decision boundaries by providing the training samples. The analyst "teaches" the classifier to recognize the informational classes.

The second method, which uses clustering, is called *unsupervised classification* because the data analyst has little control over the establishment of the decision regions. To complete the analysis, the spectral classes must eventually be converted to informational classes by identifying the ground cover which corresponds to each spectral class. Because of the analyst's reduced control, unsupervised classification is not generally so effective a method of characterizing informational classes as is supervised classification. This is particularly true when the informational classes are only marginally separable in the measurement space (e.g., two crop species which have only subtle spectral differences). Therefore, unsupervised analysis, based entirely on unsupervised classification, can be expected to produce reliable results only when the informational classes are easily discriminated in the multi-spectral data.

On the other hand, the advent of satellite-borne multispectral scanners for collecting remote sensing data has meant that for many applications purely supervised analysis is no longer feasible. The areas to be analyzed are large and heterogeneous, and it is often impractical to collect sufficient training samples to characterize every ground cover in the area. It may be too expensive or even physically impossible to do so. Furthermore, ground areas containing "pure" samples of the informational classes are difficult to locate precisely and may not contain a sufficient number of measurement vectors to adequately estimate the class statistics. To provide optimal analysis of such data, procedures have evolved which are hybrids of supervised and unsupervised analysis. One such procedure is summarized in the concluding section of this chapter (Table 3-1). Many of the analysis procedures used in the applications of Chap. 6 are also hybrid.

3-11 SUMMARY: PUTTING THE PIECES TOGETHER

Early in this chapter we observed that the job of designing a pattern-recognition system for remote sensing data analysis consists primarily of determining how best to divide the measurement space into decision regions, each region corresponding to a specific class of interest. We have since been developing the analytical tools to accomplish this on a statistically optimal basis. An appropriate conclusion to this chapter is the integration of the components into a procedure for carrying out the classifier design.

The data-analysis steps which lead to the classifier design, its use, and evaluation are shown in Table 3-1. The reader will find that the steps are given in very general terms. As we shall see in Chap. 6, the details of the analysis procedure to be used for any specific remote sensing application necessarily depend on the remote sensing instrumentation (e.g., the resolution of the sensor, the spectral bands available), the application itself (e.g., how many classes are to be discriminated and how difficult they are to discriminate), and on the resources available to implement the analysis procedure (e.g., the size and speed of the computer). See also Refs. 14 and 15.

The material presented in this chapter has been aimed primarily at analysis of the spectral content of remote sensing data. The evolution of the remote sensing technology to date has focused primarily on this aspect of the data; in other words, the feature vectors used as input to the analysis process have been almost exclusively vectors of multispectral measurements derived directly from scanners or other sources of multispectral data. However, recent research has begun to highlight the utility of both the spatial and temporal aspects of the data which were mentioned in Chap. 1. By adopting the point of view that a *pattern* is simply a *set of measurements*, regardless of how they were obtained (Sec. 3-1), the general pattern-recognition methodology can be extended to deal with spectral/spatial/temporal features as well as the purely spectral variety. The concepts presented here remain fundamental to such extensions.

Table 3-1 A typical procedure for quantitative analysis of remote sensing data

Step 1. Use *unsupervised analysis* to enhance the raw image data by deriving and displaying spectral classes.

Step 2. Use reference data to associate informational classes with the spectral classes. Select *training patterns*.

Step 3. Apply *clustering* and *separability analysis* to the training patterns to derive unimodal classes.

Step 4. Use *feature selection* or *feature extraction* to determine appropriate features for classification.

Step 5. Perform *maximum-likelihood classification* of the area of interest.

Step 6. Evaluate the results: (*a*) *rejection* rate or "thresholding," (*b*) classifier accuracy on training data, (*c*) classifier accuracy on test data.

Step 7. Refine the analysis or prepare the results for the consumer.

PROBLEMS

3-26 Give a nonmathematical, intuitive definition of clustering and illustrate two ways in which clustering can be used to aid in the analysis of multivariate data.

3-27 What are three essential elements of a mathematical definition of a cluster? Give an example illustrating the fact that clustering is not a uniquely defined operation but is dependent upon how *cluster* is defined.

3-28 State conditions under which unsupervised classification methods may be expected to work well.

3-29 State at least three practical limitations of supervised classification.

3-30 Using Table 3-1 as an outline of a typical numerical analysis procedure for remote sensing, explain the purpose or function of each step in the procedure.

3-31 Two classification procedures are described below.

Analysis procedure I
A subset of the data was formed by taking the data points corresponding to every fifth line and column. The subset of data was clustered. Initially, eight clusters were specified but examination of the cluster separabilities revealed that some clusters could be combined, resulting in five final clusters. The mean vector and covariance matrix for each cluster were computed and these were used to design a maximum-likelihood classification algorithm. The entire data set was then classified using the maximum-likelihood classifier. A unique color was assigned to each of the five classes and a color display produced showing how each data point was classified.

Analysis procedure II
Grayscale images were produced from the multispectral scanner data for a representative set of the available bands. Working from reference maps of the region, the analyst identified and selected candidate training areas for the cover types of interest. Each candidate training area was then clustered. Examination of the cluster output revealed that one of the training areas was composed of two distinct clusters while the other areas were essentially homogeneous in character. Two subclasses were defined for the nonhomogeneous training area. Statistics for the training areas were computed and used to specify the parameters of a maximum-likelihood classifier. The data set was then classified, unique colors were assigned to each training class, and a color display produced showing how each data point was classified.

    (*a*) Is procedure I supervised or unsupervised? What about procedure II?
    (*b*) Point out the significant differences between the procedures.
    (*c*) What kind of interpretation can be given to the classes produced by each procedure?
    (*d*) Under what conditions might you expect procedure I to work well?
    (*e*) What are some of the practical limitations of procedure II?

REFERENCES

1. Nilsson, N. J.: "Learning Machines," McGraw-Hill Book Company, New York, 1965.
2. Freund, J. E.: "Mathematical Statistics," Prentice-Hall, Englewood Cliffs, N.J., 1971.
3. Hadley, G.: "Linear Algebra," Addison-Wesley, Reading, Mass., 1961.
4. Anderson, T. W.: "An Introduction to Multivariate Statistical Analysis," John Wiley and Sons, New York, 1958.
5. Marill, T., and D. M. Green: On the Effectiveness of Receptors in Recognition Systems, *IEEE Trans. Information Theory*, vol. IT-9, pp. 11–17, January, 1963.
6. Swain, P. H., and R. C. King: Two Effective Feature Selection Criteria for Multispectral Remote Sensing, in *Proc. First International Joint Conference on Pattern Recognition*, IEEE Cat. no. 73 CHO 82 1-9C, IEEE Single Copy Sales, Piscataway, N.J., pp. 536–540, November, 1973.

7. Wacker, A. G.: The Minimum Distance Approach to Classification, Ph.D. Thesis, *LARS Information Note 100771*, Laboratory for Applications of Remote Sensing, and Technical Report TR-EE 71-37, School of Electrical Engineering, Purdue University, West Lafayette, Ind., 1971.

8. Kailath, T.: The Divergence and Bhattacharyya Distance Measures in Signal Selection, *IEEE Trans. Comm. Theory*, vol. COM-15, pp. 52–60, February, 1967.

9. Smedes, H. W., H. M. Spencer, and F. J. Thomson: Preprocessing of Multispectral Data and Simulation of ERTS Data Channels to Make Computer Terrain Maps of a Yellowstone National Park Test Site, *Proc. Seventh International Symposium on Remote Sensing of Environment*, vol. 3, Environmental Research Institute of Michigan, Ann Arbor, Mich., pp. 2073–2094, 1971.

10. Decell, H. P., Jr., and J. A. Quirein: An Iterative Approach to the Feature Selection Problem, in *Proc. Conference on Machine Processing of Remotely Sensed Data*, IEEE Cat. no. 73 CHO 834-2GE, IEEE Single Copy Sales, Piscataway, N.J., pp. 3B-1 to 3B-12, October, 1973.

11. Ready, P. J., and P. A. Wintz: Information Extraction, SNR Improvement, and Data Compression in Multispectral Imagery, *IEEE Trans. Communications*, vol. COM-21, pp. 1123–1131, October, 1973.

12. Ball, G. H.: Data Analysis in the Social Sciences: What about the Details?, in *Proc. Fall Joint Computer Conference*, Spartan Books, Washington, D.C., pp. 533–560, 1965.

13. Duda, R. O., and P. E. Hart: "Pattern Classification and Scene Analysis," John Wiley and Sons, New York, 1973.

14. Davis, B. J., and P. H. Swain: An Automated and Repeatable Data Analysis Procedure for Remote Sensing Applications, in *Proc. Ninth International Symposium on Remote Sensing of Environment*, Environmental Research Institute of Michigan, Ann Arbor, Mich., pp. 771–774, 1974.

15. Fleming, M. D., J. S. Berkebile, and R. M. Hoffer: Computer-Aided Analysis of Landsat-I MSS Data: A Comparison of Three Approaches, Including "Modified Clustering" Approach, in *Proc. Conference on Machine Processing of Remotely Sensed Data*, IEEE Cat. no. 75 CH 1009-O-C, IEEE Single Copy Sales, Piscataway, N.J., pp. 1B-54 to 1B-61, 1975.

FOUR

DATA-PROCESSING METHODS AND SYSTEMS

Terry L. Phillips and Philip H. Swain

In Chap. 2 we studied the central concepts of optical physics and instrumentation which must be considered in the design of systems for making quantitative remote sensing measurements. In Chap. 3 we considered a class of quantitative methods for analyzing those measurements and taking a major step toward converting them to useful information. As pointed out early in Chap. 3, these methods are most effectively implemented on a computer, which brings us to the subject of this chapter.

The potential users of the numerically oriented remote sensing technology must inevitably face the question as to what sort of data-processing system will best meet their needs. Closely related and equally important is the matter of precisely which numerical methods should be implemented by the data-processing system. These are not easy matters to resolve, in part because there is such a wide range of data-processing systems and methods to choose from. The dual purposes of this chapter, then, are (1) to make the reader aware of the spectrum of data-processing systems and methods available, and (2) to introduce some major considerations in the design of data-processing systems for remote sensing, providing a logical framework for evaluating such systems. It is beyond the scope and level of this book to provide detail sufficient to enable the reader to completely carry out the design of a data-processing system. Rather, the material presented here will provide at least a basic appreciation of how the components of such a system can contribute to the whole and a point of departure for selecting appropriate components based on the user's needs and resources. The many references in the chapter provide guidance as to where to look for greater detail.

Study objectives

After reading Sec. 4-1 you should be able to:

1. List the five entities that affect the design of a remote sensing data-processing system and describe the impact that placing a constraint on one of these entities has on the other entities.
2. Sketch a data-processing system model which allows the system designer to take into account the interfaces between the user, analyst, and data-acquisition system.
3. Defend the position that "the output subsystem is the logical place to start the system design."

4-1 A MODEL OF THE DATA-PROCESSING SYSTEM

To begin with, the reader should be aware that the purpose of the data-processing system may range from facilitating remote sensing research to producing specific products on an operational basis. Between these extremes lies every possible blend of these two purposes. The optimal design of a particular data-processing system depends on the purpose of the system probably more than on any other factor. Rather than focus on the design of systems for any specific remote sensing purpose, we shall discuss the considerations which must be universally applied regardless of the purpose and which in fact tend to lead to a design appropriate to the purpose.

Fundamental to our treatment of the remote sensing data-processing system is the interaction of five entities related to the system: the system user, the data analyst, the data-acquisition system, the data-processing hardware and software, and the system designer. It is important that we define precisely what these terms mean.

The *system user* (hereafter *user*) is defined as the consumer of the information produced by the data-processing system. In a sense, the very existence of the system depends on the user. The purpose of the system is determined by the needs of the user, which may be research results or specific products of the data-to-information conversion process. As a consumer of the system output, the user may never actually interact directly with the data-processing system.

The *data analyst*, on the other hand, may interact so intimately with the data-processing system that we actually consider the analyst a part of the system. The specific role of the analyst is determined by the purpose of the system. In a production-oriented system, the "analyst" may simply be an operator who monitors the flow of products and perhaps sees to the maintenance of the system. When research is the purpose of the system, the data analyst is likely to be the researcher himself, the system providing one of the principal tools for exploring new applications or developing new methods.

We use the term *data-acquisition system* to refer to the collection of devices and media which provides input to the data-processing system. This includes not only the primary remote sensing data-acquisition system but also any other sources which the data analyst may draw on to incorporate other forms of data (such as reference data) into the analysis process.

Data-processing hardware and software provide the core of the data-processing system in that these are tools available to the data analyst for converting input from the data-acquisition system to outputs in the form required by the user.

The *system designer*, of course, is the person or persons responsible for implementation of the data-processing system. To do the job most effectively, the system designer must develop an insightful understanding of a potential user's needs and determine how the capabilities of the data-acquisition system, the data analyst, and the data-processing methods at his or her command can best be brought to bear to meet the user's needs.

Figure 4-1 shows schematically the interrelationships of these five entities.

Although this chapter will in some respects read like a guide for the system designer, it should be clear that the material is highly relevant to the system user, the data analyst, and the designer of the data-acquisition system as well. All of these people have a potential impact on the design of the data-processing system and all must have at least a basic familiarity with the methods employed if that impact is to be maximally beneficial.

A convenient way to think of the remote sensing data-processing system and to organize the process of designing it is in terms of the model shown in Fig. 4-2. This particular way of organizing the system points up a number of very important interfaces, both among the system components and between the system and "the outside world." For example, an important aspect of the input subsystem is its interface with the data-acquisition system. The interactive subsystem provides the interface between the data-processing hardware/software and the data analyst (whom we are considering as a part of the data-processing system). And the output subsystem provides the interface with the system user.

Although we shall discuss the subsystems of our model more or less independently, it must be pointed out that the design of any of the subsystems is by

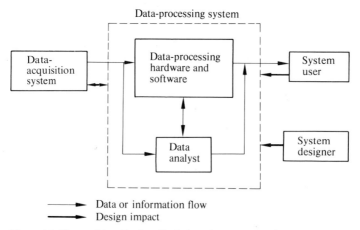

Figure 4-1 Five entities affecting the design of a remote sensing data-processing system.

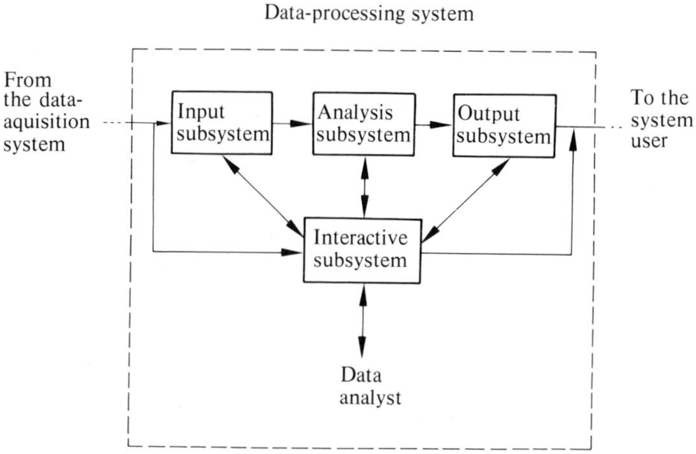

Data-processing system

Figure 4-2 A model of the data-processing system.

no means independent of the design of the others. The system design is really an iterative process so that there is opportunity to optimize trade-offs among the components, tending toward peak effectiveness of the overall system.

We observed earlier that the purpose of the data-processing system is determined by the needs of the user (or users). Since the output subsystem provides the interface with the user, this turns out to be a logical place to start the system design. Key decisions must be made jointly by the user and the system designer concerning the form of the output products to be provided by the system. In the next section of this chapter, we shall look into several important considerations which go into making these decisions. As suggested by Fig. 4-2, the design of the output subsystem must include some consideration of the data analyst's needs as well.

Attention can then be turned to the input subsystem and the forms of data required to produce the desired output. Usually the system will be required to handle several forms of input, often of widely differing characteristics and volume. Section 4-3 looks more closely at some of the design considerations resulting from this diversity.

What are the quantitative methods needed to convert the input data into output information? The system designer answers this question in specifying the analysis subsystem. As we shall see in Sec. 4-4, he has a tremendous array of methods to choose from. Again he must consider carefully the purpose of the system in order to make appropriate trade-offs among factors such as throughput, accuracy, flexibility, accessibility, and so on.

Critical to effective integration of the entire data-processing system is the design of the interactive subsystem. It has been widely observed that for many applications of quantitative remote sensing technology, the human data analyst can perform a number of functions more efficiently than they can be performed by any presently known computer implementation. In Sec. 4-5, we shall examine some considerations which contribute to effectively interfacing the man and the machine.

Finally, in Sec. 4-6 we shall introduce two approaches which can be used to evaluate a given system design or to compare a number of alternative designs.

PROBLEMS

4-1 What is the impact on the data analyst when a satellite-mounted data-collection system does not produce data with a north-south orientation? What is the impact on the system user?

4-2 Does the selection of the spectral bands used in a data-acquisition system have a greater impact on the data analyst or on the user?

4-3 Discuss the user's impact on the design of the data-processing system.

4-4 Is the analyst likely to have an impact on the hardware/software system design? Explain.

4-5 At what point in the data-processing system is the system designer likely to start his design? Why?

Study objectives
After reading Sec. 4-2 you should be able to:
1. Name the people that the output subsystem must serve.
2. Draw an output subsystem model and explain the interrelationships of the component parts.
3. Given an output subsystem and its associated user requirements, discuss the impact of changing one or more of the user requirements on the system design.
4. Read one of the application examples described in Sec. 6-3 and draw up a list of required output products, suggest hardware devices for producing them, discuss the need for analysis-results storage, and describe the type of results preparation required for each output product.

4-2 THE OUTPUT SUBSYSTEM

As suggested in Sec. 4-1, the purpose of the output subsystem is actually twofold: it must produce both the products required by the system user and some intermediate analysis results required by the data analyst. Typical outputs required by the system user include maps, tables, photographs, and digital magnetic tapes. The data analyst often needs the same products, but usually in a much less permanent form. In order to produce these products in a manner effective from the system point of view and also make them easily accessible, the organization of the output subsystem must be carefully planned. We shall take a look at some of the key characteristics of this design process.

In terms of Fig. 4-3, design of the output subsystem begins with listing the output products required, shown on the right-hand side of Fig. 4-3, and ends with detailed specification of each of the other operations and intermediate products shown. Although the design of the subsystem is generally an iterative process,

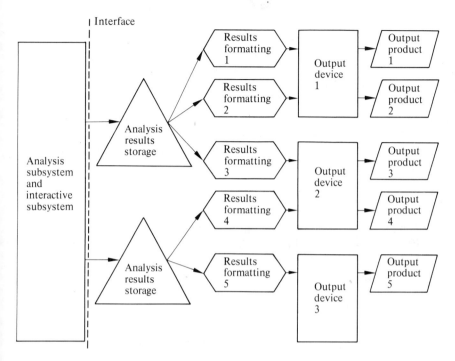

Figure 4-3 Conceptual layout of an output subsystem.

specifications tend to be developed by starting with the blocks on the right-hand side of the diagram and proceeding to the left-hand side. Once the output products have been specified, the output devices can be considered, with attention given to which products can be generated by a common device. Finally, devices for storage of the "raw" analysis results from the analysis subsystem are considered together with the software required to put the raw results into the form accepted by the output devices.

Careful attention should be given to the specification of the output products, for the system's effectiveness will be judged primarily on the ability of these products to provide the desired information in a useful format. In this chapter, we cannot discuss in detail the specifications of all possible output products; however, some sample specifications are discussed as a model for the designer.

An output product often desired is a map showing the results of classifying the remote sensing data. The specifications for such a product include:

1. The physical size, scale, and resolution of the map.
2. The means of indicating the classification of a pixel (symbol, color, etc.).
3. The method for indexing or determining the geographic positioning of any point on the map.
4. The geometric fidelity required.
5. The number of maps and the rate at which they must be produced.

6. Formatting details.
7. Limitations on the acceptable cost of producing each unit of output.

From a practical standpoint, trade-offs may be needed to arrive at an achievable set of specifications. For example, if it is necessary for the map to show the classification of each individual pixel, the scale may need to be larger and resolution coarser than would otherwise be the case. The most common trade-off, naturally, is cost versus quality and speed of production.

Another typical output product is a table or set of tables summarizing the analysis results. In this case, the specifications include:

1. The units to be used in the summarization (number of data points, acres, hectares, square miles, etc.).
2. The maximum number of classes which must be tabulated.
3. The class names to be used in the table.
4. The range of values (number of digits per value) likely to occur.
5. The boundaries of the area or areas to be tabulated.
6. Formatting details.
7. The number of copies required.

It may be necessary to implement the system with sufficient flexibility so that the user can summarize any analysis result in a multiplicity of ways. For example, he may wish to have a table for each census tract and for each transportation zone and for each county. He may wish to combine in one or more ways in the tables the classes which were defined for the classification (e.g., combine "old residential" and "new residential" into "residential"). Such flexibility can be expensive, however. Cost considerations may dictate that these manipulations be done by hand from a simple and standardized tabulation.

Once all of the output products required have been specified, attention can be turned to the hardware devices needed to produce them—although obviously this will have received considerable thought already since it bears so heavily on the cost and characteristics of the output products.

The list of output devices which could be considered is long and, as time passes, is growing steadily. In fact, the needs of remote sensing users have greatly stimulated the development of new output devices, particularly those well suited for output of maps and other imagery. Table 4-1 lists some examples, but it is by no means exhaustive.

Selection of the devices to be incorporated in the output subsystem depends heavily, of course, on the specifications for the output products required. Some additional factors which have not yet been mentioned include:

1. Permanence and stability of the products.
2. Time and labor required to produce each product.
3. The recording medium desired (paper, film, etc.).
4. Initial cost of the hardware.

Table 4-1 Some typical output devices for remote sensing data-processing systems

Primary usage/device	General characteristics			
	Throughput/ accessibility	Format versatility	Precision	Hardware cost

Hardcopy: mechanical
   Line printer
   Ink jet plotter
   X–Y plotter
   Electrostatic printer/plotter
   Thermal printer/plotter
	Low–mod	Low–mod	Low	Low–mod

Hardcopy: electronic
   Cathode-ray tube (CRT)
   Drum recorder
   Electron-beam recorder
   Laser-beam recorder
	Mod–high	Mod–high	Mod–high	Mod–high

Interactive
   Cathode-ray tube (CRT)
   Typewriter/teletype
	Low–mod	Mod–high	Low–high	Low–high

Storage
   Digital magnetic tape
   Film recorder
	Low–high	Mod–high	Low–high	Mod–high

5. Reliability of the hardware and maintenance cost.
6. Versatility (ability to produce multiple products and formats).

Again, trade-offs in performance factors and cost are inevitable in reaching a practical system design.

Many of the same considerations which go into the selection of the output devices are also relevant in determining appropriate media and hardware for storage of the raw analysis results. Thus hardware cost, reliability, and maintenance are again significant factors. Other important factors include:

1. Need for compact storage.
2. Ease and speed of results retrieval.
3. The volume of results to be stored.

In many respects, the volume of the results is the driving factor. It profoundly affects the need for compactness, which in turn influences the type of storage medium, which finally impacts the retrieval rate that is feasible. In estimating the volume, it is important to account for both the rate at which results are likely to be generated *and* the period of time it will be necessary to retain any given result in the system.

The final link in the output subsystem is the processor (hardware and/or software) required to transform the stored analysis results to a format acceptable to the output devices. If the output product is a map, this may mean a reformatting operation. If a table is required, counting operations of various complexities will be needed.

To see how these design considerations may be utilized, let us look at a simplified example of an output subsystem in the context of a specific application.

Described in Sec. 6-3 is a land-use study of the Great Lakes watershed by remote sensing which required two major output products: color-coded classification maps of land use and land-use tables organized in a number of different ways.

The user's specifications for the map products (see Fig. 6-11, between pages 36 and 37) were:

1. The classification results for each of the counties, to be printed at a scale of 1:235,000.
2. Each map to be composed of a rectangular grid of cells, each cell representing 1.75 hectares.
3. Cells to be color coded to indicate one of four land-use categories.
4. The maps to be reproducible by conventional printing processes.

The user's specifications for the tabular products (see Table 4-2) included:

1. Land use to be tabulated in units of both acres and hectares.
2. Land use to be tabulated both by county and by watershed.
3. Tables to be reproducible by conventional printing processes.

Table 4-2 Level I and level II classification results for Winnebago County, Wisconsin

Winnebago County, Wis.	Acres		Hectares		Percentage	
	Level I	Level II	Level I	Level II	Level I	Level II
Urban-commercial-industrial	26,930		10,900		7.3	
Residential		26,930		10,900		7.3
Commercial		—		—		—
Agriculture	203,360		82,320		55.0	
Row crop		134,200		54,330		36.3
Close-grown crop		23,200		9,390		6.3
Pasture		45,960		18,600		12.4
Forest	56,310	56,310	22,790	22,790	15.2	15.2
No major use	83,330		33,730		22.5	
Water		83,330		33,730		22.5
Wetland		—		—		—
Totals	369,930	369,930	149,740	149,740	100.0	100.0

The user's needs for these two types of product led to the output subsystem design shown in Fig. 4-4. The inputs to the subsystem were the classification results produced by the analysis subsystem and stored on the "results tape;" its outputs were the masters required by the conventional printing processes. The results tape actually served two purposes: (1) the data analyst used it for temporary storage of intermediate analysis results and (2) the final analysis results were stored on it prior to processing by the formatting software—the laser formatter and the tabular formatter.

The results tape contained considerable documentation concerning how the classification was obtained (training class statistics, for example). The tape was formatted to contain the classification code assigned to each 1.75-hectare cell by the analysis subsystem. The cells were arranged in "lines" running west to east and the lines sequenced so that the classified image could be reconstructed, a line at a time, north to south.

The laser formatter software was implemented with considerable flexibility to permit the data analyst to specify the portion of the results to be mapped, the scale of the map, the color to be assigned to each class, and, if necessary, combining of some of the classes defined in the analysis. The software automatically added registration indices (cross-hairs) to aid later in aligning the printing masters.

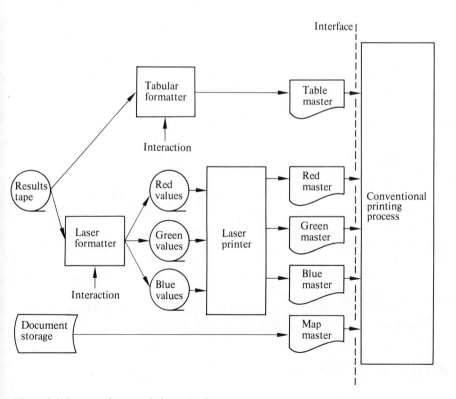

Figure 4-4 Output subsystem design example.

The output of the laser formatter was a set of three tapes from which the laser printer would produce the red, blue, and green masters for the final printing operation.

The tabular formatter software required the analyst to supply several of the same inputs required by the laser formatter: the portion of the results to be processed and any necessary combining of classes. In addition, the analyst was to specify the units to be used (acres or hectares) and the number of copies of the tables to be produced. The output of the tabular formatter was the printing master, produced directly on a computer line printer.

At every step of the design process, the design factors we introduced earlier in this section had to be accounted for in order to determine the most effective way to select or specify the subsystem components. For the sake of brevity in presenting this example, we have not discussed these considerations in detail, but readers will find it beneficial to work through them on their own. We have also elected not to complicate matters by dealing at length with the needs of the data analyst or with the possible impact of concurrent uses of this output subsystem for other remote sensing studies. The experienced system designer will account for all of these things while working toward an effective design.

PROBLEMS

4-6 Who are the people that must be served by the output subsystem?

4-7 Consider the output subsystem model shown in Fig. 4-3. Explain how storage in the system, including analysis results, could lead to a reduction in the cost of output products.

4-8 Consider the output subsystem shown in Fig. 4-4.

(*a*) How might the output subsystem change if the user had not specified geometrically corrected maps?

(*b*) In what part of the subsystem would you put the geometric correction operation?

Study objectives
After reading Sec. 4-3, you should be able to:
1. Given an input subsystem model, identify the block(s) that serve as the focal point for the subsystem design and state which person(s) the designer must keep in mind when designing the input subsystem.
2. Describe the difference between spatially scanned, spectrally scanned, and point data. Your description should show an understanding of the distinction between a difference in format and the more fundamental differences between spatial, spectral, and point data.
3. Given a list of data-collection systems and the characteristics of the data they produce, suggest devices for the input subsystem.

4-3 THE INPUT SUBSYSTEM

Having drafted a plan for the output subsystem, the system designer is ready to do the same for the input subsystem. A major goal of this step is to insure that the data-processing system will be able to accept all of the forms of input required to produce the user-specified outputs, including those needed by the data analyst (reference data, for example, if this actually needs to be handled by the data-processing system).

In order to compile a complete list of the necessary inputs, there are several sources of information which should be consulted before initiating the input sub-system design process. Two of these are, obviously, the user (and the output subsystem plan) and the data analyst. In addition, the system designer will need to study the specifications of the relevant data-collection systems and, more specifically, the characteristics of the data products created by those systems.

The design process actually gets started once the list of potential inputs is complete. Referring to the input subsystem model shown in Fig. 4-5, we may describe the design process in terms of working from one set of data media and formats produced directly by the data-collection systems to a second set of media and formats which must be very readily accessible to the analysis subsystem. The latter set is cataloged and stored in one or more system data banks, where by *data bank* we mean a collection of data stored in a single medium using a single format. Because accessibility of the typically large volumes of remote sensing data is vital to the effective overall performance of the data-processing system, the specification of the data banks in the input subsystem is really the focus of the input subsystem design.

The system designer can look for common characteristics among the different data sources to help determine the proper number and nature of the data banks required. Typically, he or she may encounter three major classes of data inputs which we shall refer to as *spatially scanned data* ("images"), *spectrally scanned data* ("spectra"), and *point data.*

The characterizing feature of spatially scanned data is that an image of the scene from which it was collected can be reconstructed from the data. The original

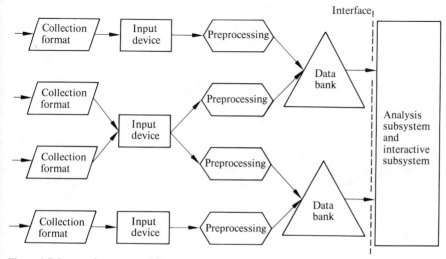

Figure 4-5 Input subsystem model.

data medium may be aerial photographs; more often in numerically oriented remote sensing systems the medium is digital or analog magnetic tapes from a multispectral scanner carried aloft by an aircraft or spacecraft.

Spectrally scanned data are most commonly collected in a research environment. These data originate from devices which make measurements over a broad, finely subdivided range of the electromagnetic spectrum. What is usually recorded is the average energy level per unit area in the field of view of the instrument as a function of wavelength; a single "scan" over the spectral range of the instrument is referred to as a *spectrum*. In numerically oriented systems, these spectra, which cannot normally be used to reconstruct an image of the scene, may also be stored on digital magnetic tapes. However, the nature of such data is fundamentally so different from the data collected by spatially scanning devices that it usually is not convenient to store it in the same format as the image data. Thus, the ensemble of spectrally scanned data will typically constitute a second data bank.

Point data, finally, consist of virtually all other forms of data which might be collected. Instruments which record individual measurements at a single instant of time (or average the measurements over a relatively small interval of time) produce point data. These measurements may also be stored on computer-compatible magnetic tape, although punched cards, paper tape, and other "low-volume" media are often used. Again, however, the appropriate format is usually sufficiently different that the point data will comprise a third distinct data bank.

A word about selecting the media for the data banks: the considerations are identical with those discussed in the preceding section for selecting the medium for storage of raw analysis results. Usually the determining factor will be the volume of data to be stored, but again the cost, reliability, and maintenance of the hardware are factors—as are the need for compact storage and the required ease and speed of retrieval. As noted earlier, spectrally and spatially scanned data are typically stored on magnetic tape, which is a very high volume, intermediate speed-of-access storage medium.

The input devices must be capable of accepting the data in the media and formats provided by the data-collection systems. Presumably these media and formats will have been selected to best accommodate the nature and volume of the data. A simple piece of hardware, such as a card reader or standard computer tape drive, may be all that is required as an input device. On the other hand, a complex subsystem such as an analog-to-digital tape converter may be necessary.

The final step of the input subsystem design process is the specification of the preprocessing functions. The purpose of each preprocessing function is to reformat the data generated by an input device into a specific format required by a data bank. Clearly, then, the key to the specification of a preprocessing function is the appropriate specification of the data bank, as discussed earlier. The preprocessing function may amount to a relatively simple reformatting task or may be a much more complex operation, depending on the nature of the input data and the data-bank specifications.

As suggested throughout this chapter, the design of the data-processing system is really an iterative process. The task of specifying what is to be included in the preprocessing functions of the input subsystem provides an outstanding example of the requirement for iteration in the design process. We shall see that several of the typical data-processing capabilities described in the next section are actually candidates for implementation as preprocessing functions in the input subsystem. The decision as to where to include them in the data-processing system depends considerably on the data analyst's need for accessing data with particular specifications. For example, if in the data-analysis process the data analyst requires several intermediate outputs all geometrically corrected to a particular map projection, then it would be appropriate to specify that the data in the data bank be preprocessed to that map projection. On the other hand, if the data analyst can interact with the system and the data without requiring that this step be included as a preprocessing function, then the total data-processing and analysis task may be completed less expensively if the correction is performed by a results formatting processor in the *output subsystem*. This is just one of many choices and trade-offs the system designer must make in coming up with the most effective plan for the overall data-processing system.

As an example, we will consider the design of an input subsystem to handle the products of a fairly typical remote sensing data-collection system designed for research. Three data products are involved: Landsat multispectral scanner (MSS) computer-compatible tapes, aircraft MSS analog tapes, and field radiometer analog tapes. A suitable system design is diagrammed in Fig. 4-6.

Since the list of inputs to the system has already been specified, we begin by considering the requirements for the system data banks. Two of the three inputs

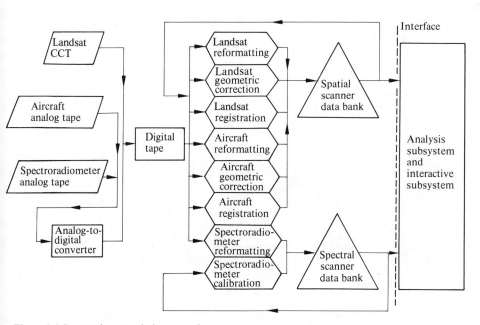

Figure 4-6 Input subsystem design example.

consist of image data (spatially scanned) and one consists of spectra (spectrally scanned). It is therefore appropriate to provide for two data banks—one for the image data and one for the spectra—since the storage and retrieval characteristics (and probably the data-analysis procedures as well) will be significantly different for these two types of input.

LARSYS, a remote sensing data-processing system developed at Purdue University's Laboratory for Applications of Remote Sensing (LARS), has its input subsystem configured in this way and is fairly typical of contemporary systems. The image data bank in LARSYS consists of digital magnetic tapes, referred to as multispectral image storage tapes. Magnetic tape was chosen as the storage medium for this data bank because:

1. It is a very reliable storage medium.
2. A very large library of multispectral image data can be stored and accessed relatively rapidly.
3. The medium is transportable, which means that the data can be made available to users of other remote sensing data-processing systems.

A multispectral image storage tape (Fig. 4-7) contains four types of records: (1) ID records; (2) data records; (3) end-of-file (EOF) records; (4) an end-of-tape record.

An ID record (Table 4-3) completely and uniquely identifies each of the runs stored as files on the tape. The ID record also contains calibration information which is used for radiometric calibration of the data in that run.

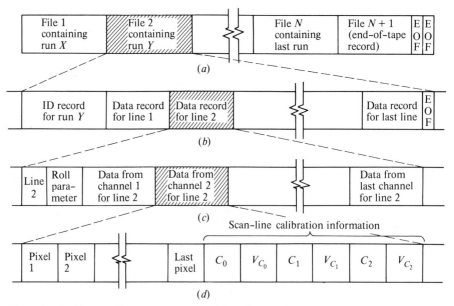

Figure 4-7 Multispectral image storage tape. (*a*) Tape format. (*b*) Run format. (*c*) Data record format. (*d*) Data from one channel.

Table 4-3 The ID record for the LARSYS multispectral image storage tape

Word	Format†	Description
ID(1)	Integer	Tape number (for example, 5, 102, etc.)
ID(2)	Integer	File number (1, 2, ...)
ID(3)	Integer	Run number (form *aabbbbcc*, where *aa* = last 2 digits of the year data were collected; *bbbb* = serial number, restarted each year; *cc* = modifier, allowing for various versions of data run *aabbbb___*)
ID(4)	Integer	Continuation code (0 means the run begins with this file; a positive integer *X* means this file contains the continuation of the run started in the last data file on tape *X*)
ID(5)	Integer	The number of data channels (spectral bands)
ID(6)	Integer	The number of data vectors per scan line, including 6 calibration vectors
ID(7–10)	Character	Alphanumeric flightline designator (16 characters)
ID(11)	Integer	Month data were taken
ID(12)	Integer	Day of month data were taken
ID(13)	Integer	Year data were taken
ID(14)	Character	Time of day data were taken (local, military)
ID(15)	Integer	Altitude of the sensor platform
ID(16)	Integer	Ground heading of the sensor platform
ID(17–19)	Character	Date data were preprocessed to generate this file (form *mmmmbdd, yyyy* where *mmmm* = month, *b* = blank, *dd* = day, *yyyy* = year)
ID(20)	Integer	The number of scan lines in this run
ID(21–50)	....	Not used; reserved for future needs
ID(51)	Real	Lower limit of first spectral band (μm)
ID(52)	Real	Upper limit of first spectral band (μm)
ID(53–55)	Real	Calibration values for the first spectral band
ID(56–200)	Real	For successive spectral bands, the same information as in ID(51) to ID(55)

† Word formats: LARSYS was developed for the IBM System/360 computers. Therefore, integer words are 32-bit two's complement signed binary numbers. Character words are extended binary-coded decimal interchange code (EBCDIC) 8-bit characters, 4 characters per word. Real words are 32-bit floating-point binary numbers in System/360 form.

A sequence of data records comprising the data run follows each ID record. Each data record corresponds to a scan line recorded by the multispectral scanner system. The scan lines are numbered sequentially, and the line number is the first piece of information recorded in each record. This is followed by a roll parameter which indicates the roll position of the scanner relative to vertical. Special values of the roll parameter are used to signal cases in which the roll is not known or no multispectral data are available for that line; in the latter case the rest of the record is just a "dummy." The data follows the roll parameter in *channel-interleaved* form as shown in Fig. 4-7(c) and (d). Channel interleaving was specified for LARSYS primarily because the characteristics of the System/360 hardware on which LARSYS was developed would thereby allow data from a subset of the multi-spectral channels to be read selectively from each data record. This was an attractive feature since as many as 30 channels were allowed for, but it was felt that most analysis would probably be done using a relatively small subset of the available channels.

Each scan line consists of a sequence of multispectral data vectors, each vector made up of values representing either the measurements made by the scanner on a ground resolution element or, at the end of the line, scan-line calibration information. Each value is recorded as an 8-bit integer. The range of possible data values is therefore 0 to 255. Through use of the calibration information in the ID record, these values can be converted, if desired, to the level of energy actually measured by the scanner in each spectral band. The scan-line calibration information, recorded following the data at the end of each line, provides information which can be used to account for any "drift" in the electronics of the data-collection system occurring during the course of a run.

Following the last scan line in a run is an end-of-file record which is simply a standard IBM end-of-file tape mark, recognized by the hardware and the system software. After the file containing the last run on the tape, an end-of-tape record is written. The end-of-tape record is identical in format to an ID record, but instead of the unique serial number identifying a data run which would appear in the ID record (Table 4-3) a zero is written, signifying "end-of-tape."

Finally, two successive end-of-file records are written immediately after the end-of-tape record.

The spectra data bank in LARSYS also consists of digital magnetic tapes, this medium having been selected for the same reasons as for the image data bank. The spectrally scanned data storage-tape format (Fig. 4-8) contains four types of records: (1) tape identification records; (2) run identification records; (3) data records; and (4) end-of-file records. The first file on the tape is a tape identification file which consists of a tape identification record and an end-of-file mark. Each tape has a unique tape identification record consisting of a tape number and a short verbal description of the type of data stored on the tape. This identification information is followed immediately by the end-of-file record.

Many data runs are stored on a single tape in separate files. For convenience, all of the runs stored on a particular tape are usually limited to those acquired by a single instrument. Each run contains a run identification record, data records, and an end-of-file record, as shown in Fig. 4-8(b).

In the LARSYS system, a large amount of information (300 bytes) is stored in the run identification record. Since these data are usually used for research purposes, every attempt is made to collect comprehensive information about the experiment, and this information is recorded in the run identification record, including: the run number, experiment number, observation number, date of data

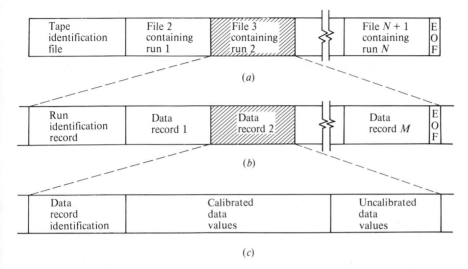

Figure 4-8 Spectrally scanned data storage-tape format. (*a*) Tape format. (*b*) Run format. (*c*) Data-record format.

collection, time of data collection, experiment name, principal investigator, scene type, location, meteorological information, calibration information, viewing angle, agronomic information, and processing parameters [see Fig. 2-36(*a*)].

The data record consists of data-record identification, calibrated data values, and uncalibrated data values. The number of data records in a run is usually determined by the number of detectors in the instrument acquiring the data. In the LARS field spectroradiometer system, the number of detectors is four; therefore, four data records are included in a run. The number of data records contained in a run is indicated in the run identification record.

The data-record identification information indicates from which detector the data in the record originated, the starting wavelength, the wavelength interval, the data-quality annotation, and the calibration values. The identification information is followed by calibrated data values, stored in 4-byte floating-point format, and uncalibrated data values, stored in 2-byte integer format. The purpose of storing the uncalibrated data is to provide the capability for recalibration should this be necessary for future work.

Finally, an additional end-of-file record is written on the end of the tape to indicate that no more data exist on the tape.

As indicated earlier in this section, the design of the data banks is the focal point of the input subsystem design, and this is why we have treated this aspect of the example in such detail.

Since the Landsat data are available in computer-compatible tape form, a digital tape drive serves as the principal input device for the sample input subsystem. The aircraft multispectral scanner data and field spectroradiometer data are recorded originally in analog tape form, so an analog-to-digital conversion system must also be included in the input subsystem. This type of conversion system was discussed in Sec. 2-7.

The final design decisions for the input subsystem shown in Fig. 4-6 concern the preprocessing functions. Three are required—for Landsat, aircraft, and spectroradiometer data reformatting—in order to convert the formats of the data available from the data-acquisition systems into the formats appropriate for the respective data banks.

In addition to these preprocessing functions, other preprocessing operations are also available in the input subsystem. These operations, which are discussed further in the next section, include geometric correction and geometric registration of the Landsat data and the aircraft data and calibration of the spectroradiometer data. The geometric correction operations have been included in the input subsystem so that all intermediate as well as final output products can be geometrically rectified. This provides the data analyst with an improved capability for interacting with the data and provides the system user with a more immediately usable output product. The registration preprocessing operations provide the data analyst with the availability of multitemporal information in the data-analysis process. Finally, the spectroradiometer calibration function improves the data analyst's facilities for extracting useful information from the data.

PROBLEMS

4-9 Refer to Fig. 4-5 and identify the system component(s) with which *people* must interact. Discuss the impact this interaction has on the subsystem design.

4-10 Consider a remote sensing research project which attempts to correlate agricultural crop yields with Landsat multispectral scanner data, laboratory spectroreflectometer data, field radiometer data, and meteorological data. In designing the input subsystem you are asked to minimize the number of data banks and formats. How many data banks do you recommend for the input subsystem? Justify your answer by describing the nature of each data source (spectral, spatial, point) and why data from different sources may (or may not) be combined in a single data bank.

4-11 For each data-collection system listed in the left-hand column below, select an input device from the right-hand column which you feel best matches the characteristics of the data-collection system.

Data-collection systems	Input devices
Landsat multispectral scanner (digital data)	Analog tape drive
Aerial camera	Card reader
Radiometer	Film scanner
Digital thermometer	Digital tape drive

4-12 Figure 4-7 shows the format of the multispectral image storage tape. The text mentions that up to 30 channels can be accommodated. Assuming that the input subsystem is to be designed for 4-channel Landsat data only, discuss the considerations which might lead you to use "point interleaving" rather than "channel interleaving" in specifying the format of the data records.

Study objectives
After reading Sec. 4-4 you should be able to:
1. State who the analysis subsystem serves.
2. Discuss how such analysis subsystem trade-offs as flexibility versus simplicity and "simple" algorithms versus "powerful" algorithms affect the data analyst and the user.
3. Select from the list of processing functions discussed those that might be good candidates for inclusion in (*a*) the input subsystem and (*b*) the output subsystem.

4-4 THE ANALYSIS SUBSYSTEM

A host of data-processing and data-analysis algorithms reported in the remote sensing literature are candidates for inclusion in the analysis subsystem. We shall mention many important algorithms and provide references to works which treat them in detail.

To organize this discussion, we shall consider various classes of processing operations in a sequence in which the operations might actually be utilized:

1. Radiometric transformations.
2. Geometric transformations.
3. Data presentation.
4. Data compression.
5. Image enhancement.
6. Statistical analysis.
7. Clustering.
8. Feature extraction.
9. Classification (supervised).
10. Results presentation.

A brief commentary on each of these classes of operations follows. Table 4-6 at the end of the chapter lists specific types of operations within each class and cites references where the reader will find more information about them.

Again we call attention to the fact that in any particular system it may be appropriate for some of these operations to reside in a subsystem other than the analysis subsystem. For instance, the system designer may choose to include the radiometric and geometric transformations in the input subsystem. Data-presentation and image-enhancement algorithms may be included in the interactive subsystem. Results-presentation operations are most likely to be included in the output subsystem. For the purposes of this discussion, however, we shall take advantage of the possibility that all of these operations may in some cases be included in the analysis subsystem, so that we can give them all a unified treatment.

Radiometric transformations are used to convert the raw multispectral data to a radiometrically consistent set of measurements. Often these transformations correct for specific types of problems in the data-collection system, such as uncompensated instability in the system electronics. Sometimes adjustments are made for environmental variations, such as changes in the illumination level over the course of a data-collection run.

As discussed in Chap. 2, the output of the data-collection system is generally translated to a range of values which is most convenient for the recording medium. A radiometric transformation may be used to absolutely calibrate the data, i.e., to reverse the translation process and recover the energy units as measured by the sensor.

Geometric transformations alter the image geometry, either to correct geometric distortions introduced by the data-collection system or to meet special requirements of the system user. Various forms of geometric distortion arise from limitations on the precision with which the attitude of the sensor can be controlled. Others may be introduced by small imperfections in the sensor optics. In either case, these distortions can be eliminated or substantially reduced by appropriate processing if data are available which characterize the sensor attitude and geometry.

"Registration" is the term used to refer to the process of geometrically aligning one set of data with another. The second set of "data" might even be a reference template, such as a map. Once this kind of alignment has been achieved, it is relatively easy to have access to all of the available aligned data corresponding to any given ground point. There is a wide range of possible types of data which might be registered. A few, in addition to data from repeated passes of the sensor over the same site, include soil type, topographical parameters, meteorological data, and land ownership.

Data presentation is a concern of the data analyst who needs to view or otherwise interact directly with the remote sensing data as part of the data-analysis process. The forms in which the data must be presented depend to a considerable extent on the nature of the application problem and the details of the analysis procedure used. The forms it is *practical* to produce depend very much on the hardware available (line printer, electrostatic printer, color CRT, etc.).

Data compression may be a particularly important processing capability because of the tremendous volume of data which remote sensing systems typically have to handle. Data compression is possible at any point in the system where the *information-carrying capability* of the data format is greater than the actual *information content*. By reducing or compressing the data format at such points, we can effectively reduce the data transmission, storage, or processing requirements on the system, a step which often results in reducing the system cost as well.

Image enhancement is a term which means different things to different people. In its narrowest usage, image enhancement refers to those processes which improve the visual quality of pictorially presented data. In its broadest usage, image enhancement can refer to any process which makes any sort of improvement on any variety of image-like data. The latter usage is so general as to encompass all of the processing functions discussed so far and many of those yet to be discussed. Having alerted the reader to this variability in usage, we shall invoke the principle that moderation in all things is virtuous and adopt a stance somewhere between the extremes, to be more precisely defined by the entries in Table 4-6.

All of the image-enhancement processes listed in the table improve, in some sense, the visual quality of the image data. But many can be useful irrespective of whether the data are actually to be viewed. For example, the filtering operations which accomplish edge enhancement might be part of a more complex procedure which locates objects in the data prior to performing an analysis of the relationships among such objects. The other operations in the table tend to remove various forms of noise, possibly introduced by the data-collection system, and thereby may be used in order to improve later classification analysis.

Statistical analysis is another very general term. We use it here to refer to a class of quantitative (numerical) tools which aid in the characterization of the data, often for the purpose of eventually training a classifier. Frequently these tools are used for other purposes, however, such as evaluating alternative data-collection plans or determining the quality of some other analysis result (such as a classification). Several statistical techniques are discussed in Chaps. 3 and 6 of this text.

Clustering, feature extraction, and *classification* are also discussed in Chap. 3. However, the material there is necessarily limited in two respects: it deals only with the spectral characteristics of the data, ignoring its spatial and temporal aspects; and it focuses principally on a single example of each processing function. The reader interested in pursuing these topics further should consult the literature cited in Table 4-6.

Results presentation, to meet the needs of the user, often involves considerably more processing capability than simply printing map-like displays or tabular summaries. This has already been suggested by our discussion of the output subsystem earlier in this chapter. For the user who needs maps, reformatting and annotation of the "raw" classification results are the minimal additional processing steps required. The production of tabular summaries may require extensive statistical analysis of the classification results to put them into the form needed for presentation.

In a number of practical cases, direct hardcopy output of the classification results is not in itself sufficient to meet the user's needs. Rather, the classification results must be stored in the computer system in such a way that the user can access and manipulate them interactively. In this instance, the *results* are really *data* to be input to other forms of systems for storage, retrieval, and manipulation.

If all of the processing capabilities listed in Table 4-6 were to be implemented in a single data-processing system, the interface between the data analyst and the system would be so complex that the analyst's effectiveness could be significantly impaired. The analyst would simply have too many choices to make! Therefore an analysis subsystem consisting of an appropriate subset of these processing capabilities is likely to provide a more effective tool for the analyst than one attempting to make all of them available. A critical aspect of the system design, then, is determining an appropriate subset. We noted at the very beginning of the chapter that the design of an optimal remote sensing data-processing system depends heavily on the purpose of the system, and here we see a major reason for that dependency. If, for example, the system is to be used for operational or production-line generation of products, it should be streamlined, with simplicity and efficiency as major considerations. If the system is to be primarily a research tool, the overriding consideration is likely to be flexibility, i.e., a broad choice of methods out of which to synthesize analysis procedures. Other aspects of "system purpose" which are factors in system design include the characteristics of the data (volume, complexity), the types of applications to which the analysis process is to be applied, and the anticipated sophistication of the users and data analysts.

These factors influence not only the selection of data-processing capabilities but the way in which they are implemented as well. A broad range of trade-offs between hardware and software is possible. If production-line operation is the goal, many key processes might be implemented, for efficiency, in special-purpose hardware. On the other hand, if flexibility for research is paramount, then software implementation on a general-purpose computer is more appropriate.

Many of these same factors influence the specification of the interface of the analysis subsystem with the data analyst. We refer to this interface as the interactive subsystem, the subject of the next section.

PROBLEMS

4-13 When designing the analysis subsystem, is it more important to consider the requirements of the user or the data analyst? In what respects?

4-14 Discuss the implications to the user if a less powerful classification algorithm is used in the analysis subsystem.

4-15 Does it follow that greater flexibility in the analysis subsystem (say, six clustering algorithms instead of one) requires less skill on the part of the analyst? Discuss.

4-16 Which of the processing functions listed at the beginning of Sec. 4-4 are candidates for inclusion in the
 (*a*) input subsystem?
 (*b*) output subsystem?

Study objectives
Upon completion of Sec. 4-5, you should be able to:
1. Define the role of the interactive subsystem in the data-processing system.
2. Discuss how the requirements of the interactive subsystem might change as the requirements of the user change.

4-5 THE INTERACTIVE SUBSYSTEM

Even those who may consider purely *automatic* processing of remote sensing data to be a long-term objective will readily concede that, in terms of present technology, the "man-in-the-loop," the data analyst, is an indispensible part of any effective contemporary remote sensing data-processing system. There are two related reasons for the unique role of human beings in the system. The first reason is their ability to efficiently acquire and maintain an awareness of "the big picture," i.e., an awareness not only of the complete remote sensing data set but also of the associated reference data, possibly in many formats and media, and of the relationships between the reference data and the remote sensing data. The second reason is their innate ability to draw on the ensemble of data available to them, making considerable use of previous training and experience to make critical decisions. In other words, they exhibit a high level of intelligent behavior in directing the analysis process.

By contrast, modern computers are capable only of relatively primitive levels of "artificial intelligence." They can take account of data only a few bits at a time, and their ability to synthesize a "big picture" from the data and apply reasoning is no match for a human being. They somewhat make up for these deficiencies by their speed and their quantitative capabilities, characteristics which we capitalize on heavily in designing the total data-processing system. But computer science will have to make some substantial breakthroughs in the artificial intelligence field before we can expect to completely eliminate the human from the remote sensing data-analysis process.

The task, then, in designing the interactive subsystem, is to provide facilities for interfacing the human being with machine and data as effectively as possible. There are seven characteristics of this subsystem which we shall focus on in turn, namely:

1. Data retrieval.
2. Analyst/data interface.
3. Analyst/data-processing system interface.
4. Analyst/results interface.
5. Distributed access.
6. Training of data analysts.
7. Hardware considerations.

As with every other aspect of the data-processing system, the relative importance of each of these seven points depends on the specific nature of the environment in which the system is to be used.

Data Retrieval

Gaining access to the data set(s) of interest can be one of the most expensive and time-consuming aspects of interacting with a remote sensing data-processing system. Convenient and rapid data retrieval deserves the attention of the system designer.

To begin with, the analyst should be able to specify the needed data by specifying a few familiar parameters such as date of collection and the location of the area of interest. The system designer would prefer to use a simple tag, e.g., a serial number, for each data set, but the analyst's job will be easier and more likely error-free if the more "conventional" parameters can be used. Implementation of this sort of access will require a data catalog in the system to link parameters to the physical location where the data are stored.

In general the analyst should not have to be concerned with the specifics of the data format. Either all data should be stored in the data library in a uniform format or the system should be capable of determining and reading, without aid from the analysts, any data format which is valid for the system. We say that the data format, medium, and input device should all be "transparent" to the data analyst.

Some forms of data which the analyst is likely to need may not be in computer-compatible form. Examples include aerial photographs and topographic maps. The analyst's job can be greatly facilitated if the data-processing system is designed to maintain a comprehensive and conveniently accessed catalog which cross-references the various forms of data.

Analyst/Data Interface

Once the analyst has located the data which need to be worked with, he or she must be able to effectively interact with and manipulate the data. To begin with, effective interaction requires that the system be capable of responding rapidly to the analyst's on-line data needs. Unfortunately, rapid response can be very expensive, with the data storage medium and input device characteristics being the principal cost-determining factors. The system designer must weigh carefully the actual needs of the analyst versus the cost of fully meeting them in the design of the inter-active subsystem.

The analyst needs to be able to view the data in image form—both the raw data and results of various processing operations. Two general categories of data-display devices should be considered: interactive and hardcopy.

Interactive displays, principally black-and-white or color CRT systems, are practically essential if the data-processing environment demands a high level of analyst participation in the analysis operations. They provide a high-quality image rapidly and, through the use of color, they can display information simultaneously from multiple images (spectral bands, registered multitemporal data, digitized reference data overlays, etc.). Usually it is desirable for the device to include a facility such as a light-pen, joy-stick, or track-ball which will enable the analyst to specify points or regions in the data to be operated on in some fashion by the data-processing system (Fig. 4-9).

The analyst may also need hardcopy displays of the data for off-line work. This is a factor which should be accounted for in designing the *output* subsystem. The characteristics of several classes of hardcopy devices are listed with our earlier discussion of the output subsystem (Sec. 4-2).

When hardcopy reference data, such as maps or photographs, must be precisely and quantitatively coordinated with numerical data in the computer, an interactive aid such as a coordinate digitizer may be required. This consists of a flat surface on which the reference data are mounted and a cursor which can be manually placed over any point in the data. Pressing a button then transmits the

Figure 4-9 Using a light-pen at the digital display to designate regions.

location of the cursor to the computer. Such a device may be considered part of the input subsystem.

Analyst/Data-Processing System Interface

The details of how the analyst actually goes about calling up the various processing capabilities can significantly impact the throughput of the entire data-processing system and hence the cost and time required for each result produced. Again the purpose of the system, with research at one end of the spectrum and production at the other, is an important consideration. In general, the research-oriented system must be designed with maximum flexibility in mind to provide the analyst with many options in specifying the steps of the data-processing procedure. On the other hand, the production-oriented system should be the essence of streamlined simplicity, designed to use a standard set of default options unless the analyst chooses to vary from a standard processing procedure.

In either case, the analyst/data-processing system interface should include facilities for recovery from both system and human errors. Every input the analyst provides the system with should be checked for intelligibility and consistency. When errors are detected or even suspected, the analyst should be given the chance to correct obvious errors and to verify unusual requests.

Some kinds of errors can only be detected by the analyst himself as processing proceeds. Efficient recovery from this type of error, either a system or an analyst error, requires "check-point" processing, which carefully preserves the intermediate results of the data-processing operations and allows the analyst to go back to an earlier step, a "check point," and reinitiate processing from that point using the saved results from preceding steps.

In order for the analyst to diagnose the type of error we have just mentioned or even to be confident that things are proceeding according to plan, the system should provide the analyst with periodic status reports. In addition to being provided on-line to the data analyst, these reports should be preserved in hardcopy form to provide an historical record of the data and procedures used and of any problems which may have been encountered in the course of the analysis. Obviously this will facilitate *post mortem* troubleshooting as well as documenting the derivation of any given set of results.

Analyst/Results Interface

It is useful and realistic to think of any analysis result as being *data* for further processing of some kind. Thus it is just as important for the data analyst to have ready access to results as it is for him or her to have ready access to the primary remote sensing data. Many of the same considerations apply: results should be uniquely identified and cataloged in the system; they should be conveniently and quickly retrievable; it should be possible to display them interactively and to selectively obtain hardcopies for off-line work.

Distributed Access

At the current state of computer technology, a good case can be made for the economies achievable through centralization of the data and the relatively large-scale computational capacity required for remote sensing data processing. However, the system designer may want to give some consideration to providing geographically distributed access to the central system. Putting the system capabilities in the analyst's office rather than making the analyst come to the system has been shown to be a very effective way to increase system utilization—which, of course, is the only way that economies of scale can actually be realized.

The design of an effective interactive subsystem for distributed access requires that the system designer be well acquainted with the currently available data-communications technology, a technology which is rapidly evolving. This topic is beyond the scope of this textbook.

Training of Data Analysts

The design of the interactive subsystem—and the entire data-processing and analysis facility, for that matter—should take account of the fact that data analysts will have to be trained to use the system. These analysts are likely to have a diversity of backgrounds including a wide range of familiarity with digital computers and their use.

The impact that these considerations may have on the system depend on:

1. The anticipated backgrounds of the trainees.
2. The data-processing environment (research versus production).

3. System resources which can be allocated for training.
4. Capital and personnel resources available for education purposes.

At one extreme, a full-blown computer-aided instructional facility, totally integrated with the data-processing and analysis system, is a possibility—but an expensive one. At a more conservative but still optional level, the system designer may elect to tailor the software to facilitate analyst training by providing for English-like control commands with elaborate diagnosis of the commands and opportunity for on-line error correction. (We have already noted the impact of error detection and recovery on system throughput and product cost.) Another training aid is a prompting capability. For example, the analyst might be permitted to enter the command HELP at any stage of on-line processing, which would result in printing a list of all control commands from which he or she may choose at that stage.

Of course, absolutely essential are off-line training facilities, including documentation of all procedures which the data analyst must know to do the job. The attention given to preparation of these materials and their format are bound to influence the time required to train the data analyst to an effective level. The level of proficiency achievable may also be affected.

Finally, one must not overlook the need to upgrade the training facilities as changes are made in the system and in the analyst's procedures. A little forethought along these lines at the time of system design can pay immense dividends in terms of the time required to acclimatize the data analyst to system changes.

Hardware Considerations

It is appropriate to summarize the functional hardware requirements of the interactive subsystem which must be considered by the system designer, realizing that the details are heavily dependent on the system environment, the available technology, and the resources which can be allocated to this component of the data-processing and analysis system. Table 4-4 contains this summary. Obviously these

Table 4-4 Key hardware elements in the interactive subsystem

Function	Typical devices
Man-to-machine communication	Keyboard (typewriter or special purpose); CRT menu and light-pen or cursor
Machine-to-man communication	Typewriter (mechanical or thermal printing); teletype; CRT
Graphics and image display	CRT (black-and-white, color); line printer
Temporary storage	*On-line*: disk, drum; *off-line*: punched cards, paper tape, magnetic tape (reel, cassette), floppy disk
Hardcopy	Line printer; film

considerations have much in common with those associated with designing the input subsystem and the output subsystem. In this case, however, greater attention must be paid to rapid response and less to permanence of the media involved.

PROBLEM

4-17 Consider the evolution of a data-processing system as it is redesigned from an initial research-oriented system to a production-oriented system. Reread Sec. 4-5 and discuss how each of the seven characteristics of the interactive subsystem might change.

Study objective
After studying Sec. 4-6 you should be able to:
1. Use two methods to compare two given data-processing systems for a given remote sensing application.

4-6 SYSTEM EVALUATION

In preceding sections, we have discussed the major considerations in the design of remote sensing data-processing systems. Early in the chapter it was pointed out that the system designer must ensure that the processing system (hardware and software) interfaces well with the system user, the data analyst, and the data-acquisition system. A fundamental approach to the system design was given in terms of a model of the data-processing system. The model included distinct subsystems providing the interfaces with the system user, the data analyst, and the data-acquisition system. In subsequent sections, major considerations in the design of each of these subsystems were discussed. We now have a foundation for the evaluation of data-processing systems for either specific or general tasks. In this section, two approaches to system evaluation will be introduced which provide us with the means for evaluating a given system design or for comparing alternative system designs.

The first approach is based on determining the cost of producing output products. This approach should be used whenever possible because it provides the most useful kind of information to the system user. The system user is able to judge the "value" of these products to his application in terms of cost effectiveness or other economic terms.

The second approach to system evaluation is based on less-precise considerations. For situations where cost is less of a concern than the evaluator's estimation of system "quality," the second approach might be preferable. This is especially true for complex situations where requirements are known only generally and the cost approach is not feasible.

The cost approach to system evaluation is simple to define but usually is not easy to implement. The outline of this approach is as follows:

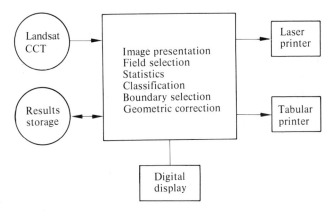

Figure 4-10 Block diagram of the system proposed by Remote Sensing Machines, Inc.

1. List the products to be produced by the system and estimate the total number of units of each product which will be required over a reasonable lifetime of the system.
2. Using the subsystem design approach previously described, estimate the cost of producing the required output products.

The system can then be evaluated on the basis of the output product cost.

As an example of the output product-cost approach, let us look at the Great Lakes watershed example introduced in Sec. 4-2. In this example the user required county maps, county tables, and watershed tables. Let us suppose two system designs were proposed to provide these output products, as described below.

> **Example** Remote Sensing Machines, Incorporated (we will abbreviate this to RSM),† proposed the system shown in Fig. 4-10, consisting of a special-purpose processor with two tape units, a digital display, a line printer, and a laser printer. RSM indicated that the Landsat computer-compatible tapes (CCTs) could be input to the special purpose processor and displayed on the digital display for the analyst to select training field boundaries. The analyst would then use these field boundaries to calculate appropriate training statistics and classify the Landsat data, storing the results on an intermediate results tape. The system would also provide the capability to display the intermediate results tape. The data analyst could then request the system to print the appropriate tables on the line printer and print the appropriate maps on the laser printer. It was suggested that through a combination of software included in the special-purpose processor and the hardware facilities of the laser printer, a geometrically corrected map of the results could be produced. The cost of the proposed system would be $600,000.

† The reader will discover that the two mythical "corporations" used in this example represent two fundamentally different ways of providing capabilities for remote sensing data analysis.

On the other hand, Remote Sensing Services, Incorporated (RSS), proposed to provide a specialized service to the data analyst via a dial-up terminal. It was indicated that if the data analyst would order imagery of the Landsat data and draw the county and watershed boundaries on the Landsat imagery, RSS would then obtain the Landsat tapes, digitize the county and watershed boundaries, and geometrically correct the Landsat data to provide the data analyst with access to a data base containing the counties (see Fig. 4-11). Through access to RSS's computer via a remote terminal, the data analyst could call up and display appropriate subsets of the county data base. He would then control the calculation of statistics by the system, followed by classification of the data and creation of a results tape. The output subsystem at RSS, shown in Fig. 4-4, would produce the appropriate tables and maps under control of the data analyst via the dial-up terminal. RSS indicated that the data analyst would have to provide the terminal link to their system and pay for the services used by the data analyst. It was estimated that the cost of these services would be: dial-up terminal, $3000; data-input preparation, $500 per county; data analysis, $600 per county; and output production, $200 per county. These quotes were for counties averaging 40 × 40 km, at a resolution of 80 m, and up to 16 classes per classification.

The system designer discussed both of these proposals with the data analyst. Based on a typical county which was processed on both systems to provide a benchmark, the data analyst felt the needed results could be produced using either of these systems. Both systems were found to have offsetting advantages and disadvantages and were judged to be about equal in capability.

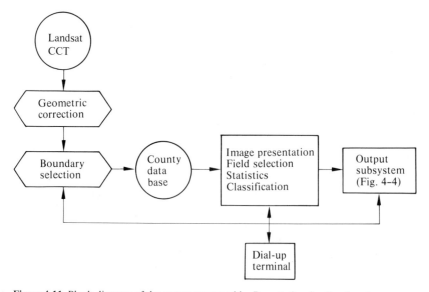

Figure 4-11 Block diagram of the system proposed by Remote Sensing Services, Inc.

The system designer then discussed both systems with the user who wanted the Landsat data-analysis results. Initially the user felt that both systems produced products containing the information desired. However, it was felt that the number of times it would be necessary to have the counties analyzed would not offset the initial investment to acquire the system proposed by RSM plus the cost of maintaining and upgrading the system over its expected lifetime. The system designer pointed out that there were other applications for which similar output products were required. The system user felt that the decision between the two systems depended primarily on whether it was desirable to establish a business of producing these products, and the final decision was made on this basis.

The second approach to system evaluation is based on Table 4-5. In this approach, the system designer assigns a weight factor to each of the capabilities of the system. Then these capabilities are evaluated, rating each on a scale of 1 to 10 according to its ability to satisfy a system requirement. This evaluation is multiplied by the associated weight, and the sum of the resulting products for all system capabilities gives the system designer a quantitative measure of the overall effectiveness of the system, which can then be used for decision-making purposes.

Table 4-5 System-evaluation format

System capabilities	Weight (a)	System 1 evaluation (b)	System 1 $(a \cdot b)$	System 2 evaluation (b)	System 2 $(a \cdot b)$
Output subsystem (25)					
Product 1					
Product 2					
Product 3					
⋮					
Input subsystem (20)					
Input 1					
Input 2					
Input 3					
⋮					
Analysis subsystem (20)					
Capability 1					
Capability 2					
Capability 3					
⋮					
Interactive subsystem (35)					
Access to inputs					
Access to processing capabilities					
Access to output product					
⋮					
Overall system effectiveness (totals of a · b columns)					

The system designer may approach this type of evaluation by first discussing the output subsystem with the system user. Together they can assign the appropriate weight to the output products according to the value of each output product to the user.

Next, the system designer may discuss the input subsystem, the analysis subsystem, and the interactive subsystem with the data analyst. The data analyst first assigns appropriate weights to the various data inputs which he or she feels will be required to produce the types of output product desired by the system user. Then weights are assigned to the processing capabilities, drawing on the experience of the data analyst. Finally, weights are assigned to the interactive subsystem capabilities according to the needs of the anticipated analysis operations.

Typical weights for entire subsystems are indicated in Table 4-5. Notice that the interactive subsystem has been assigned the largest total weight of 35 out of 100. This is because the data analyst's ability to interact with the system is considered critical to the effectiveness of the entire system. The next largest weight of 25 out of 100 is assigned to the output subsystem capabilities because of the absolute necessity to provide the user with the products which are required. Weights of 20 are assigned to the input subsystem and the analysis subsystem.

This evaluation method provides a quantitative approach to the qualitative evaluation of a system but is not intended to serve as the sole criterion for selecting a system. Rather, the information contained in the evaluation table will provide both the data analyst and the system user with useful information which together with cost and other pertinent considerations constitute a rational basis for selecting the most effective alternative system for the remote sensing application at hand.

PROBLEMS

4-18 Consider the Great Lakes watershed example discussed in Secs. 4-2 and 4-6. Assume that it costs $150,000 per year to maintain and upgrade the Remote Sensing Machines, Incorporated (RSM) system (Fig. 4-10) and that the system has an expected lifetime of five years.

(*a*) How many counties could you analyze using the Remote Sensing Services, Incorporated (RSS) dial-up terminal before you exceeded the cost of the RSM system?

(*b*) What would the lifetime of the RSM system have to be in order to be cost effective relative to the RSS system, assuming all 191 counties in the Great Lakes watershed example were required to be mapped once per year?

4-19 Evaluate the RSS and RSM systems using the method of Table 4-5. (*Hint*: you may want to review previous sections in this chapter in order to decide what output products, inputs, and processing capabilities are required.)

Table 4-6 Typical data-processing operations for remote sensing

(References appear at the end of the table. For a broad treatment of many of the subjects, the reader may wish to consult Refs. 1, 2, 3, 4.)

Processing function	Purpose	References
1. Radiometric transformations	**Correct radiometric abnormalities due to sensor and environmental characteristics**	5
Bias and gain adjustment	Correct for electronic "drift," detector differences	6, 7, 8
System noise removal	Correct for noise introduced by the detection, recording, and transmission systems	9, 10
Radiometric calibration	Convert data values to physical units	11, 12
Scan/illumination angle correction	Remove variable effects due to the changing relationship between view angle and illumination angle	13
Scene/environment variation corrections	Correct for variations in illumination, topography, atmospheric effects	10, 14, 15, 16
2. Geometric transformations	**Alter image geometry**	17, 18, 19
Registration	Achieve congruency of multiple sets of data from the same scene to each other or to another data base	17, 20, 21, 22, 23
Scaling	Adjust scale to meet user requirements	24
Projection conversion	Change to map projection required by user	
Correction for systematic distortions	Remove distortions due to sensor or platform characteristics	17, 24, 25, 26, 27, 28
3. Data presentation	**Data interface with analyst**	
Plotting (1-, 2-, or 3-dimensional)	Visualization of measurement space relationships	
Level slicing (gray-level or color display)	Observe data relationships in image format	29, 30
Photographic display of data	Observe data relationships in image format	31
4. Data compression	**Improve efficiency of transmitting, storing, or manipulating data**	32, 33, 34, 35
Source encoding		
Axis rotation		
5. Image enhancement	**Accentuate a characteristic of the data, often for visual display**	36
High-pass filtering	Edge enhancement	36
Low-pass filtering	Smooth local "noise"	10, 37
Laplacian filtering	Edge enhancement	36
Data restoration	Adjust for sensor or data system intermittency	
Finite aperture correction (MTF roll-off correction)	Reduce blurring due to finite optics	29, 38

Table 4-6 (*continued*)

Processing function	Purpose	References
6. Statistical analysis	**Characterize classes, select training samples, design classifier**	
Histogram	Observe intrachannel frequencies	
Correlation analysis	Observe interchannel relationships	39
Compute mean vectors, covariance matrices	Obtain first- and second-order sample statistics and Gaussian approximation of density functions	
Eigenvector transformation	Factor analysis, feature extraction	40
Sampling design	Selection of training and test data	41, 42, 43
7. Clustering	**Determine spectrally similar data classes; unsupervised classification**	30, 44, 45, 46
8. Feature extraction	**Dimensionality reduction to improve computational efficiency; enhancement of discriminating features**	47
Feature selection	Dimensionality reduction	30, 48
Linear feature combinations	Dimensionality reduction	40, 49
Ratios	Enhance discriminability	10, 50
Image segmentation	Enhance discriminability	51, 52
Texture analysis	Enhance discriminability	53
Spatial frequency analysis (Fourier analysis)	Enhance discriminability	17, 54
9. Classification—supervised	**Categorize (classify) data into ground-cover types**	47, 55, 56, 57
Linear		58
Quadratic (Gaussian)		30, 56
Nonparametric		59
Sample (per field)		30, 52
10. Classification— unsupervised	(see 7. Clustering)	
11. Results presentation	**Format results to meet user or analyst requirements**	3, 60
Image format	Map-like presentation for analysis,	
Line printer	evaluation or use as end-product	30, 61
CRT		
Film		
Plotter		
Annotation		
Tabular format	Quantitative summarization for analysis, evaluation or user end-product	
Areal measurements		30, 62
Areal estimation		43
Accuracy assessment		63
Analysis of variance		
(Interactive) storage and display	(On-line) availability for user interrogation or manipulation	64, 65

REFERENCES

1. Reeves, R. G. (Ed.): "Manual of Remote Sensing," 2 vols., American Society of Photogrammetry, Falls Church, Va., 1975.
2. Haralick, R. M.: Glossary and Index to Remotely Sensed Image Pattern Recognition Concepts, *Pattern Recognition*, vol. 5, pp. 391–403, 1973.
3. Lintz, J., Jr., and D. S. Simonett (Eds.): "Remote Sensing of Environment," Addison-Wesley, Reading, Mass., 1976.
4. Duda, R. O., and P. E. Hart: "Pattern Classification and Scene Analysis," John Wiley and Sons, New York, 1973.
5. Nalepka, R. F., and J. P. Morgenstern: Signature Extension Techniques Applied to Multispectral Scanner Data, in *Proc. Eighth International Symposium on Remote Sensing of Environment*, Environmental Research Institute of Michigan, Ann Arbor, Mich., pp. 881–893, October, 1972.
6. National Aeronautics and Space Administration: "ERTS Data User's Handbook," General Electric Document no. 71SD4249, Goddard Space Flight Center, Greenbelt, Md., 1972.
7. Hoffer, R. M., and F. E. Goodrick: Variables in Automatic Classification over Extended Test Sites, in *Proc. Seventh International Symposium on Remote Sensing of Environment*, Environmental Research Institute of Michigan, Ann Arbor, Mich., pp. 1967–1981, 1971.
8. Phillips, T. L.: Calibration of Scanner Data for Operational Processing Programs at LARS, *LARS Information Note 071069*, Laboratory for Applications of Remote Sensing, Purdue University, West Lafayette, Ind., 1969.
9. Rindfleish, T. C., J. A. Dunne, H. J. Frieden, W. D. Stromberg, and R. M. Ruiz: Digital Processing of the Mariner 6 and 7 Pictures, *Journal of Geophysical Research*, vol. 76, no. 2, pp. 394–417, 1971.
10. Maxwell, E. L.: Multivariate System Analysis of Multispectral Imagery, *Photogrammetric Engineering and Remote Sensing*, vol. 42, no. 9, pp. 1173–1186, 1976.
11. Cochran, J. K., and R. E. Bailey: Computer-Aided Extension of Digitized Remotely Sensed Water Surface Temperatures into the Third Dimension, *LARS Information Note 070576*, Laboratory for Applications of Remote Sensing, Purdue University, West Lafayette, Ind., 1976.
12. Bartolucci, L. A., P. H. Swain, and C. L. Wu: Selective Radiant Temperature Mapping Using a Layered Classifier, *IEEE Trans. Geoscience Electronics*, vol. GE-14, no. 2, pp. 101–106, 1976.
13. Landgrebe, D. A.: Data Processing II: Advancements in Large-Scale Data Processing Systems for Remote Sensing, in *Proc. Fourth Annual Earth Resources Program Review*, NASA Manned Spacecraft Center, Houston, Tex., pp. 51–1 to 51–31, 1972.
14. Parsons, C. L., and G. M. Jurica: Correction of Earth Resources Technology Satellite Multi-spectral Scanner Data for the Effect of Atmosphere, *LARS Information Note 061875*, Laboratory for Applications of Remote Sensing, Purdue University, West Lafayette, Ind., 1975.
15. Crane, R. B.: Preprocessing Techniques to Remove Atmospheric and Sensor Variability in Multi-spectral Scanner Data, in *Proc. Seventh International Symposium on Remote Sensing of Environment*, Environmental Research Institute of Michigan, Ann Arbor, Mich., pp. 1345–1355, 1971.
16. Rogers, R. H., K. Peacock, and N. J. Shah: A Technique for Correcting ERTS Data for Solar and Atmospheric Effects, in *Proc. Third Earth Resources Technology Satellite Symposium*, vol. I(B), NASA Goddard Space Flight Center, Greenbelt, Md., pp. 1787–1804, 1973.
17. Emmert, R. A., and C. D. McGillem: Multitemporal Geometric Distortion Correction Using the Affine Transformation, in *Proc. Symposium on Machine Processing of Remotely Sensed Data*, IEEE Cat. no. 73 CHO 834-2 GE. IEEE Single Copy Sales, Piscataway, N.J., pp. 1B–24 to 1B–32, 1973.
18. Kingston, P., and V. LaGarde: Spatial Data Massaging Techniques, in R. F. Tomlinson (Ed.), "Geographical Data Handling," vol. 2, International Geographical Union Commission, Ottawa, Canada, 1972.
19. Van Wie, P., and M. Stein: A Landsat Digital Image Rectification System, in *Proc. Symposium on Machine Processing of Remotely Sensed Data*, IEEE Cat. no. 76 CH 1103-1 MPRSD, IEEE Single Copy Sales, Piscataway, N.J., pp. 4A–18 to 4A–26, 1976.

20. Anuta, P. E.: Spatial Registration of Multispectral and Multitemporal Digital Imagery Using Fast Fourier Transform Techniques, *IEEE Trans. Geoscience Electronics*, vol. GE-8, no. 4, pp. 353–368, 1970.

21. Svedlow, M. C., C. D. McGillem, and P. E. Anuta: Analytical and Experimental Design and Analysis of an Optimal Processor for Image Registration, *LARS Information Note 090776*, Laboratory for Applications of Remote Sensing, Purdue University, West Lafayette, Ind., 1976.

22. Kaneko, T.: Evaluation of Landsat Image Registration Accuracy, *Photogrammetric Engineering and Remote Sensing*, vol. 42, no. 10, pp. 1285–1299, 1976.

23. Barnea, D. K., and H. F. Silverman: A Class of Algorithms for Fast Digital Image Registration, *IEEE Trans. Computers*, vol. C-21, no. 2, pp. 179–186, February, 1972.

24. Anuta, P. E.: Geometric Correction of ERTS-1 Digital Multispectral Scanner Data, *LARS Information Note 103073*, Laboratory for Applications of Remote Sensing, Purdue University, West Lafayette, Ind., 1973.

25. Baker, J. R., and E. M. Mikhail: Geometric Analysis and Restitution of Digital Multispectral Scanner Data Arrays, *LARS Information Note 052875*, Laboratory for Applications of Remote Sensing, Purdue University, West Lafayette, Ind., 1975.

26. Anuta, P. E.: Spline Function Approximation Techniques for Image Geometric Distortion Representation, *LARS Information Note 103174*, Laboratory for Applications of Remote Sensing, Purdue University, West Lafayette, Ind., 1974.

27. Wong, K. W.: Geometric and Cartographic Accuracy of ERTS-1 Imagery, *Photogrammetric Engineering and Remote Sensing*, vol. 41, no. 5, pp. 621–635, 1975.

28. Puccinelli, E. F.: Ground Location of Satellite Scanner Data, *Photogrammetric Engineering and Remote Sensing*, vol. 42, no. 4, pp. 537–543, 1976.

29. Billingsley, F. C.: Digital Image Processing for Photoreconnaissance Applications, in *Proc. Conference on Parallel Image Processing for Earth Observational Systems*, X-711-72-308, NASA Goddard Space Flight Center, Greenbelt, Md., pp. 41–57, March, 1972.

30. Phillips, T. L. (Ed.): "LARSYS Version 3 User's Manual," Laboratory for Applications of Remote Sensing Purdue University, West Lafayette, Ind., 1973.

31. Jones, A. D.: Photographic Data Extraction from Landsat Images, *Photogrammetric Engineering and Remote Sensing*, vol. 42, no. 11, pp. 1423–1426, 1976.

32. Duan, J. R., and P. A. Wintz: Information-Preserving Coding for Multispectral Data, in *Proc. Conference on Machine Processing of Remotely Sensed Data*, IEEE Cat. no. 73 CHO 834-2 GE, IEEE Single Copy Sales, Piscataway, N.J., pp. 4A–28 to 4A–35, 1973.

33. Andrews, H. C.: "Computer Techniques in Image Processing," Academic Press, New York, 1970.

34. Wintz, P. A.: Transform Picture Coding, *Proc. IEEE*, vol. 60, no. 7, pp. 809–820, 1972.

35. Ready, P. J., and P. A. Wintz: "Multispectral Data Compression Through Transform Coding and Block Quantization," Technical Report TR-EE 72-2, School of Electrical Engineering, Purdue University, West Lafayette, Ind., 1972.

36. Rosenfeld, A.: "Picture Processing by Computer," Academic Press, New York, 1969.

37. Brown, W. L.: Reduced Variance in Remotely Sensed Multispectral Data—A Pragmatic Approach, in F. Shahroki (Ed.), "Remote Sensing of Earth Resources," vol. 1, University of Tennessee, Tullahoma, Tenn., pp. 525–537, 1972.

38. Riemer, T. E., and C. D. McGillem: Optimum Constrained Image Restoration Filters, *LARS Information Note 091974*, Laboratory for Applications of Remote Sensing, also Technical Report TR-EE 74-37, School of Electrical Engineering, Purdue University, West Lafayette, Ind., 1974.

39. Basu, J. P., and P. Odell: Effects of Intraclass Correlation Among Training Samples on the Misclassification Probabilities of Bayes' Procedures, *Pattern Recognition*, vol. 6, pp. 13–16, 1974.

40. Fukunaga, K., and W. L. G. Koontz: Application of Karhunen-Loeve Expansion to Feature Selection and Ordering, *IEEE Trans. Computers*, vol. C-19, no. 4, pp. 311–318, April, 1970.

41. VonSteen, D. H., and W. H. Wigton: Crop Identification and Acreage Measurement Utilizing ERTS Imagery, in *Proc. Third Earth Resources Technology Satellite Symposium*, vol. 1, NASA Document SP-351, Goddard Space Flight Center, Greenbelt, Md., pp. 87–92, 1973.

42. Langley, P. G., J. Van Roessel, and S. Wert: "Investigation to Develop a Multistage Forest Sampling

Inventory System Using ERTS-1 Imagery," Technical Report NAS5-21853, NASA Goddard Space Flight Center, Greenbelt, Md., 1974.

43. Wigton, W. H.: Use of Landsat Technology by Statistical Reporting Service, in *Proc. Symposium on Machine Processing of Remotely Sensed Data,* IEEE Cat. no. 76 CH 1103-1 MPRSD, IEEE Single Copy Sales, Piscataway, N.J., pp. PB-6 to PB-10, 1976.

44. Ball, G. H., and D. J. Hall: "ISODATA: A Novel Method of Data Analysis and Pattern Classification," Technical Report, Stanford Research Institute, Menlo Park, Calif., 1965.

45. Schell, J. A.: A Comparison of Two Approaches for Category Identification and Classification Analysis from an Agricultural Scene, in F. Shahroki (Ed.), "Remote Sensing of Earth Resources," vol. 1, University of Tennessee, Tullahoma, Tenn., pp. 374–394, 1972.

46. Ball, G. H.: Data Analysis in the Social Sciences: What About the Details?, in *Proc. IEEE Fall Joint Computer Conference,* Spartan Books, Washington, D.C., pp. 533–560, 1965.

47. Fu, K. S., D. A. Landgrebe, and T. L. Phillips: Information Processing of Remotely Sensed Agricultural Data, *Proc. IEEE,* vol. 57, no. 4, pp. 639–653, April, 1969.

48. Swain, P. H., and R. C. King: Two Effective Feature Selection Criteria for Multispectral Remote Sensing, in *Proc. First International Joint Conference on Pattern Recognition,* IEEE Cat. no. 73 CHO 821-9C, IEEE Single Copy Sales, Piscataway, N.J., pp. 536–540, 1973.

49. Wheeler, S. G., and P. N. Misra: Linear Dimensionality of Landsat Agricultural Data with Implications for Classification, in *Proc. Symposium on Machine Processing of Remotely Sensed Data,* IEEE Cat. no. 76 CH 1103-1 MPRSD, IEEE Single Copy Sales, Piscataway, N.J., pp. 2A–1 to 2A–9, 1976.

50. Smedes, H. W., M. M. Spencer, and F. J. Thomson: Processing of Multispectral Data and Simulation of ERTS Data Channels to Make Computer Terrain Maps of a Yellowstone National Park Test Site, in *Proc. Seventh International Symposium on Remote Sensing of Environment,* Environmental Research Institute of Michigan, Ann Arbor, Mich., pp. 2073–2094, 1971.

51. Robertson, T. V., and K. S. Fu: "Multispectral Image Partitioning," Technical Report TR-EE 73-26, School of Electrical Engineering, Purdue University, West Lafayette, Ind., 1973.

52. Kettig, R. L., and D. A. Landgrebe: Classification of Multispectral Image Data by Extraction and Classification of Homogeneous Objects, *IEEE Trans. Geoscience Electronics,* vol. GE-14, no. 1, pp. 19–26, January, 1976.

53. Haralick, R. M., K. Shanmugam, and I. Dinstein: Textural Features for Image Classification, *IEEE Trans. Systems, Man, and Cybernetics,* vol. SMC-3, pp. 610–621, 1973.

54. Higgins, J. L., and E. S. Deutsch: The Effects of Picture Operations in the Fourier Domain and Vice Versa, in F. Shashroki (Ed.), "Remote Sensing of Earth Resources," vol. 1, University of Tennessee, Tullahoma, Tenn., pp. 460–480, 1972.

55. Fukunaga, K.: "Introduction to Statistical Pattern Recognition," Academic Press, New York, 1972.

56. Nilsson, N. J.: "Learning Machines," McGraw-Hill Book Company, New York, 1965.

57. Siegal, B. S., and M. J. Abrams: Geologic Mapping Using Landsat Data, *Photogrammetric Engineering and Remote Sensing,* vol. 42, no. 3, pp. 325–337, 1976.

58. Crane, R. B., and W. Richardson: Rapid Processing of Multispectral Scanner Data Using Linear Techniques, in F. Shahroki (Ed.), "Remote Sensing of Earth Resources," vol. 1, University of Tennessee, Tullahoma, Tenn., pp. 581–595, 1972.

59. Haralick, R. M.: Automatic Remote Sensor Image Processing, in A. Rosenfeld (Ed.), "Digital Picture Analysis," Springer-Verlag, Heidelberg, pp. 5–63, 1976.

60. Tomlinson, R. D. (Ed.): "Geographical Data Handling," Symposium ed., International Geophysical Union Commission, Ottawa, Canada, 1972.

61. Wilson, L. L.: Purdue/LARS Digital Display User's Guide, *LARS Information Note 022675,* Laboratory for Applications of Remote Sensing, Purdue University, West Lafayette, Ind., 1975.

62. Swain, P. H.: Land Use Classification and Mapping by Machine-Assisted Analysis of Landsat Multispectral Scanner Data, *LARS Information Note 111276,* Laboratory for Applications of Remote Sensing, Purdue University, West Lafayette, Ind., 1976.

63. Hoffer, R. M.: "Natural Resource Mapping in Mountainous Terrain by Computer Analysis of

ERTS-1 Satellite Data," Research Bulletin 919, Agricultural Experiment Station, Purdue University, West Lafayette, Ind., 1975.

64. Bryant, N. A., and A. L. Zobrist: IBIS: A Geographic Information System Based on Digital Image Processing and Image Raster Datatype, in *Proc. Symposium on Machine Processing of Remotely Sensed Data*, IEEE Cat. no. 76 CH 1103-1 MPRSD, IEEE Single Copy Sales, Piscataway, N.J., pp. 1A-1 to 1A-7, 1976.

65. Hitchcock, H. C., T. L. Cox, F. P. Baxter, and C. W. Smart: Soil and Land Cover Overlay Analyses, *Photogrammetric Engineering and Remote Sensing*, vol. 41, no. 12, pp. 1519-1524, 1975.

CHAPTER FIVE

BIOLOGICAL AND PHYSICAL CONSIDERATIONS IN APPLYING COMPUTER-AIDED ANALYSIS TECHNIQUES TO REMOTE SENSOR DATA

Roger M. Hoffer

Many aspects of the instrumentation systems utilized in remote sensing have been discussed thus far, as well as the theory and procedures involved in quantitatively analyzing data from such systems. However, we have not yet considered the relationships that exist among the instruments, the data-processing techniques and the earth surface features of interest. In applying pattern-recognition theory to data obtained from multispectral scanner systems for purposes of identifying and mapping various earth surface features, a basic underlying premise is that the cover types of interest are indeed spectrally separable. Experience has shown that this premise is often valid and that many earth surface features can be identified and mapped on the basis of their spectral characteristics, but we also know that some features of interest cannot be spectrally separated and identified. Thus, effective utilization of remote sensor data requires thorough knowledge and understanding of the spectral characteristics of the various earth surface features and the factors that influence these spectral characteristics. It is the purpose of this chapter to discuss the fundamental energy-matter interactions that control and influence the spectral characteristics of vegetation, soil, water, and snow features and to discuss the temporal and spatial effects on the spectral characteristics of earth surface features. The last major section of the chapter, Sec. 5-6, shows how we can *apply* our knowledge of the spectral characteristics of earth surface features to more effectively interpret multispectral remote sensor data.

Study objectives

After studying Sec. 5-1, you should be able to:

1. Define the two primary factors that cause the spectral response of major cover types, e.g., vegetation, soil, or water, to be dynamic rather than static in nature.
2. Describe at least two examples that illustrate the way in which the spectral properties of earth surface features of interest might change as a function of time for a specific geographic location, and as a function of geographic location at a specific time during the year.

5-1 SPECTRAL, TEMPORAL, AND SPATIAL CHARACTERISTICS OF THE SCENE

In considering the relationships between sensor systems and data-processing techniques, it is important to remember that most remote sensing data-acquisition systems simply record, in selected wavelength bands, variations in the amount of energy being reflected or emitted by objects on the surface of the earth. For photographic sensors, the data are recorded directly onto a two-dimensional array (photographic film). With scanner systems, the energy is usually recorded on magnetic tape and later may be displayed in a two-dimensional array to obtain an image format.

After the data are displayed in a two-dimensional array one can usually interpret the spatial features in the data, such as size, shape, texture, or linear features. From an image-interpretation standpoint, the spatial features are often vitally important to achieve proper identification of the object. However, at present, the application of pattern-recognition algorithms to earth resources data has not advanced operationally to the point where spatial features can be utilized as effectively as spectral features. Thus, for the purpose of applying computer-aided analysis techniques to remotely sensed data, we are primarily interested in the spectral characteristics of the data. For this reason, a knowledge of the spectral characteristics of the various vegetation, soil, water, and other earth surface features of interest is a basic and very essential ingredient for proper analysis and interpretation of remote sensor data.

If we pause for a moment to consider the biological system with which we are working, it becomes obvious that the spectral characteristics of various earth surface features do not remain static—they change with geographic location and time. For example, one often finds that on a particular date a certain crop species will have reached various stages of maturity in different geographic areas. Wheat may be mature and in the process of being harvested in the Southern Great Plains, whereas on the same date in the Northern Plains States it is still green and therefore has a distinctly different spectral response. Or, consider the spectral changes that take place with time in one geographic location. A forest canopy that is green in July may become red, yellow, or brown in late September, depending on the species involved. Urban or suburban areas that have a large number of deciduous trees or large lawns will look very different (spectrally) in autumn, after the leaves have fallen from the trees and the grass has become brown, than they did a few months earlier. Thus, seasonal changes or temporal variability of the biological system must be taken into account in order to determine the best time of the year to obtain remote sensor data in order to meet a particular information objective. In considering temporal changes, we recognize immediately that their effect is basically a variation in the spectral characteristics of the materials which we are sensing. Study of the spectral characteristics of earth surface features must therefore include consideration of the temporal variations of these spectral characteristics.

Temporal changes in spectral response can be either natural, such as the seasonal changes in tree foliage, or they can be caused by human beings, such as the effect in spectral response caused by the farmers who plow under the stubble in their fields or the construction crew who, at the start of a construction project, must remove the surface vegetation and expose the soil below. In remote sensing, "change-detection" techniques can be used to monitor these temporal changes; data are obtained over the same geographic area at different times of the year or at the same time in different years and analyzed to determine where and how extensive certain types of change in spectral response have been. Such change-detection procedures have been used with satellite data for a variety of applications, such as determining the extent of flooding of large rivers, mapping the areal extent of snow cover, assessing urban expansion, and monitoring forest logging operations.

In addition to the spectral and related temporal considerations, the spatial characteristics of the biological system with which we are involved must be considered. For example, if a certain spectral pattern is found in large rectangular blocks on the ground, we would infer that we are in an agricultural area and we are looking at a group of agricultural fields and possibly some woodlots. Such a pattern would be quite different from one found in a wildland area dominated by forest and rangeland. Such spatial characteristics are apparent because of differences in the spectral characteristics of the various materials the remote sensor system is "looking at." That is, it is the contrasts in spectral response (tone or color) between an object and the surrounding objects which allow one to determine the size and shape (spatial characteristics) of the target object. For example, a concrete highway can be easily seen and identified on satellite imagery in the visible wavelengths if the highway is wide enough and is surrounded by green vegetation or water (as in the case of a bridge). This is possible because of the large amount of contrast in spectral response in the visible wavelengths between the highly reflective concrete and the low reflectance of green vegetation or water in these wavelengths. However, if there were sandy soil (as in a desert area) beside the highway, the contrast between the highly reflective concrete and the highly reflective soil is not large, and detecting the highway would be much more difficult.

Another spatial characteristic of importance in some remote sensing situations is texture. In essence, texture is simply a repeated variation in tone (spectral response) over relatively small areas. It is often the textural differences on aerial photos that allow one to accurately identify forest cover or various agricultural crops. Frequently, however, texture is not as apparent on multispectral scanner data because of the system's lower spatial resolution. From satellite altitudes, texture is often not discernable and one is much more dependent upon tonal variations (spectral response) and very gross spatial features to identify the various objects or cover types of interest. Therefore, when utilizing satellite multispectral scanner data, the identification of many earth surface features is primarily a function of the spectral response of these features.

In summary, multispectral scanner systems provide a very effective method for obtaining spectral data in a quantitative format over a broad range of wavelengths, and, as discussed in Chaps. 1 and 3, computer-aided analysis techniques provide a very powerful method for analyzing such quantitative spectral data. The critical point here is that the essence of both the data collection and the analysis systems are focused upon the spectral characteristics of the various earth surface features. Therefore, it is vitally important that we thoroughly understand these spectral characteristics and examine them in the light of the temporal and spatial variations which affect them. The following three sections of this chapter will examine the spectral characteristics of basic, naturally occurring materials on the earth's surface—vegetation, soil, water and snow.

PROBLEMS

5-1 As a part of a hydrological study, an airborne multispectral scanner system was used to obtain reflectance measurements of a major river, following the river from its source in a forested area, through an industrial area, and stopping in a flood plain area. The first mission was flown in March during spring flooding; the second in June. Table P5-1 describes the relative reflectance of the water observed in the visible wavelengths and the associated water-quality conditions for each river segment for both dates.

Table P5-1

Area	March		June		
	Relative reflectance of water	Water quality	Relative reflectance of water	Water quality	
Forest	Low	Clear, clean	Low	Clear, clean	|
Industrial	Medium	Some industrial pollution	Medium	Some industrial pollution	Factor 2
Flood plain	High	Suspended soils	Medium	Some industrial pollution	↓

———————— Factor 1 ————————→

Recall the two primary factors that cause the spectral response of major cover types to be dynamic rather than static in nature, and identify which one is operating as Factor 1 in the example above and which as Factor 2. Describe the spectral changes related to each factor, as outlined in the table.

5-2 A sparsely wooded area is being converted to a suburban residential area. Describe the factors or events that you would expect to affect the spectral response of this area as the project progresses from start to finish. (You need not describe the spectral responses in any detail.)

5-3 Select a ground-cover type of interest (agricultural crop, natural vegetation, water, urban areas, etc.) and describe factors which might affect the spectral response over a wide geographic area at some fixed time.

5-4 Describe the basic temporal changes in a deciduous forest stand which would affect its spectral response over a period of one year. (Your answer need not describe more than four such time-related conditions.)

Study objectives

After completing Sec. 5-2 you should be able to:
1. Write the energy-balance equation and give a physical interpretation of each term in the equation.
2. Describe the nature of the energy-matter interactions that take place when energy in the (*a*) visible, (*b*) near-infrared, and (*c*) middle-infrared wavelength bands strikes a leaf.
3. State the approximate wavelength (in micrometers) where absorption by chlorophyll, xanthophyll, and anthocyanin pigments (green, yellow, and red, respectively) each dominate the spectral reflectance properties of vegetation.
4. Calculate the amount of reflectance expected from a canopy composed of several layers of leaves when given the appropriate components of the energy-balance equation for a single leaf.
5. Identify the approximate spectral location of the five primary absorption bands found in green vegetation between 0.4 and 2.6 μm, and describe the cause of absorption at each of these wavelengths.
6. Describe the changes in the reflectance properties of vegetation as a function of moisture content.

5-2 SPECTRAL CHARACTERISTICS OF VEGETATION

The spectral reflectance of green vegetation is distinctive and quite variable with wavelength. Figure 5-1 shows a typical spectral reflectance curve for green vegetation and identifies the spectral response regions of major significance. In the visible wavelengths, pigmentation dominates the spectral response of plants; chlorophyll is especially important, although we shall see later that other plant pigments also play a significant role in this portion of the spectrum. In the near-infrared region, the reflectance rises noticeably because the green leaf absorbs very little energy in this region. In the middle-infrared region, water absorbs energy strongly in particular wavelengths, and, because green leaves have very high

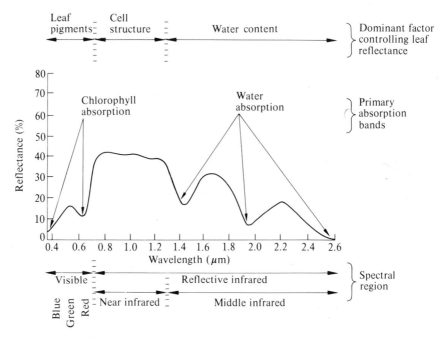

Figure 5-1 Significant spectral response characteristics of green vegetation.

moisture content, these water-absorption bands dominate the spectral response in this region.

We are going to take a closer look at the physical and biological mechanisms which influence the spectral response of vegetation. In doing so, it will be useful to have firmly in mind the fact that plant leaves reflect, absorb, and transmit incident radiation in a manner that, as Gates[1] said so well, "is uniquely characteristic of pigmented cells containing water solutions." These interrelationships, as a function of wavelength λ, are shown by the energy balance equation:

$$I_\lambda = R_\lambda + A_\lambda + T_\lambda \tag{5-1}$$

where I denotes incident energy, R denotes energy reflected, A denotes energy absorbed, and T denotes energy transmitted. Since most remote sensing instruments which operate in the 0.3 to 3.0-μm portion of the spectrum measure only reflected energy, it is often more useful to think of this relationship in the form:

$$R_\lambda = I_\lambda - (A_\lambda + T_\lambda) \tag{5-2}$$

i.e., the reflected energy is equal to the incident energy reduced by the energy which is either absorbed or transmitted.

Considering again the details of green-leaf reflectance shown in Fig. 5-1, we note first that there is very low reflectance in the blue and red regions of the visible spectrum. This low reflectance corresponds to the two chlorophyll-absorption bands, so called because the chlorophyll present in the leaf absorbs most of the incident energy in these wavelength bands centered at approximately 0.45 and 0.65 μm.[2] In the visible wavelengths most of the energy striking a green leaf is absorbed and very little is transmitted through the leaf. Therefore, the energy-balance equation indicates that in the visible portion of the spectrum the incident energy which is not absorbed will largely be reflected. A relative lack of absorption in the wavelengths between the two chlorophyll-absorption bands allows a reflectance peak to occur at approximately 0.54 μm, which is the green-wavelength region. Thus, we could say that it is essentially the relatively low absorption in the green wavelengths that causes normal, healthy foliage to appear green to our eyes. When a plant is under stress and chlorophyll production is decreased, the lack of chlorophyll pigmentation causes plants to absorb less in the chlorophyll-absorption bands. Such plants will have a much higher reflectance, particularly in the red portion of the spectrum, and therefore appear yellowish or "chlorotic."

Other pigments of interest include the carotenes and xanthophylls (both yellow pigments) and the anthocyanins (red pigments). Carotenes and xanthophylls are frequently present in the green leaves but have an absorption band only in the blue portion of the spectrum (at approximately 0.45 μm). Since chlorophyll, which is usually also present, also absorbs in the blue, it normally masks these yellow pigments. However, as a plant undergoes senescence, the chlorophyll frequently disappears, allowing the carotenes and xanthophylls to be dominant. This is the primary cause for the yellow coloration of tree foliage in the fall season. Also in the fall, as chlorophyll production decreases, some tree species produce anthocyanin in great quantities, giving the leaves a bright red appearance.

The effects of differences in pigmentation on leaf spectra are seen in Figs. 5-2, 5-3, and 5-4 (between pages 36 and 37). Figure 5-2(a) shows the striking differences in the visible portion of the spectrum for a variegated coleus leaf with different pigmentation characteristics, including a white area on the leaf where there were no pigments, a red area having anthocyanin but no chlorophyll, a normal green-leaf area having a high chlorophyll content, and a very deep reddish-purple portion of the leaf containing both anthocyanin and chlorophyll. Color and color infrared photos of the leaf from which these curves were obtained are shown in Fig. 5-3 (between pages 36 and 37). In spite of the differences in reflectance in the visible portion of the spectrum caused by the differences in pigmentation, the energy-matter interactions are such that there are no significant differences in reflectance in the near- or middle-infrared portions of the spectrum, as seen in Fig. 5-2(a). This same phenomenon appears in the curves of the red and green maple leaves, Fig. 5-2(b). Again we see a significant difference in reflectance in the visible region due to the difference in pigmentation of these leaves, but there is no significant difference in the near and middle infrared. Figure 5-4 (between pages 36 and 37) shows the color and color infrared photos of the foliage from which the curves in Fig. 5-2(b) were obtained. Because color infrared film is sensitive to the green and red visible wavelengths as well as to the 0.7 to 0.9-μm wavelengths in the near infrared, pigmentation differences will cause a distinct difference on color infrared film, even though such pigmentation differences do not cause significant differences in near-infrared reflectance.

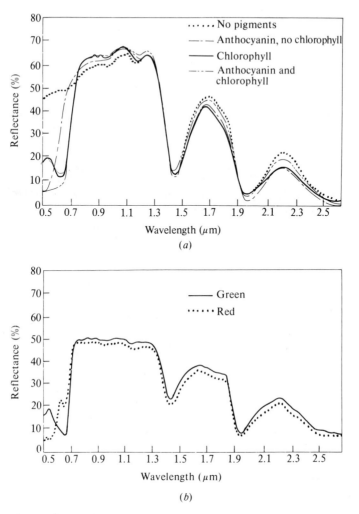

Figure 5-2 The effect of pigmentation on leaf reflectance. (*a*) Coleus leaf. (*b*) Maple leaves. (*After Hoffer and Johannsen.*[3])

Turning to an examination of the near-infrared portion of the spectrum, note in the figures shown so far the very marked increase in reflectance in passing from the visible to the near infrared, at approximately 0.7 μm. In the near infrared, healthy green vegetation is characterized by very high reflectance, very high transmittance, and very low absorptance, as compared to the visible wavelengths. In fact, for most types of vegetation, we will find approximately 45 to 50 percent reflectance, 45 to 50 percent transmittance, and less than 5 percent absorptance in the near-infrared wavelengths.

The internal structure of plant leaves is very complex, and it is this internal structure that largely controls the reflectance in the near-infrared region.[2,5] As shown in Fig. 5-5, the internal structure of corn and soybean leaves is very different. It is this difference in internal structure that accounts for the difference in near-infrared reflectance of corn and soybean leaves, as shown in Fig. 5-6. Although Fig. 5-6 indicates that little, if any, difference in reflectance exists in the visible wavelengths, the difference that is evident throughout the near- and middle-infrared wavelengths is significant. Many other plant species also produce similar distinct differences in reflectance in the near infrared, even though differences in reflectance in the visible wavelengths are often negligible.

It is important to note that in comparison to the reflectance from a single leaf, multiple-leaf layers can cause an even higher reflectance (up to 85 percent) in the near-infrared portion of the spectrum.[8] This is due to additive reflectance: energy transmitted through the first (uppermost) layer of leaves and reflected from a second layer is partially transmitted back through the first layer. For example, consider a leaf which reflects approximately 50 percent and transmits approximately 50 percent of the near-infrared energy incident on it. As indicated in Fig. 5-7, the transmitted energy falls on the second leaf layer where half of it again is transmitted (25 percent of the original) and half reflected. The reflected energy then passes back through the top leaf layer which allows half of that energy (or 12.5 percent of the

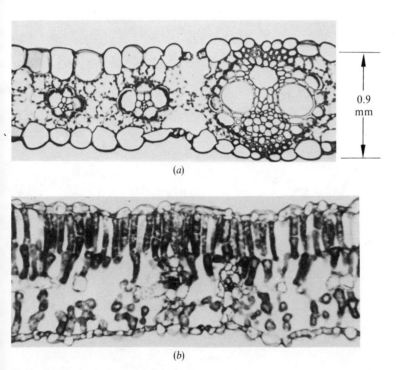

(a)

(b)

Figure 5-5 Cross sections of portions of corn and soybean leaves, showing the complexity of the internal structure of vegetation as well as the differences in leaf structure that exist between these two species. (a) Corn,[6] (b) Soybean[7] (Reproduced from *Agronomy Journal*, vol. 63, pp. 864–868, 1971, by permission of the American Society of Agronomy.)

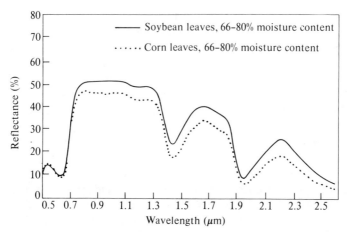

Figure 5-6 Reflectance curves for corn and soybean leaves with comparable levels of moisture content. (*After Hoffer and Johannsen.*[3])

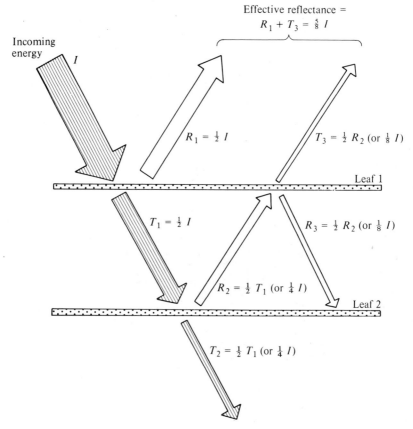

Figure 5-7 Simplified sketch of the effect of multiple-leaf layers on vegetative reflectance. $I =$ incoming energy; $T =$ transmitted energy; $R =$ reflected energy.

original) to be transmitted and half reflected. The resulting total energy coming from the top leaf layer in this two-layer example is 62.5 percent of the incoming energy. Modeling of multiple-leaf layers shows significant increases in near-infrared reflectance as more leaf layers are added, up to about six layers.[8] A good example of this is shown in Fig. 5-8. The impact of this effect on overall spectral reflectance is that a row of cotton or soybeans, for example, will appear to have a much higher near-infrared reflectance in the center of the row where the multiple-leaf layer effect is greater than at the edge of the row where there are not as many leaf layers and more energy is transmitted through the leaves and absorbed by the soil below.

Next, let us consider vegetative reflectance in the middle-infrared portion of the spectrum. In these wavelengths, the spectral response of green vegetation is dominated by strong water-absorption bands which occur near 1.4, 1.9, and 2.7 μm (as shown previously in Fig. 5-1). The absorption band at 2.7 μm is a major one, referred to as a fundamental vibrational water-absorption band. In the strictest

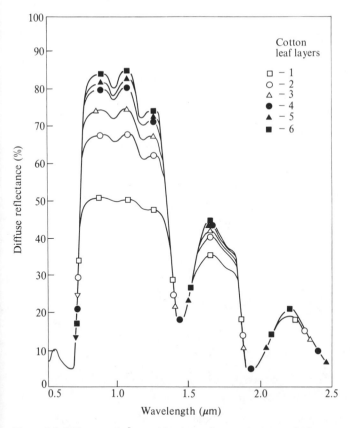

Figure 5-8 Reflectance from combinations of cotton leaves stacked on one another, up to six deep. (*After Myers.*[8])

sense this band is actually a combination of two fundamental vibrational water-absorption bands located at 2.66 and 2.73 μm. (There is another such fundamental vibrational water-absorption band at 6.27 μm.) Bands at 1.9, 1.4, 1.1, and 0.96 μm are all overtone and combination bands and are successively weaker absorption bands than the fundamental water-absorption bands.[1] The absorption bands at 1.9 and 1.4 μm dominate the spectral reflectance of leaves in the middle-infrared wavelengths. However, even the very minor water-absorption bands at 1.1 and 0.96 μm have a significant impact on reflectance, particularly for multiple layers of leaves, as shown in Fig. 5-8.

In these middle-infrared wavelengths, reflectance peaks occur at about 1.6 and 2.2 μm, between the water-absorption bands. Figure 5-9 shows that leaf reflectance in the middle infrared is approximately inversely related to the absorptance of a layer of water approximately 1 mm in depth. The degree to which incident solar energy in the middle-infrared portion of the spectrum is absorbed by vegetation is a function of the total amount of water present in the leaf, and that, in turn, is a function of both the percentage of moisture content of the leaf and the leaf thickness. As the moisture content of leaves decreases, reflectance in this middle-infrared wavelength region increases markedly. This is shown in Fig. 5-10, in which the average of a number of spectral curves for corn leaves has been obtained for four different groupings of moisture content. As indicated by this figure, the decrease in moisture content does not cause significant spectral differences until the moisture content of the plants has become very low (e.g., below about 54 percent). However,

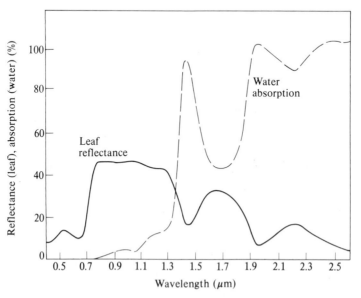

Figure 5-9 The inverse relationship between leaf reflectance and water absorption. The water-absorption curve represents the amount of absorption caused by a layer of water 1 mm deep. (*After Hoffer and Johannsen.*[3])

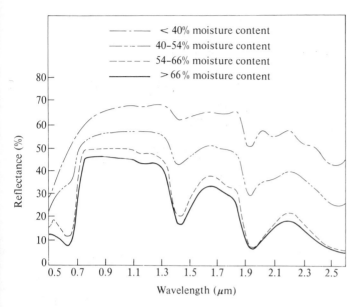

Figure 5-10 Effect of moisture content on reflectance of corn leaves. (*After Hoffer and Johannsen.*[3])

at the very low levels of moisture content associated with the top two curves, the plants were dying or dead, the leaves had lost most of their chlorophyll, and the increase in reflectance was substantial throughout the reflective region of the spectrum. As you might expect, the increased reflectance in the water-absorption bands was very large. For remote sensing purposes, however, we are primarily interested in the effect on reflectance in the "atmospheric window" wavelength bands in which multispectral scanner systems can operate: approximately 0.3 to 1.3, 1.5 to 1.8, and 2.0 to 2.6 μm in the reflective portion of the spectrum. (See Sec. 2-2 for a discussion and complete listing of atmospheric windows.) As Fig. 5-10 indicates, there are substantial changes in reflectance in the wavelength bands in which scanner systems can function, as well as in the water-absorption bands. This increased reflectance in the wavelengths between the water-absorption bands could be referred to as a "carryover effect," whereby the increased reflectance in the water-absorption bands influences or "carries over" to the wavelengths between the water-absorption bands, thus resulting in an increased reflectance throughout the middle-infrared region.

It is possible that a similar "carryover effect" is responsible for at least part of the increased reflectance in the near-infrared region observed in Fig. 5-10, due in this case to the large increase in reflectance in the 0.66-μm chlorophyll-absorption band coupled with the increase in reflectance in the 1.4-μm water-absorption band. In addition, many other complex energy-matter interactions occur in these near-infrared wavelengths. As leaves lose moisture, drastic changes which occur in their internal structure affect the near-infrared reflectance.[5,7] Much remains to be learned about energy-matter interactions in these infrared wavelengths.

To summarize, distinct differences in reflectance are found among the visible, near-infrared, and middle-infrared portions of the spectrum. In the visible wavelengths, the pigmentation of the leaves is the dominating factor. Most of the incident energy is absorbed and the remainder is reflected. The internal structure of the leaves controls the level of reflectance of the near infrared, where about half of the incident energy is reflected, nearly half is transmitted, and very little is absorbed by the leaf. The total moisture content of the vegetation controls the middle-infrared reflectance, with much of the incident energy being absorbed by the water in the leaf, the remainder being reflected. Much remains to be learned about the spectral characteristics of many different species of vegetation and the factors that influence the spectral response patterns that are measured by remote sensor systems.

PROBLEMS

5-5 (*a*) All energy striking a leaf must be either _____ or _____ or _____.
  (*b*) Write the energy-balance equation and give a brief physical interpretation of each term.
5-6 Indicate in Table P5-6 whether a leaf will have relatively "high," "medium," or "low" reflectance in the four wavelength bands indicated if the various pigment combinations cited are present.

Table P5-6

	0.4–0.5 μm (blue)	0.5–0.6 μm (green)	0.6–0.7 μm (red)	0.7–0.9 μm (near infrared)
Chlorophyll (green)				
Xanthophyll (yellow), no chlorophyll				
Anthocyanin (red), no chlorophyll				
Anthocyanin (red) + chlorophyll (green)				
No pigments				

5-7 Compute the amount of energy reflected from two leaf layers if 45 percent of the energy incident on a single leaf is reflected, 5 percent is absorbed, and 50 percent is transmitted through the leaf.

5-8 If two different types of vegetation appear in different tones or colors on color infrared film, is it safe to assume that the color difference is caused by a difference in infrared reflectance for the two types of vegetation? Why?

5-9 Most types of healthy, green vegetation have similarly shaped reflectance curves. Explain this fact by discussing the energy-matter interactions which take place in the visible, near-infrared and middle-infrared portions of the spectrum.

5-10 A decrease in moisture content in a leaf will cause the reflectance in the middle-infrared wave-lengths to _____.

Study objectives

Upon completion of Sec. 5-3 you should be able to:

1. List five major properties which affect the reflectance characteristics of soils. State whether these properties are inherent to the soil or not.
2. State whether an *increase* in each of the major soil properties listed above will cause an increase or decrease in reflectance in the 0.6 to 0.7-μm wavelength band.
3. Describe the differences in behavior of the reflectance characteristics of soils with different particle sizes (e.g., sand and clay) as a function of moisture content.
4. Describe two different approaches involving multispectral scanner data that could be used to determine whether a soil with a dark tone in the reflective wavelengths is dark because of high organic matter content or high moisture content.

5-3 SPECTRAL CHARACTERISTICS OF SOIL

Spectral reflectance curves from most soil materials are generally less complex in appearance than those from vegetation. Figure 5-11, for example, shows typical spectral reflectance curves for three different types of soil in an air-dried condition. As these curves indicate, one of the most outstanding reflectance characteristics of dry soils is a generally increasing level of reflectance with increasing wavelength, particularly in the visible and near-infrared portions of the spectrum. Energy-matter interactions are perhaps less complicated for soils, as compared to vegetation, since all incoming energy will be either absorbed or reflected, and one need not be concerned about energy being transmitted through the material, as is the case when dealing with vegetation. However, the soil itself is a complex mixture of materials having various physical and chemical properties which can affect the absorptance and reflectance characteristics of the soil. Therefore, although the reflectance curves shown in Fig. 5-11 are similar in their general shape, there are a number of interrelated soil properties that must be considered when discussing the reasons for the differences in amplitude of the curves. The moisture content, the amount of organic matter, the amount of iron oxide, the relative percentages of clay, silt, and sand, and the roughness characteristics of the soil surface all significantly influence the spectral reflectance of soils, as we shall see.

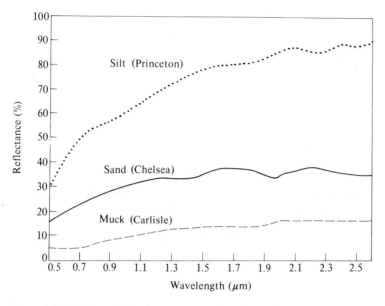

Figure 5-11 DK-2 spectral reflectance curves for three soil types at low moisture contents. (*After Hoffer.*[9])

Perhaps the soil characteristic that should be examined first is soil texture, which refers to the relative proportions of clay, silt, and sand particles present in a mass of soil. Soil particles less than 0.002 mm in diameter are defined as clay, 0.002 to 0.05 mm as silt, and 0.05 to 2.0 mm as sand.[10][†] The relative amount of these different components determines the textural name (e.g., sandy clay, silt loam, etc.) that is given to a particular soil, as illustrated by Fig. 5-12. Very small particle sizes, such as found in clay soils, will enable the particles to be packed very closely together with only very minute spaces between the soil particles. On the other hand, relatively large sand particles allow larger air spaces between the particles, thereby enabling more air or water movement through sandy soils as compared to clay soils. When moisture is present, each soil particle is covered by a very thin layer or film of water, and water will also occupy at least some of the air space between the soil particles. Although the film of water that can coat a soil particle is very thin at best, millions of small, tightly packed clay particles, each covered by this thin coating of water, can hold a significant amount of water. In addition, since the spaces between the particles of clay soils are so minute, water will not drain from or even be evaporated from a clay soil as easily as from sandy soils, which are composed of larger particles and have larger air spaces between the particles. This relationship between size of the soil particles and the moisture content of the soil

---

† The International System defines 0.02 mm as the division between silt and sand rather than 0.05 mm as indicated in this scheme, which is used by the U.S. Department of Agriculture and most soil scientists in the United States.

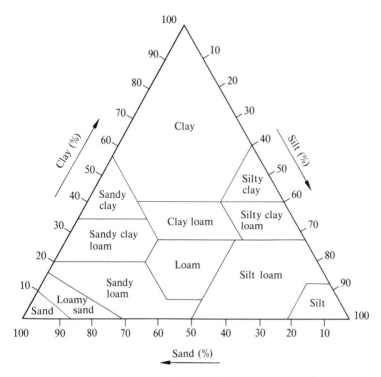

Figure 5-12 Soil texture triangle, showing the relative percentages of clay particles (below 0.002 mm), silt (0.002 to 0.05 mm), and sand (0.05 to 2.0 mm) present in the various soil textural classes. (*After USDA Soil Survey Staff.*[10])

has a significant impact on the spectral response of soils, as is illustrated in Figs. 5-13, 5-14, and 5-15.

Figure 5-13 shows typical reflectance curves for a sandy soil at three different levels of moisture content. The relatively large particle size of the sand allows most of the moisture to be removed from the sand when it is air dried. Consequently, as Fig. 5-13 demonstrates, there are no significant decreases in reflectance in the water-absorption bands of air-dried sandy soils. However, the sandy soils that have not been subjected to air drying and which contain significant amounts of water have distinct decreases in reflectance in the absorption bands at approximately 1.4, 1.9, and 2.7 μm. These are the same water-absorption bands that were discussed in Sec. 5-2, where we examined the relationship between reflectance and moisture content of vegetation. Just as we found for vegetation, as the moisture content of soil increases, the reflectance decreases, particularly in the water-absorption bands. The reason for this, of course, is that the incoming radiation is strongly absorbed by water at these particular wavelengths, regardless of whether the material containing the water is vegetation or soil. As previously noted, the 1.4- and 1.9-μm water-absorption bands represent overtones of the fundamental frequencies (2.66, 2.73, and 6.27 μm) at which water molecules vibrate. The decreased reflectance in

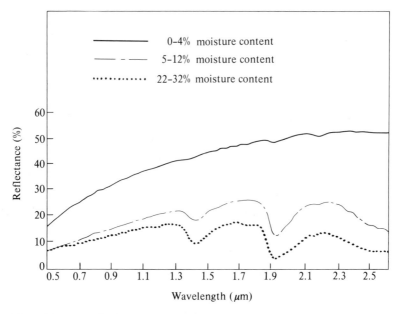

Figure 5-13 Spectral reflectance curves for Chelsea sand in three moisture-content groupings. (*After Hoffer and Johannsen.*[3])

the water-absorption bands causes a decreased reflectance in the wavelengths between the water-absorption bands.

In the visible portion of the spectrum there is also a distinct decrease in reflectance for moist soil as compared to dry soil. Almost everyone has observed this phenomenon many times in life—as rain drops fall on the dry sand at the beach, on the soil in a garden, or even on the sidewalk, the wet areas become darker in color due to a decreased reflectance. You may have also noticed that if the soil is moist initially, increasing the moisture content further will not always cause a proportional decrease in reflectance (i.e., darkening in color). The similarity in reflectance in the shorter wavelengths of the two curves for moist sand in Fig. 5-13 illustrates this phenomenon.

An outstanding example of the relationship between increasing moisture content and decreasing level of reflectance is seen in Fig. 5-14. In this illustration, even the curve for air-dried silt has decreased levels of reflectance in the water-absorption bands. This is because the small size of silt particles, as compared to sand, enables a significant amount of water to adhere to the soil particles even when the soil is in an air-dried condition.

Figure 5-15 shows reflectance curves for a Pembroke clay soil in two moisture-content groupings. Again, in addition to the overall decreased level of reflectance with increased moisture content, it is apparent that because of the extremely small size of clay particles, soil samples that have been air dried still retain some moisture, as evidenced by the decreased reflectance in the water-absorption bands. These

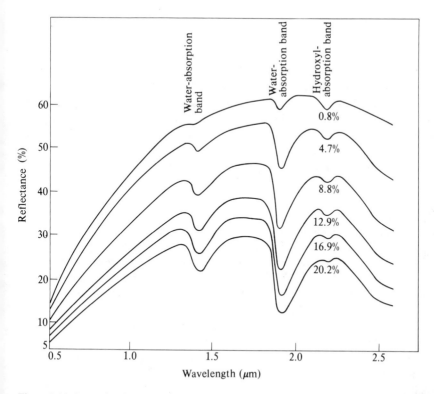

Figure 5-14 Spectral reflectance curves for Newtonia silt loam at various moisture contents. (*After Bowers and Hanks,*[11] © Copyright 1965, The Williams and Wilkens Co., Baltimore, Md. Adapted with permission of the author.)

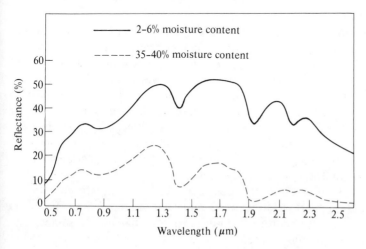

Figure 5-15 Spectral reflectance curves for a typical clay soil (Pembroke) at two moisture contents. (*After Hoffer and Johannsen.*[3])

water-absorption bands are evident in the reflectance curves of all clay soils in either field or air-dried condition. In addition to the water-absorption bands, most clay soils have rather strong hydroxyl absorption bands at approximately 1.4 and 2.2 μm.[12] This gives a rather distinctive shape to spectral reflectance curves for soils containing moderate-to-large percentages of clay. Notice, for example, that the reflectance curves in Fig. 5-14 for the Newtonia silt loam displayed a decreased reflectance at 2.2 μm, due to the absorption by the clay in this soil type. (Silt loam soils have up to 27 percent clay content, as indicated in Fig. 5-12.) It should also be noted that the particular type of clay present in a soil (e.g., Kaolinite, montmorillonite, etc.) will influence the strength of the absorption and therefore affect the characteristics of the spectral curves of the various soils containing these different types of clay.

The texture of soil affects the spectral reflectance of the soil both because of its influence on the moisture-holding capacity and because the size of soil particles per se strongly influences the reflectance. If other factors are constant, as particle size decreases, the soil surface becomes smoother and more incoming energy is reflected. For example, Bowers and Hanks[11] calculated that an increase in particle size from 0.022 to 2.65 mm would cause an increase in absorption of incoming solar radiation of at least 14 percent. Montgomery[13] found that the amount of silt present was the major factor in explaining the level of reflectance in both the visible and reflective infrared portions of the spectrum for the soils with which he was working. Since silt particles are of a relatively small size, the relationship was directly proportional (i.e., an increase in the amount of silt present caused an increase in the level of reflectance).

The organic matter content is another soil property that significantly influences the reflectance characteristics of a soil. The organic matter content of soil is extremely important to agriculturalists since it determines the amount and form of nitrogen in a soil. Although the level of organic matter found in most temperate-zone soils ranges only from about 0.5 to 5 percent, a soil with 5 percent organic matter will usually appear quite dark brown or black in color, whereas lower amounts of organic matter content will result in lighter brown or gray tones in the soil. The degree of decomposition of the organic matter will also influence the color to a great extent. For example, raw peat is brown in color but the well-decomposed and more fertile organic soil produced from peat is black or nearly so.[10] The relationship between organic matter content and reflectance throughout the visible wavelengths has been shown to be curvilinear in character, as indicated in Fig. 5-16. A very similar curvilinear relationship was also obtained for reflectance data obtained from the air, by an airborne multispectral scanner operating in the 0.62 to 0.66-μm wavelength band.[15]

It must be pointed out that soils developed under different climatic conditions may not show the same relationship between color and organic matter. In areas of relatively high annual temperature, well-drained soils having the same high organic matter content as soils in cooler regions tend to be brown rather than black. Also, it has been found that in tropical and warm-temperate climates, the dark clays seldom contain as much as 3 percent organic matter; yet these soils

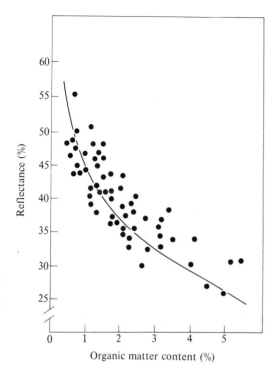

Figure 5-16 Relationship between organic matter content and reflectance (*After Page,*[14] adapted from *Agronomy Journal,* vol. 66, pp. 625–653, 1974, by permission of the American Society of Agronomy.)

include some of the blackest soils of the world.[10] Thus, the climatic region and the drainage conditions must be taken into account when considering the relationships between spectral reflectance and the organic matter content of soils.

Iron oxide can also have a significant influence on the spectral reflectance characteristics of soil. The red colors of many soils are generally related to unhydrated iron oxide, although partially hydrated iron oxide and manganese dioxide can also cause this red coloration.[10] An increase in iron oxide can cause a significant decrease in reflectance, at least in the visible wavelengths. Figure 5-17 is an excellent example of the inverse relationship between reflectance in the visible portion of the spectrum and the percentage of iron oxide present in the soil. This figure indicates that the iron oxide content of the soil can cause a difference in reflectance of as much as 40 percent! Figure 5-18 illustrates that removal of the iron oxide from a soil will cause a marked increase in reflectance throughout the 0.5 to 1.1-μm wavelength region, but the reflectance above 1.1 μm is not particularly affected. This figure also shows that the removal of the organic matter from a soil will cause a similar marked increase in reflectance over about the same range of wavelengths.

To review this discussion of soil reflectance, it has been found that increased moisture will cause decreased reflectance throughout the reflective portion of the spectrum; that texture of the soil will cause an increased reflectance with decreased particle size; that a decrease in surface roughness will cause an increased level of

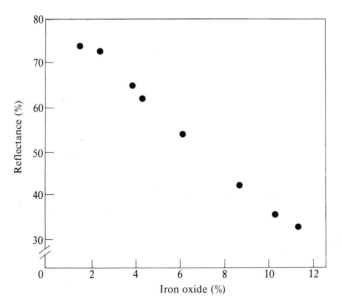

Figure 5-17 Relationship between iron oxide and soil reflectance in the 0.50- to 0.64-μm wavelength band. (*After Obukhov and Orlov.*[16])

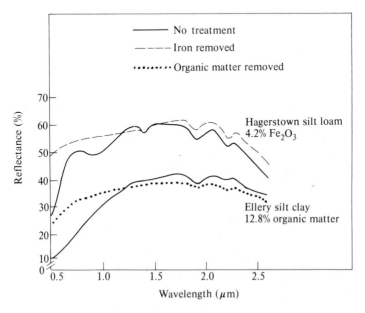

Figure 5-18 Spectral reflectance curves illustrating the effect of removal of iron oxide and organic matter from the soil. (*After Matthews.*[17])

reflectance; that an increase in organic matter content will cause a decrease in reflectance; and that an increase in iron oxide can cause a decrease in reflectance. However, many of these factors are very closely interrelated. Therefore, these guidelines for interpreting soil spectra will often hold true only for certain ranges of conditions. For example, Montgomery[13] indicated that the relationship between increasing organic matter content and decreasing reflectance exists primarily for soils with more than 1.5 percent organic matter content. For soils with less than 1.5 percent organic matter, the iron content of the soil exerted a significant influence on the level of reflectance. Furthermore, for those soils having over 1.5 percent organic matter, the relationship of decreasing reflectance with increasing organic matter content held true only for soils in which the percentage of silt was approximately the same. Conversely, if two soils had the same organic matter content but differed in silt content, the soil having the greatest amount of silt had the highest level of reflectance.

The interrelationships between the iron oxide and organic matter content of a soil and the influence of these soil properties on the reflectance characteristics of soil is important in interpreting data from different geographic regions. Because of the relationship between climate and soil development, one finds, for example, that in the southern portion of the United States the amount of organic matter present in the soil is generally relatively low, whereas the level of iron oxide present is usually fairly high, thereby creating the typical red color of the soils in the southeastern United States. In the more northern states, climatic conditions are such that the organic matter content of soils generally has a more dominant effect on soil color than does the iron oxide content. Also, in the more northern regions, the iron oxides tend to be more yellow than red in color, and therefore the organic matter in these soils often masks the level of iron oxide content. Thus, many brown soils can contain relatively high levels of iron oxide as well as organic matter. However, because the yellow-colored iron oxides tend to be masked, it is the organic matter content as well as moisture content, texture, and surface roughness that are the dominant factors in determining the reflectance characteristics of the soils in these more northern climatic regions. In the tropical and warm-temperate regions, the reflectance is more closely related to the iron oxide content than to the organic matter content, while moisture content, texture, and surface roughness continue to be important factors.

The ability to characterize the moisture content of a soil by remote sensing would have many practical advantages. However, as we have seen, even though an increased moisture content in the soil will cause a decreased reflectance, a similar decrease in reflectance can also be caused by an increase in organic matter content. Therefore, measurements obtained only in the reflective portion of the spectrum cannot be utilized to reliably discriminate between dark tones caused by a relatively high moisture content and those caused by a relatively high organic matter content. An example of such a situation is shown in Fig. 5-19. It is clear that there is a decreased level of reflectance in the area around the letter A, as compared to the area at the letter B, but the reason for this difference is not apparent when looking at the reflective imagery alone. It could be caused by differences in either moisture

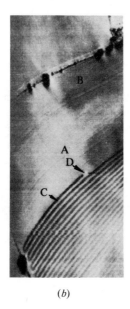

(a)   (b)

Figure 5-19 Reflected and emitted (thermal infrared) imagery of dry and moist soils. Dry, dark-toned soil (letter A) becomes relatively light in tone on thermal infrared imagery, whereas dry, light-toned soil (letter B) becomes relatively dark in tone on the thermal infrared image. Areas of moist soil (exemplified by the strips of freshly cultivated soil at the letter C) are relatively dark in tone on imagery obtained in both portions of the spectrum. The relatively hot tractor engine causes a white "blip" on the thermal infrared image at the letter D. (a) Reflected image. (b) Emitted (thermal) image.

content or organic matter content. If available, imagery from the reflective portion of the spectrum obtained on two or more dates could be compared; an obvious change from a dark to light tone for the areas of interest would lead you to conclude that the dark tone on the original set of imagery was due to a temporary influence (i.e., moisture content) rather than a more permanent characteristic of the soil such as the amount of organic matter present. However, even if the soil areas of interest display a dark tone on the imagery from both dates, you still cannot be absolutely certain that it is due to a permanent soil characteristic. Use of thermal infrared data provides a second and more effective approach.

Whether a soil is relatively light or dark in color, if it is moist, it is usually relatively cool due to evaporation. A dry soil, on the other hand, will absorb solar energy but will not be cooled by evaporation and will therefore be much warmer and will appear relatively light in tone on thermal infrared imagery. Thus, a soil that is relatively dark in the reflective portion of the spectrum and remains relatively dark in the thermal portion of the spectrum can be interpreted as having a relatively high moisture content. (The recently tilled strips in Fig. 5-19 provide a good example of this situation.) On the other hand, a soil showing a reversal between the reflective and thermal regions, such that it is relatively dark in the reflective portion of the spectrum but relatively light in the thermal portion of the spectrum, can be assumed to be fairly high in organic matter content but relatively dry. (Area A in Fig. 5-19 provides an example of this situation.)

To summarize, we can say that a dark area of bare soil that exhibits no reversal between the reflective and thermal portions of the spectrum (i.e., dark in both the reflective and emissive wavelengths) is indicative of a relatively moist soil; an area that is light in the reflective as well as the thermal portions of the spectrum is indicative of a relatively light-colored, dry soil; finally, an area that is relatively dark in the reflective region and light in the thermal portion of the spectrum indicates a relatively dry soil, perhaps having a high organic matter content.

In this section, we have seen that the spectral reflectance of soil can be dominated by several different but often closely interrelated soil properties. Soil texture, moisture content, surface roughness characteristics, organic matter, and iron oxide are among the most important soil parameters involved. Both the reflective and emissive portions of the spectrum can provide useful information concerning soil features, and data from both of these spectral regions must be utilized to effectively interpret the soil characteristics.

One of the most interesting and significant facts concerning soil reflectance characteristics is that while there are significant differences in the amplitude of reflectance among various soils and soil conditions, these differences are relatively consistent throughout the various wavelength regions. With vegetation, we emphasized the need to interpret relative differences in reflectance as a function of wavelength, but this is not the case for interpreting the reflectance characteristics of soils. In the next section, dealing with the reflectance characteristics of water, we will see that the different spectral regions are important but that most of the information about the water characteristics is obtained from visible wavelength data.

PROBLEMS

5-11 List (in a column) five major factors influencing the reflectance characteristics of surface soils. Put an "I" next to the factors which are inherent soil characteristics and a "T" next to those which are temporary.

5-12 Using the same list you developed in answer to Prob. 5-11, indicate whether an *increase* in each of these factors would *increase* or *decrease* the reflectance in the 0.62 to 0.68-μm wavelength band.

5-13 The spectral reflectance characteristics of a moist sandy soil are shown in Fig. P5-13. Sketch a curve on the graph that would be representative of the reflectance characteristics of the same soil after it had been air dried.

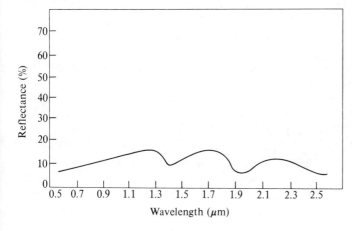

Figure P5-13

5-14 Repeat Prob. 5-13 assuming the soil had a high clay content.

5-15 Why do the reflectance characteristics of air-dried clay soils exhibit distinctive water-absorption bands while those of air-dried sandy soils do not?

5-16 Two students were discussing an experimental procedure for using remote sensing to measure the moisture content of soil. Student A (who had thoroughly studied Sec. 5-3 but not Sec. 2-2 of this book) suggested making reflectance measurements in the visible region of the spectrum and in a band centered at 1.4 μm. Student B (who had thoroughly studied both Secs. 5-3 and 2-2 of this book) agreed that the visible wavelength-band measurement would be useful, but said that measurements around 1.4 μm would be useless. What is the basis of this statement?

5-17 Consider how the various soils listed in Table P5-17 would appear in reflective and thermal imagery (see Fig. 5-19). Fill in the table, indicating whether the soils would appear light or dark on the two images.

Table P5-17

	Appearance in the	
Soil condition	Reflective image	Thermal image
Low organic content—dry		
Low organic content—wet		
High organic content—dry		
High organic content—wet		

5-18 Suppose an area of exposed soil was dark in tone on a visible wavelength image. Describe two approaches involving multispectral scanner data that could be used to determine whether the dark tone was primarily the result of high organic matter content or high moisture content.

Study objectives

After finishing Sec. 5-4 you should be able to:

1. Identify those wavelength bands which are best suited for distinguishing water bodies from surrounding land areas and those wavelength bands which are best suited for assessing water conditions.
2. Sketch the general shape of the reflectance characteristics of both clear and turbid water.
3. Identify factors that can significantly affect the reflectance characteristics of water.
4. State which wavelength bands are most useful for differentiating clouds from snow cover.

5-4 SPECTRAL CHARACTERISTICS OF WATER AND SNOW

As with vegetation and soil, the spectral response of water varies with wavelength, according to the energy-matter interactions taking place. For water bodies, the interactions are a result of the nature of the water itself and are further affected by various conditions of the water. As we shall see, locating and delineating water bodies by remote sensing can be done most easily in the near-infrared wavelengths, while some aspects of the condition of the water can best be assessed using data obtained in the visible wavelengths. Mapping the extent and condition of snow cover in order to obtain better predictions of runoff from the snowpack can best be accomplished using the middle-infrared bands.

Let us first look at the near- and middle-infrared wavelengths where, as you will recall from Fig. 5-9, even a very thin layer of liquid water displays several distinct, strong absorption bands. In a natural setting, water bodies absorb nearly all incident energy in both the near-infrared and middle-infrared wavelengths, even when the water is very shallow. Therefore, since water absorbs energy in the near- and middle-infrared wavelengths so effectively, there is very little energy available to be reflected at these wavelengths. This is very advantageous for remote sensing purposes, since it causes water features to have a significant and distinctly lower reflectance than either vegetation or soil throughout the reflective infrared portion of the spectrum, as is indicated in Fig. 5-20. Such distinct differences in infrared reflectance allow water bodies to be easily identified and mapped. For example, on black-and-white infrared photography or on multispectral scanner imagery obtained in the reflective infrared portion of the spectrum, as shown in Fig. 5-21, water bodies are black and stand out in stark contrast to surrounding vegetative and soil features. People working with Landsat data have found that even relatively small ponds or lakes of 2 to 4 hectares can be located quickly and easily.[19,20]

In the visible portion of the spectrum, the energy-matter interactions for water bodies become more complex. Although it is still helpful to think of these

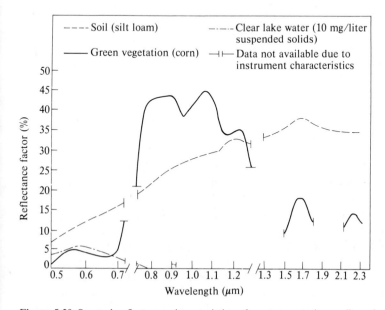

Figure 5-20 Spectral reflectance characteristics of green vegetation, soil, and clear water (*After Bartolucci*,[18] © copyright 1977, American Society of Photogrammetry. Used with permission.)

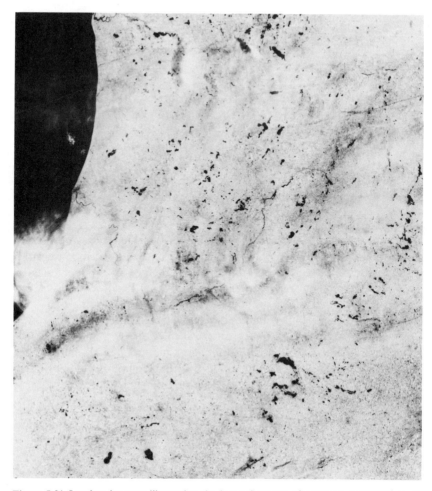

Figure 5-21 Landsat imagery illustrating the low reflectance of water as compared to vegetation or soil in the reflective infrared portion of the spectrum.

interactions in terms of the energy-balance equation, $I_\lambda = R_\lambda + T_\lambda + A_\lambda$, the components of the different terms are not easily determined. The reflectance term can involve reflectance from the surface of the water, from the bottom materials, or from suspended materials within the water body. The absorption and transmission characteristics are not only functions of the water per se, but are also significantly influenced by the various types and sizes of materials in the water—both organic and inorganic.

In order to determine what information about the condition and quality of water bodies can be obtained from visible reflectance measurements, studies have been made on the spectral characteristics of water bodies having a wide range of conditions. Considering first the absorption and transmission characteristics of clear water, you will notice from Fig. 5-22 that distilled water absorbs very little incoming energy in the visible wavelengths below 0.6 μm. Conversely, the transmission of incoming radiation in the shorter visible wavelength region is very high for distilled water, as seen in Fig. 5-23. The transmission characteristics of very clear ocean and lake water are similar to those of distilled water.[21,23] Note in Fig. 5-23, however, that the transmittance for natural water decreases rather significantly as the level of turbidity in the water increases and that the wavelength for maximum transmittance shifts somewhat toward the longer wavelengths. The high transmission and low absorption characteristics of clear water indicates that where the water bodies are relatively shallow and the water is very clear, the reflected energy that is recorded by remote sensor systems operating in the shorter visible wavelengths must largely be a function of the sand, muck, rock, or whatever is on the bottom. (A major exception to this statement would be in situations where specular reflection occurs at the surface of the water, creating what is often called "sun glint.")

As might be anticipated from the data shown in Fig. 5-23, the most accurate depth measurements in clear water can be obtained in the portion of the spectrum where transmittance is highest, at approximately 0.48 μm in the blue-green portion

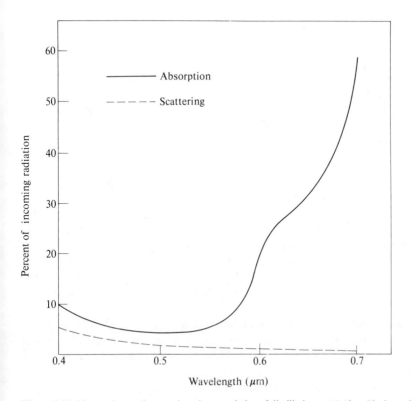

Figure 5-22 Absorption and scattering characteristics of distilled water (*After Clarke and James.*[21])

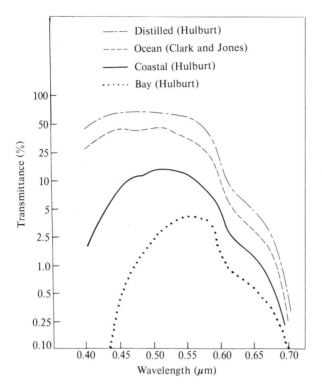

Figure 5-23 Spectral transmittance through ten meters of water of various types (*After Specht et al.,*[22] © copyright 1973, American Society of Photogrammetry. Used with permission.)

of the spectrum.[22,24] An indication of the relationship between wavelength and the potential for determining water depth from satellite altitudes has been shown in some of the results of work with Landsat 1 data. Lepley, Foster, and Everett[25] estimated the depth of penetration for clear water to be 10 m for the 0.5 to 0.6-μm band, 3 m for the 0.6 to 0.7-μm band, 1 m for the 0.7 to 0.8-μm band, and only 10 cm for the 0.8 to 1.1-μm band. Other studies[26,27] also determined that the 0.5 to 0.6-μm band of Landsat data was best for obtaining depth measurements, but also found that depth measurements using Landsat data were successful only for relatively shallow depths of less than 5 to 15 m. If Landsat data were available in wavelengths below 0.5 μm, it is possible that depth information could be obtained in deeper water since maximum penetration occurs in the wavelengths below 0.5 μm.

In the natural world, water bodies are usually not clear but contain a variety of organic and inorganic materials, some of which are in suspension. These materials cause scattering and absorption of incident energy and consequently cause significant variations in the transmission of energy through water, as seen in Fig. 5-23. Thus, the interactions between scattering, absorption, and transmission control the spectral reflectance that can be measured by remote sensing techniques.

These interactions have been the subject of a great many research studies involving a wide variety of water conditions.

Turbidity caused by suspended sediments is one of the major factors affecting the spectral response of water bodies. Bartolucci, Robinson, and Silva,[18] measuring the spectral response of turbid and clear water under natural conditions, showed that turbid water has a significantly higher reflectance than clear water and that the peak reflectance for turbid water is at a longer wavelength than for clear water (Fig. 5-24). They also found that for water bodies having turbidity levels of 100 mg/liter of suspended solids and where the depth of the water was at least 30 cm, the reflectance measured from above was due to the water itself and was not a function of the bottom characteristics.

Even from satellite altitudes, reflectance measurements have been shown to be highly correlated with the level of turbidity in the water. In one study, it was shown that reflectance in the 0.6 to 0.7-μm band was almost linearly related to the level of turbidity.[28] This was true for both the Landsat data and the reference measurements obtained just above the surface of the water (which had been obtained simultaneously with the Landsat data), as shown in Fig. 5-25.

The concentration of chlorophyll in water also affects the spectral response.[29,30] As shown in Fig. 5-26, as the chlorophyll concentration increases, there is a significant decrease in the relative amount of energy reflected in the blue wavelengths but an increase in the green wavelengths. This relationship between chlorophyll concentration and spectral response is significant in that chlorophyll levels are useful indices of both primary productivity and eutrophication of water bodies. Various remote sensing techniques have been utilized to study the potential for monitoring algal presence and concentrations, and several investigators have reported promising results.[31,32] However, since some species of algae grow in relatively deep water, remotely sensed reflectance measurements of surface or near-surface conditions are not necessarily indicative of all algal growth. Also, since algal bloom cycles are sometimes quite brief, they may occur between satellite data-

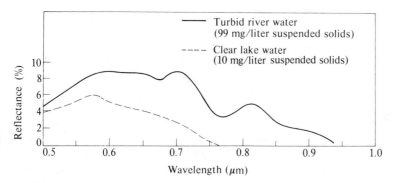

Figure 5-24 Spectral reflectance characteristics of turbid and clear water in the 0.5–1.0 μm wavelength interval (*After Bartolucci et al.,*[18] © copyright 1977, American Society of Photogrammetry. Used with permission.)

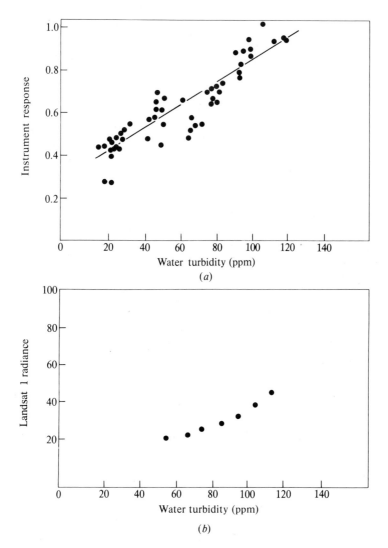

Figure 5-25 Reflectance in the 0.6- to 0.7-μm band in relation to differences in water turbidity. (*a*) Reference data obtained by hand-held instrument. (*b*) Landsat 1 radiance levels. (*After Weisblatt, Zaitzeff, and Reeves.*[28])

collection dates, e.g., between the 18-day cycle of Landsat data collection. Therefore, although it was mentioned earlier in this chapter, it must be emphasized that knowledge of the biological characteristics of the targets of interest is essential before remote sensing technology can be effectively utilized.

In addition to turbidity caused by inorganic matter in suspension and chlorophyll concentrations, many other natural and synthetic substances can cause variations in spectral response of water bodies. For example, the distinct brownish-yellow color of the water in many rivers in the northern United States is due to the

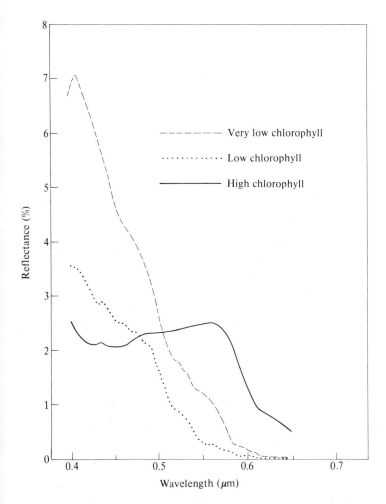

Figure 5-26 Spectral reflectance from ocean water having different concentrations of chlorophyll (*After Clarke et al.,*[29] © copyright 1970 by the American Association for the Advancement of Science. Used with permission.)

high concentrations of tannin from the Eastern hemlock (*Tsuga canadensis*) and various other species of trees and plants growing in bogs in these areas. Along this same line, it has been pointed out that these types of dissolved and suspended organic matter (i.e., tannin dyes from the adjacent land vegetation) which are found in fresh water are different in nature and perhaps in spectral response from the yellow gelbstoff from chlorophyll-bearing phytoplankton found in sea water.[33] In a study of pollutants from pulp mills, Scherz, Graff, and Boyle[34] found distinct differences in spectral response which were related to differences in the types and concentrations of suspended matter in the streams. Another environmental pollution problem that has been investigated using remote sensing techniques involves oil spills, which have been successfully delineated using a variety of photographic and scanner systems.[35] Munday et al.[36] reported that thin films of oil and the edge of an oil slick tended to be defined best using the ultraviolet and blue wavelengths, that the green wavelengths allowed the best delineation of the thicker portions of oil slicks, and that the thermal infrared region also could be utilized to detect oil slicks.

Some water characteristics of interest do not produce observable differences in spectral response. For example, no spectral differences are observable when gases (e.g., oxygen, nitrogen, carbon dioxide) or inorganic salts (e.g., sodium chloride, sodium sulfate) are dissolved in distilled water.[37] Similarly, significant differences in relative acidity found in the waters of abandoned strip mines (pH levels ranging from 3 to 7) could not be reliably correlated with differences in spectral response.[38]

Even though the desired information may not be directly obtainable through differences in spectral response, correlation of the information of interest with the ground features that are observable through remote sensing is often a useful and powerful method for obtaining the desired information. For example, in the strip mine example mentioned above, one could locate areas of *potentially* acid water through correlation with areas identifiable as strip mines on aerial photographs. Pollution caused by soil erosion from non-point sources can be effectively evaluated by studying land-use patterns of the areas adjacent to the water bodies. In one study, land-use data for the areas around different lakes were correlated with coliform counts made on samples of the lake water.[39] Only seven lakes were utilized in this study, but it was found that if 20 percent or less of the land surrounding the lake was classified in the urban category, the water quality (as defined by coliform standard) was well maintained. However, lake zones having over 20 percent urban category had water that was in a questionable state, and such water should be monitored frequently by health departments, according to the authors of this study. Furthermore, if 40 percent or more of the lake-shore area was in the urban category, they found that the water would very likely be unsafe for swimming or other human contact. Thus, even though the desired information about water quality may not be obtained directly through spectral measurements, correlation techniques can often be utilized in such a way that the desired information may still be obtained.

In summary, in almost all cases involving differences in water condition, there are many complex interrelationships between the spectral response measured by remote sensing systems and the particular water characteristics of interest. Unless one has appropriate reference data, it is often difficult to correctly interpret remotely obtained reflectance measurements from such areas. However, one must not overlook the fact that remote sensing techniques may be very beneficial for purposes of simply detecting areas of different spectral response and quantitatively delineating and mapping such areas, even if the specific reason for the differences cannot be determined from the spectral response measurements alone. As Wezernak and Polcyn[40] pointed out:

In addition to color originating from pollution sources, decaying vegetation, or other natural sources, color changes may occur as a result of changes in phytoplankton ecology and frequently are indicative of environmental changes in a body of water. Therefore, the occurrence of colored water deserves to be delineated and documented even if the observed phenomena are not immediately interpretable.

Let us now turn our attention to the spectral characteristics of water in the frozen state, i.e., snow! The ability to obtain reflectance data from satellite altitudes has been of particular interest to hydrologists involved in predicting runoff from the snowpack in many of the mountainous regions of the world. The potential for satellites to obtain data rapidly over large geographic areas and at very frequent intervals offers a capability that was previously unavailable. In the past, runoff forecasting has been done by ground-based measurements of the depth and water content of the snowpack at many individual locations or by making observation flights in light aircraft to estimate the areal extent of snow cover. However, there has been no economical, timely method for obtaining accurate estimates of the area of the snowpack. This task is ideally suited for satellite data-collection systems and computer-processing techniques.

Some of the early work with satellite data indicated that snow could not be reliably mapped because of the similarity in spectral response between snow and clouds. This similarity was particularly evident to investigators working with data from Landsat 1, who found that the spectral response of clouds and snow was so similar throughout the 0.5 to 1.1-μm sensitivity range of the Landsat detectors that the areas of snow cover could not be spectrally differentiated from the clouds.[41] It was also found that in many cases the reflectance from either the snow or clouds was so high that the detectors on the satellite were saturated, so, even if differences in spectral response actually existed, they could not be measured. To illustrate this problem, Table 5-1 shows spectral response data for both clouds and snow, as measured from Landsat data obtained on three different dates over the San Juan Mountains in southwestern Colorado.

Clearly, it is difficult to spectrally separate snow from clouds in the 0.5 to 1.1-μm portion of the spectrum. However, multispectral scanner data obtained from the Skylab satellite showed for the first time that snow and clouds could be easily differentiated in the middle-infrared portion of the spectrum, particularly in

Table 5-1 Spectral response of snow and clouds (mean $\pm$ 1σ)† [41]

	Wavelength band, μm			
	0.5–0.6	0.6–0.7	0.7–0.8	0.8–1.1
Snow	126 $\pm$ 2.6	127 $\pm$ 1.5	126 $\pm$ 2.3	125 $\pm$ 4.2
Clouds	127 $\pm$ 1.8	127 $\pm$ 0.9	125 $\pm$ 3.7	126 $\pm$ 3.1

† The saturation level for Landsat data is 128. The original response values in the 0.8 to 1.1-μm band have been doubled, for ease of comparison with the other three wavelength bands.

the 1.55 to 1.75 and 2.10 to 2.35-μm wavelength bands.[42] In these wavelengths, the clouds have a very high reflectance and appear white on the imagery, while the snow has a very low reflectance and appears black on the imagery. Figure 5-27 shows an example of this phenomenon; in the visible, near-infrared, and thermal infrared images, spectral discrimination between snow and clouds is not possible, while in the middle infrared it is.

Spectral reflectance curves obtained by O'Brien and Munis[43] show why the snow has this type of response on the Skylab imagery. Figure 5-28 shows the reflectance curve for nearly fresh snow and for two-day-old snow throughout the 0.6 to 2.5-μm portion of the spectrum. The decrease in reflectance of the snow from almost 100 percent at wavelengths less than about 0.8 μm to nearly 0 percent at 1.5 μm is unparalleled in other naturally occurring earth surface materials. The influence of the water-absorption bands discussed previously is also clearly seen in

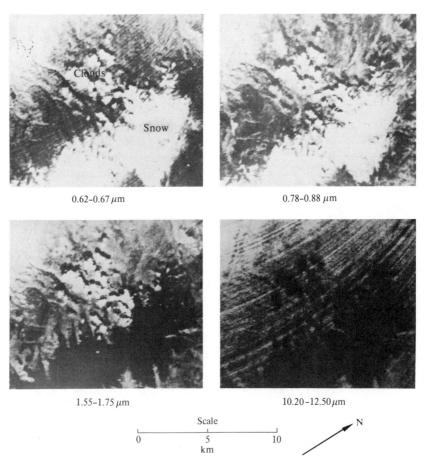

0.62–0.67 μm   0.78–0.88 μm

1.55–1.75 μm   10.20–12.50 μm

Scale

0   5   10
km

N

Figure 5-27 Snow and clouds as seen on Skylab imagery in four spectral regions: visible, near infrared, middle infrared, and thermal infrared. (*After Hoffer and Staff.*[42])

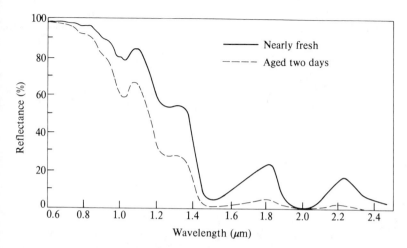

Figure 5-28 Spectral reflectance characteristics of snow. (*After O'Brien and Munis.*[43])

these reflectance curves for snow. It should also be noted from this figure that in the visible portion of the spectrum shown there is no difference in reflectance between the fresh and the two-day-old snow. In the wavelengths greater than 0.8 µm, however, the fresh snow has a higher reflectance, indicating that as a snowpack ages, the reflectance decreases in the infrared though not in the visible wavelengths. This would suggest that the reflective infrared wavelengths are potentially useful for assessing the age and condition of snow cover.

In the case of clouds, sunlight is nonselectively scattered by the clouds, so there is a high, uniform spectral response throughout the reflective portion of the spectrum (0.3 to 3.0 µm).[20] Therefore, in the middle-infrared wavelengths, the clouds have a very high reflectance in contrast to the very low reflectance of snow, as was so evident on the Skylab imagery in Fig. 5-27. In the thermal infrared wavelength region, the cloud tops and snow are often in the same temperature range, so spectral differentiation in this wavelength region is not reliable.

Figure 5-29 shows a partially snow-covered region in all 13 wavelength bands of Skylab data. Even though some of the data is of relatively poor quality, the relationship between the reflectance curves shown in Fig. 5-28 and the spectral response measured by the Skylab multispectral scanner system is clearly demonstrated. Figure 5-28 showed that as the wavelength increases, the reflectance of snow generally decreases throughout the reflective infrared portions of the spectrum. This relationship is enhanced in these Skylab data because at the time the data were obtained, portions of the snowpack at lower elevations were starting to melt, and the increase in amount of liquid water near the outer edge of the snowpack (i.e., at the lower elevations) caused a decrease in reflectance in the infrared wavelengths. Thus, as shown in Fig. 5-29, the snowpack appears to diminish in size throughout the near-infrared wavelengths. The relationship of decreasing reflectance with increasing wavelength throughout the reflective infrared portion

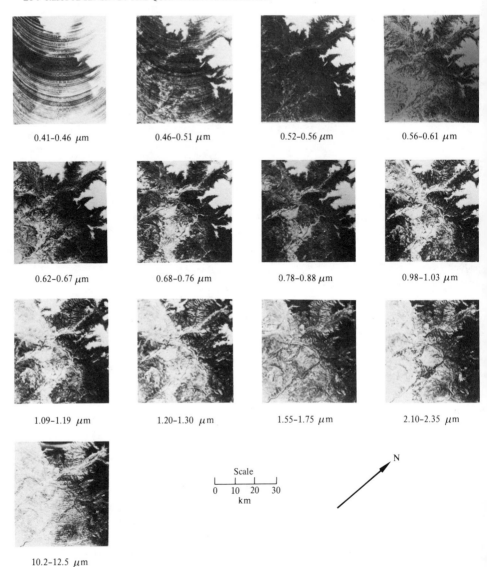

Figure 5-29 Multispectral scanner images from Skylab 2 over the San Juan Mountains, Colorado, June 5, 1973.[42]

of the spectrum was also confirmed by quantitative measurements of the reflectance from various areas within the snowpack on this Skylab data.

Thus, we have seen that the middle-infrared wavelengths are most useful for mapping snow cover and that the condition of the snow may be assessed by comparison of the data from different bands within the near-infrared region.

PROBLEMS

5-19 Match the various applications with the most appropriate wavelength region:

Locating water bodies	Visible
Assessing water quality	Near infrared
Mapping snow cover	Middle infrared

5-20 If the transmittance of very clear water is high, how is it possible for measured reflectance values from water bodies to be very low in some cases and very high in others?

5-21 Of the items listed below, which ones affect the reflectance characteristics of water?

(*a*) Oxygen (*e*) pH level
(*b*) Inorganic turbidity (*f*) Sodium chloride
(*c*) Chlorophyll (*g*) Bottom characteristics
(*d*) Carbon dioxide

5-22 If water turbidity due to suspended soil particles increases, reflectance in the 0.6 to 0.7-μm wavelength band will _____ (increase, decrease, be unaffected).

5-23(*a*) If water turbidity due to chlorophyll increases, the reflectance in the 0.5 to 0.6-μm wavelength band will _____ (increase, decrease, be unaffected).

 (*b*) How would you have answered (*a*) if the 0.4 to 0.5-μm wavelength band had been designated?

Study objectives

After reading Sec. 5-5 you should be able to:

1. List and discuss a number of factors which cause either temporal or spatial variations in spectral response for agricultural, forest, and rangeland cover types.
2. Define four approaches for collecting remote sensing data that will help to minimize the spectral variability normally encountered within agricultural crops.
3. Discuss the need for defining the cover types of interest in terms other than simply the names of the species involved.
4. Discuss the capabilities and limitations of remote sensing systems for obtaining land-use maps.

5-5 TEMPORAL AND SPATIAL EFFECTS ON THE SPECTRAL CHARACTERISTICS OF EARTH SURFACE FEATURES

One of the greatest challenges to effective interpretation of remote sensor data is understanding the temporal and spatial variations in the spectral characteristics of the natural resources or cover types being observed. Interpretation of large quantities of multispectral scanner data, color and color infrared photography, as well as of other types of remote sensor data, has clearly shown that the various earth surface features of interest, particularly vegetation, are often quite variable in their spectral response characteristics.

First, let us clarify the use of the terms "temporal effects" and "spatial effects." Variations in spectral response due to *temporal effects* involve situations where the spectral characteristics of the vegetation or other cover type in a given location change over time. In some cases, the time period may involve only a few hours, but more often the period of reference involves days, weeks, or even months. When, at a given point in time, the spectral responses are different for a single type of vegetation growing in different geographic locations, we refer to this as a *spatial effect*. In some cases the locations involved may be only a short distance apart (e.g., a few meters), but in other instances the distance involved may range from a few to hundreds of kilometers.

From a temporal standpoint, the spectral characteristics of many species of vegetation are in a nearly continual state of change and, in addition, different species often change at different times throughout the annual growing cycle. For instance, some trees leaf out earlier in the spring than others, and various crop species undergo senescence at different times. Also, as leaves mature, they change from a very light green to a darker green, and such changes are accompanied by increases in chlorophyll content and changes in the internal leaf structure.[1] As discussed in Sec. 5-2, such physiological changes in the vegetation also cause changes in its spectral characteristics. Many crop species exhibit very rapid changes in spectral characteristics throughout the growing season. For example, in central Indiana in late May, winter wheat presents a fairly solid canopy of lush, green vegetation to the remote sensor instruments, but by late June the same winter wheat is golden brown and nearing maturity. Two weeks later, it has probably been harvested, and remote sensor instruments measure only the reflectance of the yellow straw stubble. In many instances, another two weeks allows weeds and green vegetation to mix with the straw, and in remote sensor data the fields produce a response that is very much like grazed pasture or, perhaps, hay. These types of rapid change emphasize the importance of understanding seasonal changes of crops or other earth surface features, if optimal utilization is to be made of remote sensor data that may be available for several different dates over an area of interest.

From the standpoint of spatial effects on spectral response, two different types of effects must be considered. The first involves the impact on spectral response of spatial variations over a limited area, such as the effects of the spacing of row crops or of changes in plant morphology. For example, differences in the row spacing of agricultural crops can influence the relative vegetation/soil mixture "seen" by the remote sensor system at any one time during the growing season (especially early in the season), thereby causing differences in spectral response from field to field. Diseases or stress conditions often cause differences in plant size and morphology that influence the measured spectral responses. Again, such differences in spectral responses are often due, in part, to differences in the vegetation/soil mixture seen from above by the remote sensing system. The second type of spatial effect, spectral variation over relatively large geographic areas, is caused by such geographic variables as different weather conditions, soil types, or cultural practices. These differences in spectral response over large areas are sometimes referred to as "geographic effects" on spectral response.

One cannot ignore the impact of the temporal and geographic (or broad-area spatial) effects on spectral response. The fact that they exist in varying degrees for different types of vegetation has sometimes caused a great deal of difficulty in situations where a limited set of spectral response data from a single geographic location and a single date has been utilized as training data for mapping a particular species of vegetation over a large geographic area. Recognition and understanding of the temporal and geographic effects on spectral response is the first step, and a very necessary one, in developing effective procedures for analyzing remote sensor data obtained over large geographic areas. We will come back to this point later in this section.

From a very practical standpoint, information concerning the annual growing cycle of crop species or other cover types of interest and the appearance of these crops at each phase of their growing cycle and in different geographic regions of a country or the world will allow the user of remote sensor data to schedule data-collection missions much more effectively than might otherwise be the case. There may be certain times during the year when it would be rather easy to identify a certain crop species and other times when identification of the same species would be extremely difficult. For example, even though corn and soybean plants are completely different in their morphology and appearance, entire fields of these crops are spectrally quite similar at some times of the year but significantly different at other times. Thus, there are periods of several weeks early in the growing season when it is difficult to reliably separate these crops using very small-scale photography or scanner data if *only* the spectral differences are utilized. Later in the season after the corn has tasseled or when other spectral differences related to the maturity of the crops have developed, these two species can be readily separated. It becomes very important for people working with remote sensor data to be knowledgeable about these seasonal changes and the spectral responses associated with them. One must be familiar not only with the seasonal changes of the particular crop species of interest but also with the seasonal changes of other associated crops that might be confused with the species of interest.

The above comments point out that proper interpretation of remote sensor data requires a thorough understanding of the temporal and spatial characteristics inherent in the vegetative cover types present and of the related changes in spectral response. Some of the more important causes of variation in spectral response for vegetative cover types are summarized below. Agricultural crops are emphasized ; in some instances there is overlap between the causes cited, due to the interrelated nature of these factors.

1. Differences in the *amount of ground cover* (*due to cultural practices*). In crops, differences in the amount of fertilizer (particularly at the low fertility levels), planting dates, and planting procedures which cause differences in crop geometry (such as differences in row width) can cause significant differences in the amount of ground cover.
2. Variations in the *amount of ground cover* (*due to natural causes*). Spatially, such variations might be due to differences in soil type, soil moisture, uneven germination, and disease or insect conditions which result in small, stunted plants. Temporally, of course, data collected over the same area at different times throughout the growing season normally display differences in the amount of ground cover, particularly early in the growing season. Forest and rangelands show significant differences in ground cover or density for many of the same reasons as agricultural crops, except for planting-date variables.
3. Variations in *maturity*. In agricultural crops these can be caused by differences in

variety, planting date, soil types, or soil moisture conditions, as well as the normal seasonal changes. In forest and rangelands, soil type and moisture conditions as well as differences in microclimate and normal seasonal change are often the cause of such variations.

4. *Diseases, moisture stress, and insect infestations.* These factors can cause changes in the reflection and emission characteristics of the vegetation. Such stress conditions also can cause distinct variations in the amount of ground cover and plant geometry, thereby accentuating the changes in reflection and emission from the vegetation itself.

5. *Geometric configuration* of the ground cover. In crops, significant spectral variability could be due to lodging (blowing down) of the plants in addition to such things as differences in plant morphology or even row direction. In forested areas, stand density and size of the trees can cause significant differences in spectral response, as will topographic variables such as slope and aspect.

6. *Environmental variables.* Atmospheric conditions, wind, angle of reflection in relation to the angle of solar incidence, and soil moisture conditions (as affected by the amount of previous rainfall), as well as the length of time and weather conditions since the last rainfall, are some of the possible causes of spectral variations.

Figures 5-30 through 5-37 (between pages 36 and 37) show some examples of different types of variation in spectral response. It is quite apparent that variations in spectral response may cause difficulties in the analysis of remote sensor data *if* the objective is to identify the species of interest. However, it is important to recognize that, in many cases, it is actually these variations in spectral response *within the species* that are of interest, e.g., identifying healthy versus diseased or insect-infested vegetation, or differences in crop density that can be associated with differences in yield. Therefore, it must be emphasized that effective understanding of the natural scene is necessary, not only to minimize unwanted spectral variability (when attempting to identify and map different species) but also to maximize the spectral differences that may exist within a species when the information needed involves the density and/or condition of the species.

In considering seasonal changes of agricultural crops and variations in maturity, a problem involving definitions is often encountered. Suppose you are interested in utilizing remote sensing to map the acreage of soybeans throughout the state of Illinois. At what stage of development are you going to define a particular agricultural field as being "soybeans"? Do you call field X a field of soybeans after the beans have been planted? Or after emergence? Or when the beans are four inches high? Ten inches? Or is it not until the soybeans are covering 25 percent of the ground surface? Or 50 percent? Work with multispectral scanner data from aircraft altitudes has shown, for example, that early in the growing season the ability to spectrally differentiate between corn and soybeans is highly correlated with the percentage of ground cover, rather than with any distinctive differences between the spectral characteristics of the soybean and corn vegetation per se. Early-planted corn fields could be reliably distinguished from soybean fields because of the higher percentage of ground cover and smaller percentage of exposed soil measured by the scanner system in each resolution element. However, corn planted later in the season was found to occupy about the same percentage of ground area as the soybeans did. Therefore, this essentially involved a situation where 50 percent of the ground surface was bare soil and the other 50 percent was green vegetation, and there did not seem to be a significant difference as to whether the green-vegetation component of the scene consisted of soybeans or corn. The small spectral differences that exist between soybean and corn canopies could not be distinguished because of the more powerful influence of the soil-vegetation mixture that was being measured by the scanner system at this time during the growing season. This type of interaction of spectral responses has sometimes caused difficulties for analysts working with multispectral scanner data from agricultural crops, particularly if the data had been obtained early in the growing season. Use of multitemporal data and careful selection of the time of year when such interaction can be avoided seems to be one of the most practical solutions to such situations.

When considering forested areas, a similar question involving definitions is often encountered, although in this case the question is one of spatial variability rather than temporal variability. In essence, if one observes a fairly dense forest canopy, it is not too difficult to identify it as a "forested" area. However, how do you categorize an area when there is only a 10 to 20 percent crown closure? Suppose, for example, that some areas of forestland within the region of interest have a very low stand density (e.g., 20 percent crown closure), and other areas within this study region are used as pastureland but contain some scattered trees to shade the cattle (also 20 percent crown closure). From aircraft or satellite altitudes, such areas might be very similar in spectral response since the actual cover type could be the same in both instances. However, the land-use categories to be identified and mapped would be quite different (i.e., forest versus pastureland). Therefore, we sometimes face situations in which the amount of crown closure required to define an area as "forestland" is specified somewhat arbitrarily. For example, forestlands might include only those areas having 30 percent or more crown closure, even though this will mean that some true forestlands may be erroneously classified into the pastureland category.

Questions involving the definition of classes of interest also arise when we seek to utilize multispectral scanner data for land-use mapping. Many different disciplines need current and accurate land-use maps for various areas of interest. However, the term "land use" is normally used in an interrelational context involving both land cover types and actual uses of the land, as opposed to the potential use of the land or land suitability. For example, although the cover type of an area might be forest, this does not indicate whether the *use* of this forested area involves timber production, forest recreation, or perhaps management of the area for wildlife or watershed purposes. Thus it should be understood that many actual uses of the land can only be inferred and not directly determined from remote sensing data collected from *any* altitude. To put it another way, it is possible to identify and classify various surface features or vegetative cover types from remotely sensed data, but the specific activity involving people's *use* of that land must often be inferred or verified by other means.

This basic difference between cover type and land use becomes especially apparent when we compare manual interpretation techniques and computer-aided analysis techniques. Manual interpretation techniques are often used in identifying and classifying various earth surface features and vegetative cover types, and then the land-use activities of the area involved are immediately inferred by the interpreter, thereby enabling land-use maps to be generated. On the other hand, with computer-aided analysis techniques, cover-type maps are obtained through identifying, delineating, and displaying only the various surface features and vegetative cover types; techniques involving only pattern-recognition algorithms to classify the data do not enable the computer to *infer* land use. If a land-use map is required, one must go through a second step in which the cover-type information initially produced by the computer is merged with ancillary or reference data from other sources. For example, a computer classification might show an area as having forest, water, and grassland cover types, but it would require additional information such as the location of state parks to be able to categorize this particular area as having a primary *land use* of recreation. In some situations, the additional information that allows land use to be inferred can best be obtained by such techniques as manual interpretation of aerial photos, while in other situations the additional data (such as land ownership) must be obtained from some existing data source and then possibly added to the cover-type information by computerized techniques.

One other term that has frequently been used (and misused) in remote sensing and is closely related to our discussion of spectral variability is that of "spectral signature." As often used, this term implies a unique, well-defined, and characteristic spectral pattern by means of which a particular earth surface feature can be positively identified. However, as the previous discussion concerning the spectral characteristics of vegetation indicates, all green vegetation has rather similar, basic spectral characteristics, and there are many factors that may cause variations in the spectral behavior of any one cover type or even species at any point in time. Thus, it should be recognized that *unique, unchanging spectral signatures do not exist in the natural world*. Rather, at any point in time, in a particular geographic area, there may exist measurable spectral response patterns from the various vegetation types of interest that are combinations of the reflectance and emittance from the vegetation per se, the soil, shadow effects due to differences in density or planting patterns, etc., which in toto are distinctive enough to allow the various vegetation types of concern to be identified. Temporal variations in spectral response are particularly important in this regard, and having data from two dates often makes possible the identification of some kinds of cover type which would be impossible using data from any one date alone. For example, in the fall season after winter wheat has been planted, it is not possible to identify it because the fields planted to winter wheat as well as many other fields are all spectrally represented as bare soil. In the early spring, the winter wheat has a spectral response that is characteristic of dense green vegetation, but forage crops may have a similar spectral response. However, only the fields of winter wheat have the *combination* of spectral characteristics represented by bare soil in the fall and green vegetation in the spring!

In summary, then, the term "spectral signature" tends to imply a degree of uniformity and lack of variability, as well as absolute calibration of the data that are usually not present in multispectral systems. This sometimes leads to questions such as, "Have you defined *the* spectral signature for wheat (corn, exposed soil, turbid water, etc.) yet, and what does it look like?" Experience indicates that there is no such thing as "the" spectral signature for a particular earth surface feature. Therefore, to avoid the misconceptions which the term "spectral signature" tends to convey, the term "spectral response pattern" is preferred and is used throughout this discussion to indicate a quantitative but relative set of measurements corresponding to a specific cover type on a particular set of multispectral scanner data.

This discussion concerning the spectral variability of earth surface features may appear to indicate serious problems in identifying and mapping earth surface features of interest. Indeed, it does cause problems, and such spectral variability must not be overlooked or disregarded by those attempting to utilize computer analysis techniques and multispectral scanner data. However, much of the variation that is normally encountered can often be eliminated or overcome through proper consideration of the conditions under which the remote sensor data are collected. In general, there appear to be at least four possibilities that should be considered in attempting to overcome this problem of spectral variability of vegetative cover. These include the following:

1. Collect data, if possible, at the times during the growing season when the cover type or feature of interest has a spectral response pattern that is significantly different from any other cover type (e.g., when wheat is a mature, golden-yellow color and all other crops are various tones of green).
2. Obtain remote sensor data when the variations for a given species of interest are at a minimum (e.g., the middle of the growing season for corn or soybeans, after the crop has reached maximum canopy coverage but before senescence has started for any variety of that crop).
3. Collect data at intervals throughout the growing season, since no single time period will be optimal for all species or physiognomic groups.
4. Collect data under restricted environmental conditions, such as at a minimum specified sun angle, with less than 10 percent cloud cover, or after a certain number of days since the last rainfall.

Careful consideration of the temporal variability of many earth surface features will certainly allow some of the variability to be minimized or at least accounted for. However, there is nearly always a need for so-called "ground truth data" in remote sensing. "Ground truth" involves the collection of measurements and observations about the type, size, condition, and any other physical or chemical properties believed to be of importance concerning the materials on the earth's surface that are being sensed remotely. Lately the term "ground truth" has fallen into disfavor for several reasons. Sometimes, errors in data collection have caused "ground *truth*" actually to be incorrect data; similarly there are often so many variables involved that one wonders what the "truth" of the situation really is! Furthermore, if you are obtaining data through interpretation of large-scale photos collected from the air, or if you are obtaining measurements of the temperature of a water body, should such data really be referred to as "*ground truth*"? Therefore, it seems more logical to call such data "reference data" or some similar term. "Reference data" is the term most commonly used in this book.

Regardless of the name used, the *type* of data collected and the *procedures* involved in the collection of "reference data" should be carefully defined in order to meet the objectives of the project. For example, if one is interested in identification of crop species, detailed information concerning the micrometeorological conditions within the crop canopy are not needed. However, if one is interested in the use of thermal infrared systems for disease detection, such micrometeorological data could be essential. If one is working with remote sensing instruments such as photographic systems that operate only in the reflective portion of the spectrum, soil moisture information might be required for the soil surface, but soil moisture measurements throughout the profile would be of much less importance (except as they influence the surface soil conditions). Thus, there does not appear to be a single answer to the question of exactly what types of reference data are required. We can only say that the types of reference data required must be closely related to the objectives of the remote sensing activity and the problems involved. For general mapping of cover types or vegetative species using satellite data, it has often been found that good-quality aerial photos, obtained over relatively small sample areas, are a very effective form of reference data. Such photos can be used to assist in defining training statistics and to positively identify the cover types in test areas.

In summary, there is a significant amount of spectral, temporal, and spatial variability which must be understood and taken into account in order to properly analyze remote sensor data. Thorough knowledge of the discipline area involved in any particular application and effective utilization of remote sensing capabilities and supporting reference data will allow many variabilities to be minimized and the desired information to be obtained.

PROBLEMS

5-24 List 10 agricultural factors that would cause spectral variations in a particular crop species. For each of the factors listed, indicate whether the spectral variation will be temporal or spatial in character. If spatial, indicate whether it would be most likely to occur within a particular field or between fields of the same crop species.

5-25 Define four methods that can be used to minimize the amount of spectral variability that is frequently encountered when trying to use remote sensing techniques to map and identify particular crop species.

5-26 Discuss the problems that may be encountered when only the name of a crop species is utilized to identify its spectral characteristics.

5-27 Can a satellite multispectral scanner system be used effectively to map "land use"?

Study objectives

After studying Sec. 5-6 you should be able to do the following:
1. Based on relative differences in spectral response, identify bare soil, green vegetation, and water on a set of multispectral scanner imagery, coincident spectral plots, or a set of histograms.
2. State one reason for examining histograms of each wavelength band for each class of training data to be used in a computer-aided classification.

5-6 COMPARATIVE INTERPRETATION OF REMOTE SENSOR DATA

Effective interpretation of multispectral remote sensor data is dependent upon knowledge of the spectral characteristics of earth surface features. Thus far, we have treated the various cover types rather independently, but now we shall compare the spectral reflectance data for the different cover types and relate this reflectance data to multispectral imagery.

Figure 5-38 shows spectral reflectance curves for the average of a large number of samples of typical green vegetation and of relatively dry loam soils of a medium gray-brown color. In the visible portion of the spectrum, the soil reflects

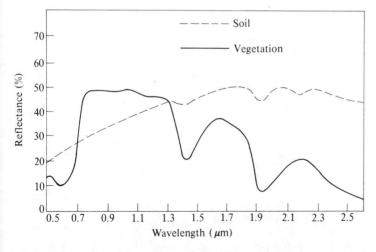

Figure 5-38 Spectral reflectance curves for healthy green vegetation and air-dried soils. These curves represent averages of 240 spectra from vegetation and 154 spectra from air-dried soils. The relative differences in reflectance in the visible (0.4 to 0.7 μm), near-infrared (0.7 to 1.3 μm), and middle-infrared (1.3 to 3.0 μm) portions of the spectrum are clearly shown by this data. (*After Hoffer.*[49])

more than the vegetation, but in the near infrared the vegetation reflects much more than the soil. Another such reversal occurs at about 1.3 μm, making the soil much more reflective in the middle-infrared portions of the spectrum than the vegetation. This type of interaction between *relative* reflectance levels for vegetation and bare soil is important in interpreting multispectral scanner imagery. In Fig. 5-39 a large field of bare soil near the center of the images appears nearly white in tone, due to high reflectance in all seven of the visible wavelength images (0.40 to 0.70 μm). The vegetation (e.g., the forest cover) reflects very little and therefore has a dark appearance in the visible-band images. In the near-infrared wavelengths, however, the situation is reversed; here the soil is relatively dark in tone and the vegetation is relatively light. These relative reflectance levels are easily seen in all three images obtained in the near-infrared range (0.72 to 0.92, 0.82 to 1.1, and 1.0 to 1.4 μm). In addition, the reversal predicted from the spectral reflectance curves for the middle-infrared portion of the spectrum is seen in the images obtained from the 1.5 to 1.8 and 2.0 to 2.6-μm wavelength bands, where the soil once again reflects much more highly than the vegetation. Water, which has a high infrared absorption, becomes very black on multispectral scanner imagery *throughout* the reflective infrared portion of the spectrum (0.72 to 2.6-μm images).

The typical vegetation and soil curves displayed in Fig. 5-38 were very useful for beginning to learn how to interpret multispectral scanner imagery as a function of spectral reflectance. However, to see more clearly the interrelationships in spectral reflectance among basic earth surface cover types, turn to Fig. 5-40, which displays field spectral data for vegetation, soil (a very light, highly reflecting soil and a very dark, low-reflecting soil), and both clear and turbid water. Examination of these reflectance curves indicates that the visible wavelength region is not as definitive as the reflective infrared region for spectrally differentiating among the basic cover types, i.e., vegetation, soil, and water. For example, in both the near- and middle-infrared regions, both clear and turbid water have little if any reflectance and so can be easily separated from any soil or vegetative cover type. In considering only the near-infrared wavelengths, we see that vegetation is more reflective or at least as highly reflective as soil, whereas in the middle-infrared wavelengths the soil is usually more reflective or at least as reflective as the vegetation.

Examination of the spectra shown in Fig. 5-40 indicates the advantage of utilizing more than one wavelength band to differentiate the various cover types of interest. For example, a very light soil tends to have a reflectance similar to that of vegetation in the near-infrared portion of the spectrum (0.7 to 1.3 μm); therefore, vegetation and light soil often cannot be easily differentiated on imagery obtained in these near-infrared wavelengths. However, the vegetation could be easily differentiated from the soil using data obtained in either the visible or the middle-infrared regions since the light soil will have a much higher reflectance than the low-reflecting, dark-toned vegetation. Similarly, a dark soil often cannot be easily differentiated from vegetation in only the visible or the middle-infrared wavelengths, since both cover types will have a relatively low reflectance. However,

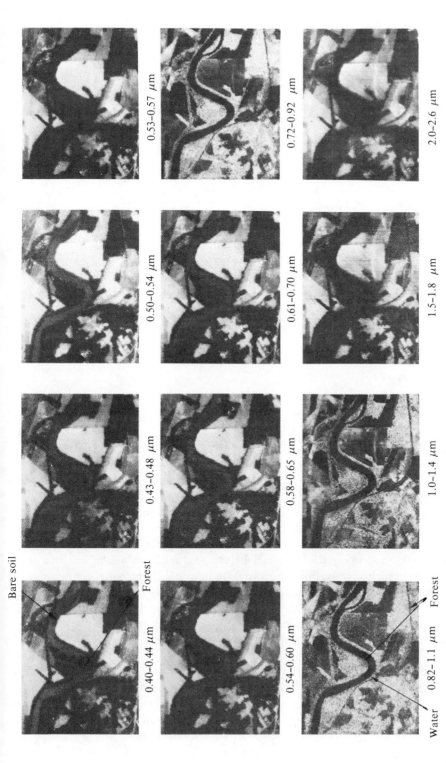

Bare soil

Forest

0.40–0.44 μm

0.43–0.48 μm

0.50–0.54 μm

0.53–0.57 μm

0.54–0.60 μm

0.58–0.65 μm

0.61–0.70 μm

0.72–0.92 μm

Water   0.82–1.1 μm   Forest

1.0–1.4 μm

1.5–1.8 μm

2.0–2.6 μm

Figure 5-39 Twelve wavelength bands of multispectral imagery in the visible, near-infrared, and middle-infrared wavelength regions.[50]

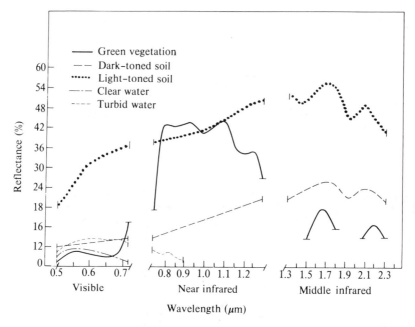

Figure 5-40 Spectral reflectance curves for green vegetation, light and dark soils, and clear and turbid water.

in the near-infrared portion of the spectrum the vegetation will have a much higher reflectance than the dark soil, and so can be easily distinguished and mapped.

A considerable amount of work with multispectral scanner data from both aircraft and satellite altitudes, obtained over a variety of geographic areas and cover types, has proven the value of the data that can be obtained in each of the four major portions of the optical spectrum: visible, near infrared, middle infrared, and thermal infrared. Water can be best separated from the other cover types using the reflective infrared wavelengths (either near or middle infrared), and then the thermal infrared region can be utilized to define differences in water temperature such as those found downstream from discharge outlets of nuclear or fossil-fuel power plants or other industrial plants.[51] Only the middle infrared can effectively distinguish between snow and clouds, whereas the near-infrared wavelengths are effective for showing areas of snow cover with relative levels of melting, and the visible wavelengths are best for defining relative amounts of snow and forest-canopy cover present, particularly in the case of coarse-resolution satellite data.[20] The near-infrared portion of the spectrum, and to a lesser extent the middle-infrared region, has been shown to be of greatest value for differentiating among various vegetative species and conditions, both from aircraft and satellite altitudes.[42,52,53] Combinations of data from the reflective and thermal wavelengths allow differentiation between dark-toned, exposed soil areas with high organic matter content and those with a high moisture content (see Sec. 5-3). Thermal data have also been effective in several situations for differentiating vegetative conditions and cover types that could not be reliably separated using only data obtained in the reflective wavelengths.[52,53,54]

Figure 5-41 illustrates the relative response for areas of bare soil, green vegetation, and dead vegetation in the visible, near-infrared, and thermal infrared wavelengths. This is the same data that we saw previously in Fig. 1-7, but now we can study it in more detail to investigate how a multispectral interpretation approach can be utilized to identify the various cover types present. Field A, for example, is dense, green, healthy alfalfa, and therefore is relatively dark in tone in the visible wavelength image, highly reflective and light in tone in the near-infrared image, and, since it is relatively cool due to transpiration, it appears dark in tone on the thermal infrared image. Field B is bare soil and appears somewhat mottled in the visible band, relatively dark-toned due to low reflectance in the near-infrared band, and, since the soil was relatively warm when these multispectral data were obtained, it appears very light in tone on the thermal infrared image. Field C is dry, dead corn which has a tan color and thus gives a medium response in the visible image, whereas in the near infrared, dead vegetation (e.g., corn) has a much lower response than living vegetation (e.g., alfalfa) but is still more reflective than the bare soil, resulting in a response that is between that of the alfalfa and soil. In the thermal infrared, the corn field has a medium relative response; although the dry vegetation is not transpiring and therefore is not as cool as the living vegetation, neither does it heat up as much as dry soil. Field D is dry wheat stubble and therefore has the same relative pattern of response as the mature corn in all three images.

It should be apparent from this imagery and the above discussion that the energy-matter interactions which take place in the natural world are quite different in the various regions of the electromagnetic spectrum. Therefore, interpretation of multispectral imagery for the purpose of understanding these energy-matter interactions must be carried out independently in each portion of the spectrum, always taking into account the energy-matter interactions peculiar to that wavelength region. In processing multispectral data by computer, the machine simply utilizes the measurements that were obtained in the different wavelength bands without understanding why differences in spectral response values exist. It is for this reason that an effective analyst/machine interface is needed. *The machine is a powerful and effective tool, but to be used most effectively, it requires a person who understands the spectral characteristics of the scene being analyzed.*

Although the previous example involved interpretation of data in an image format, these same relationships between spectral response and the various cover types of interest are present when the reflectance data are in computer-tape format. In this format, the analyst can examine and compare spectral characteristics of the data in a more quantitative manner. A particularly useful aid in doing this is the multispectral response graph, or the so-called "coincident spectral plot," an example of which is shown in Fig. 5-42. In such a plot, the mean spectral response for each cover type of interest is plotted for each wavelength

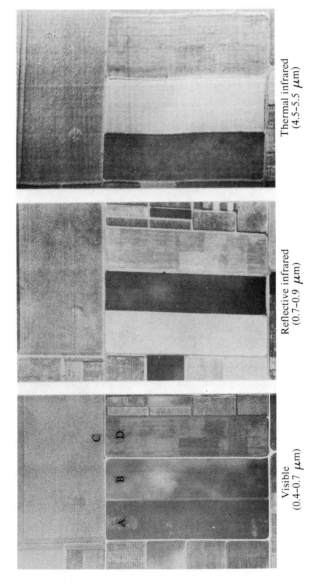

Figure 5-41 Multispectral characteristics of agricultural cover types in three wavelength bands. The data were obtained on September 30 at 1000 hours. Dense, green alfalfa is in the field labeled A, while B is a field of bare soil, C is brown mature corn, and D is dry wheat stubble.[44]

band and designated with a letter. The standard deviation (σ) from the mean is also calculated, and in this figure the $\pm 1\sigma$ is shown as a line of asterisks on either side of the mean. Thus, the longer the line, the greater the variance of the data; conversely, cover types and wavelength bands with little variance will be shown as relatively short lines.

Notice in Fig. 5-42 that water (class C) has a very low response in the near-infrared and middle-infrared wavelength bands (0.72 to 0.92 and 1.50 to 1.80 μm, respectively). In the 0.52 to 0.57-μm band, the corn (E) and soybeans (F) have a rather similar response, but in the 0.61 to 0.70-μm band the soybeans have a lower response than the corn, while in the 0.72 to 0.92- and 1.50 to 1.80-μm reflective infrared bands the soybeans have higher reflectance than corn, just as would be predicted on the basis of the DK-2 spectrophotometer reflectance data discussed earlier (i.e., the dicotyledonous species, soybeans, have higher reflectance in the near infrared than the monocotyledonous species, corn (Sec. 5-2). In the thermal infrared band, both corn and soybeans are similar in response, whereas the forest cover types (both deciduous and coniferous) have a significantly lower response, and water and forage (pasture lands) have a relatively high response. Because forage areas often have considerable amounts of bare soil exposed through the relatively short, dry stubble present, these areas often have a relatively high

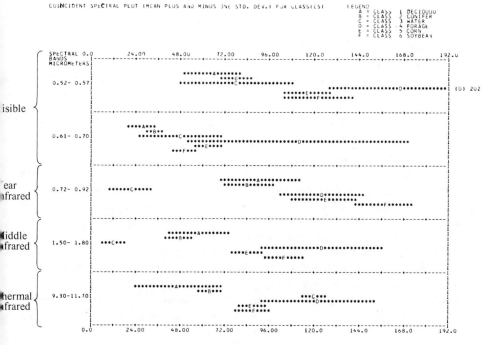

Figure 5-42 Multispectral response graph depicting relative reflectance and emittance in five wavelength bands for six major cover types. (*After Coggeshall and Hoffer.*[53])

response in the thermal infrared wavelength data. Such a condition would also explain the relatively high response for forage in the visible wavelengths.

In examining such coincident spectral plots, the analyst should be particularly alert for reversals in relative spectral response from one part of the spectrum to another. Even though there may not be any single wavelength band that will allow clear spectral separation between two cover types, combinations of wavelength bands which display these relative reversals in response may enable spectral discrimination to be achieved. For example, in Fig. 5-42, deciduous and coniferous forest cover cannot be completely discriminated in any single wavelength band, but several reversals can be seen: the deciduous have a somewhat lower response than the coniferous in the visible wavelength bands, a higher response in the reflective infrared bands, and a lower response again in the thermal infrared band. The value of this relative reversal in spectral response can be illustrated through the use of histograms of the data, as shown in Fig. 5-43. Here we see the same type of situation involving overlapping data values that was discussed in Chap. 1 (see Fig. 1-17). In essence, as Fig. 5-43 shows, the portion of class A that spectrally overlaps class B in wavelength band 1 can be easily separated in wavelength band 2. Thus, such a combination of wavelength bands allows virtually complete spectral discrimination, even though such discrimination is not possible in any single wavelength band.

Although the utilization of many wavelength bands is often an effective way to obtain maximum classification accuracy, it has been found that in practice, as more and more wavelength bands are utilized, the improvement in classification accuracy is not always commensurate with the increase in computer time required by the more complex calculations.[53] The coincident spectral plot frequently provides the analyst with valuable insights as to which wavelengths are likely to be most effective in maximizing classification accuracy, thus providing an effective tool to help in minimizing the number of wavelength bands required. However, where a large number of classes and wavelength bands are present and the interactions between wavelength bands are important for accurate discrimination, the analyst must resort to more powerful quantitative tools, such as the feature-selection methods discussed in Sec. 3-8.

Through most of this discussion, the assumption has been made that the data utilized have a Gaussian or normal distribution. This has generally been a reasonable assumption for most earth surface features; however, this is not always the case, and the analyst should check the data for normality when the methods used depend on this assumption. Experience has shown that the data may be normally distributed in many wavelength bands, yet in a certain spectral region it will have a bimodal distribution, indicating that in this case the data should be divided into two spectral classes for analysis. Figures 5-44 and 5-45 show examples of this. In Fig. 5-44 the data for corn appear to demonstrate fairly Gaussian distributions for all channels except the near-infrared 0.72 to 0.92-μm band (channel 8) where it is distinctly bimodal. In Fig. 5-45, the forest cover shows a reasonable Gaussian distribution in the visible and near-infrared portions of the spectrum, but bimodal distributions exist in both the middle-infrared and

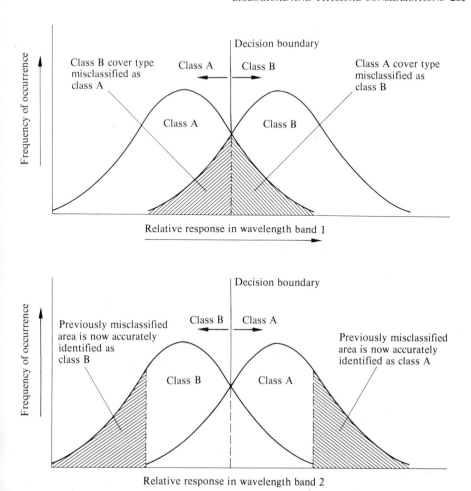

Figure 5-43 Class separability through use of spectral reversals. Even though neither class can be completely separated in either wavelength band alone, areas of spectral overlap in wavelength band 1 are clearly distinguished in wavelength band 2 due to the spectral reversal present.

thermal infrared bands (channels 10 and 12). Such bimodal distributions of reflectance values from vegetative cover can occur, rather unexpectedly, in different portions of the spectrum. Histogram plots can aid the analyst in checking the characteristics of potential training data sets and thus provide a very useful supplement to the coincident spectral plots. Since bimodal distributions of the data are often not apparent in the feature selection output or the coincident spectral plots, it is only through at least a brief examination of the histogram data that such bimodal distributions are found. Failure to separate a bimodal training class into two classes may cause classification errors which could have been avoided.

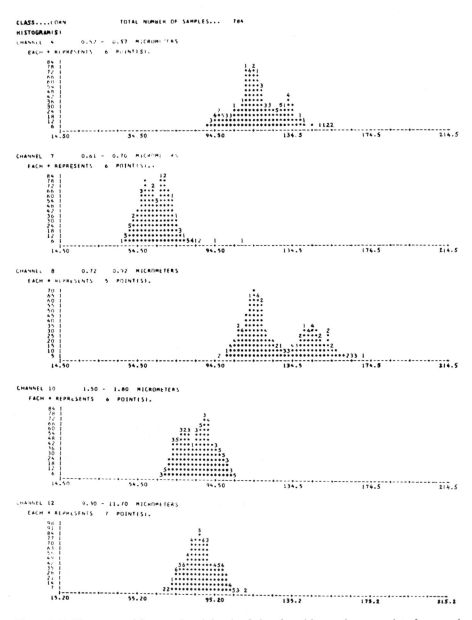

Figure 5-44 Histograms of five wavelength bands of aircraft multispectral scanner data for several different corn fields.

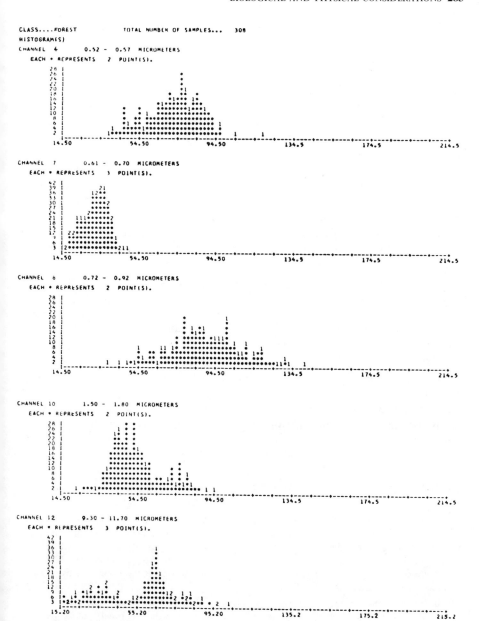

Figure 5-45 Histograms of five wavelength bands of aircraft multispectral scanner data for several blocks of deciduous forest cover.

As indicated by the histograms of reflectance in Figs. 5-44 and 5-45, various species and cover types of interest cannot always be represented by a single spectral class. Often, analysts working with multispectral scanner data have found that a particular category of interest (i.e., a species or cover type) is represented by several distinct spectral classes. The reverse may also occur; i.e., more than one category of interest may be present in a single spectral class. This situation requires really effective analyst/machine/data interaction and also indicates that the analyst must be knowledgeable about the spectral characteristics of the cover types with which he or she is working.

There is often a tendency to define the categories of interest that we want to map and then try to force the spectral classes to match the categories of interest. This can cause problems in the classification and result in classification errors. An effective alternative has often been to cluster the data, obtain "spectral maps" in which the spectrally distinct classes are displayed, and then use reference data to identify the cover types and their characteristics within each of the spectral classes. (See Sec. 3.10 for the discussion of "unsupervised analysis.") In this way, the analyst can easily determine if some categories of interest are spectrally very similar or, conversely, if several spectral classes actually represent a single category of interest. In the latter case, the different spectral classes sometimes represent important differences in the condition of the category of interest, such as differences in severity of a crop disease or differences in density of forest cover.

Thus, in summary, a key to increasing the effectiveness of utilizing computer-aided analysis techniques to analyze multispectral scanner data involves the proper interpretation of the spectral characteristics of the data and the skill to relate the spectral data to the information requirements that have been defined by the user.

PROBLEMS

5-28 Coincident spectral plots of bare soil, green vegetation, and water are shown in Fig. P5-28. Based on your knowledge of the relative reflectance of each of these materials in different wavelength bands, determine which symbol (A, B, or C) should be associated with water, which with green vegetation, and which with bare soil. (*Note*: the data have not been calibrated, so differences in response values are relative *within* each wavelength band; differences in response between wavelength bands are not significant.)

5-29 Based upon the data shown in the coincident spectral plot for Prob. 5-28, which single wavelength region (visible, near infrared, middle infrared, or thermal infrared) allows the best discrimination among the three cover types designated?

5-30 Which spectral region(s) does not allow water to be differentiated from vegetation?

5-31 In which spectral region(s) does soil have the highest relative response among the cover types shown?

5-32 In what way does the near infrared differ from the middle infrared in terms of the spectral responses of these cover types?

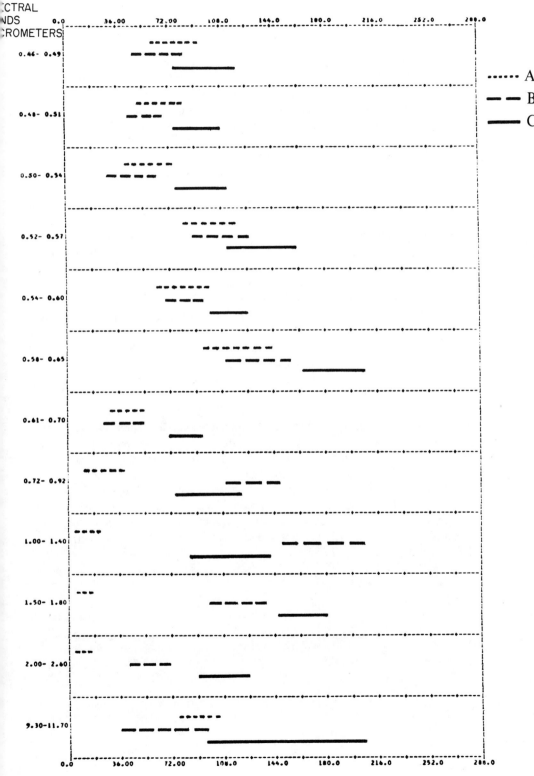

Figure P5-28

285

5-7 SUMMARY AND CONCLUSIONS

Effective utilization of computer-aided analysis techniques must be based on a solid understanding of both the techniques involved and the spectral characteristics of the materials. It has been the purpose of this chapter to describe the second of these two aspects of data analysis, and, minimally, to create a basic awareness of some of the fundamental energy-matter interactions involved in interpreting spectral characteristics of earth surface features.

We have discussed the spectral characteristics of major earth surface features, including vegetation, soil, water, and snow, and the fundamental energy-matter interactions which govern the spectral responses from these materials. The temporal and spatial effects on spectral response are often very significant, particularly among vegetative cover types, and must be carefully considered when interpreting multispectral scanner data. Interpretation of data from the various cover types indicates the value of each of the four major wavelength regions: the visible (0.4 to 0.7 μm), near infrared (0.7 to 1.3 μm), and middle infrared (1.3 to 3.0 μm) in the reflective portion of the spectrum, and the thermal infrared (8 to 14 μm) in the emissive wavelength region.

We have emphasized in this chapter that truly effective analysis of multispectral scanner data depends on in-depth understanding of the spectral characteristics of the cover types involved in addition to a thorough knowledge of the computer analysis techniques used. Without an understanding of the spectral characteristics of the cover types, the analyst soon finds himself "playing games" with the computer and does not really understand what the data do or do not represent. The analyst has a number of techniques available which allow interaction with the data and thus provide a better understanding, in a quantitative manner, of the spectral characteristics in the data. Without proper knowledge of the analysis techniques involved, the analyst cannot effectively utilize the power of the computer and the analysis capabilities that are available.

In the next chapter, we will see how the analyst combines understanding of the spectral characteristics of the earth surface features and knowledge of the computer-aided analysis techniques discussed in Chap. 3 for analyzing multispectral scanner data. An extended case study is presented along with several shorter ones, which together represent a variety of applications of the technology.

REFERENCES

1. Gates, D. M.: Physical and Physiological Properties of Plants, in "Remote Sensing with Special Reference to Agriculture and Forestry," National Academy of Sciences, Washington, D.C., pp. 224–252, 1970.
2. Gates, D. M., H. J. Keegan, J. C. Schleter, and V. R. Weidner: Spectral Properties of Plants, *Applied Optics*, vol. 4, no. 1, pp. 11–22, 1965.
3. Hoffer, R. M., and C. J. Johannsen: Ecological Potentials in Spectral Signature Analysis, in P. L. Johnson (Ed.), "Remote Sensing in Ecology," University of Georgia Press, Athens, Ga., 1969.

4. Hoffer, R. M.: Spectral Reflectance Characteristics of Vegetation, in "Fundamentals of Remote Sensing," Mini course Series, Purdue University, West Lafayette, Ind., 1976.
5. Knipling, E. B.: Physical and Physiological Basis for Differences in Reflectance of Healthy and Diseased Plants, in "Workshop on Infrared Color Photography in the Plant Sciences," Florida Department of Agriculture, Winter Haven, Fla., 1967.
6. Laboratory for Agricultural Remote Sensing: "Remote Multispectral Sensing in Agriculture," vol. 3, Research Bulletin no. 844, Agricultural Experiment Station, Purdue University, West Lafayette, Ind., 1968.
7. Sinclair, T. R., R. M. Hoffer, and M. M. Schreiber: Reflectance and Internal Structure of Leaves from Several Crops during a Growing Season, *Agronomy Journal*, vol. 63, pp. 864–868, 1971.
8. Myers, V. I.: Soil, Water, and Plant Relations, in "Remote Sensing with Special Reference to Agriculture and Forestry," National Academy of Sciences, Washington, D.C., pp. 253–297, 1970.
9. Hoffer, R. M.: Spectral Reflectance Characteristics of Earth Surface Features, in "Fundamentals of Remote Sensing," Minicourse Series, Purdue University, West Lafayette, Ind., 1976.
10. Soils Survey Staff, Bureau of Plant Industry: "Soil Survey Manual," Handbook no. 18, U.S. Department of Agriculture, Washington, D.C., 1951.
11. Bowers, S. A., and R. J. Hanks: Reflection of Radiant Energy from Soil, *Soil Science*, vol. 100, pp. 130–138, 1965.
12. Hunt, G. R., and J. W. Salisbury: Visible and Near-Infrared Spectra of Minerals and Rocks, *Modern Geology*, vol. 1, pp. 283–300, 1970.
13. Montgomery, O. L.: "An Investigation of the Relationship between Spectral Reflectance and the Chemical, Physical and Genetic Characteristics of Soils," Ph.D. Dissertation, Department of Agronomy, Purdue University, West Lafayette, Ind., 1976.
14. Page, W. R.: Estimation of Organic Matter in Atlantic Coastal Plain Soils with a Color-Difference Meter, *Agronomy Journal*, vol. 66, pp. 652–653, 1974.
15. Baumgardner, M. F., S. J. Kristof, C. J. Johannsen, and A. L. Zachary: Effects of Organic Matter on the Multispectral Properties of Soils, *Proc. Indiana Academy of Science*, vol. 79, pp. 413–422, 1970.
16. Obukhov, A. E., and D. S. Orlov: Spectral Reflectivity of Major Soil Groups and Possibility of Using Diffuse Reflections in Soil Investigations, *Soviet Soil Science*, vol. 1, pp. 174–184, 1964.
17. Matthews, H. L.: "Application of Multispectral Remote Sensing and Spectral Reflectance Patterns to Soil Survey Research," Ph.D. Dissertation, Pennsylvania State University, College Station, Pa., 1972.
18. Bartolucci, L. A., B. F. Robinson, and L. F. Silva: Field Measurements of the Spectral Response of Natural Waters, *Photogrammetric Engineering and Remote Sensing*, vol. XLIII, no. 5, pp. 595–598, 1977.
19. Work, E. A., and D. S. Gilmer: Utilization of Satellite Data for Inventorying Prairie Ponds and Lakes, *Photogrammetric Engineering and Remote Sensing*, vol. XLII, no. 5, pp. 685–694, 1976.
20. Bartolucci, L. A.: "Digital Processing of Satellite Multispectral Scanner Data for Hydrologic Applications," Ph.D. Dissertation, Department of Geosciences, Purdue University, West Lafayette, Ind., 1976.
21. Clarke, G. L., and H. R. James: Laboratory Analysis of the Selective Absorption of Light by Sea Water, *Journal of the Optical Society of America*, vol. 29, pp. 43–55, 1939.
22. Specht, M. R., D. Needler, and N. L. Fritz: New Color Film for Water-Photography Penetration, *Photogrammetric Engineering*, vol. XXXIX, no. 4, pp. 359–369, 1973.
23. Hulbert, E. O.: Optics of Distilled and Natural Water, *Journal of the Optical Society of America*, vol. 35, no. 11, pp. 698–705, 1945.
24. Duntley, S. Q.: Light in the Sea, *Journal of the Optical Society of America*, vol. 53, pp. 214–233, 1963.
25. Lepley, L. K., K. E. Foster, and L. G. Everett: Water Quality Monitoring of Reservoirs on the Colorado River Utilizing ERTS-1 Imagery, in K. P. B. Thomson, R. L. Lane, and S. C. Csallany (Eds.), Remote Sensing and Water Resources Management, *Proc. American Water Resources Association*, no. 17, Urbana, Ill., pp. 105–111, 1975.

26. Polcyn, F. C., and D. R. Lyzenga: Calculations of Water Depth from ERTS MSS Data, *Proc. Symposium on Significant Results Obtained from the Earth Resources Technology Satellite*, vol. 1, sec. B, Goddard Space Flight Center, New Carrollton, Md., pp. 1433–1441, 1973.

27. Ross, D. S.: Water Depth Estimation with ERTS-1 Imagery, in *Proc. Symposium on Significant Results Obtained from the Earth Resources Technology Satellite*, vol. 1, sec. B, Goddard Space Flight Center, New Carrollton, Md., pp. 1423–1432, 1973.

28. Weisblatt, E. A., J. B. Zaitzeff, and C. A. Reeves: Classification of Turbidity Levels in the Texas Marine Coastal Zone, in *Proc. Symposium on Machine Processing of Remotely Sensed Data*, IEEE Catalog no. 73 CHO 834-2GE, IEEE Single Copy Sales, Piscataway, N.J., pp. 3A-42 to 3A-59, 1973.

29. Clarke, G. L., G. C. Ewing, and C. J. Lorenzen: Spectra of Back Scattered Light from the Sea Obtained from Aircraft as a Measure of Chlorophyll Concentrations, *Science*, vol. 167, pp. 1119–1121, 1970.

30. Wezernak, C. T.: The Use of Remote Sensing in Limnological Studies, in *Proc. Ninth International Symposium on Remote Sensing of Environment*, vol. II, Environmental Research Institute of Michigan, Ann Arbor, Mich., pp. 963–980, 1974.

31. Egan, W. G.: Boundaries of ERTS and Aircraft Data Within Which Useful Water Quality Information Can Be Obtained, in *Proc. Ninth International Symposium on Remote Sensing of Environment*, vol. II, Environmental Research Institute of Michigan, Ann Arbor, Mich., pp. 1319–1343, 1974.

32. Szekielda, K. H., and R. J. Curran: Biomass in the Upwelling Areas Along the Northwest Coast of Africa as Viewed with ERTS-1, in *Proc. Symposium on Significant Results Obtained from the Earth Resources Technology Satellite*, vol. 1, sec. B, Goddard Space Flight Center, New Carrollton, Md., pp. 1385–1401, 1973.

33. Fok, Y. S., and W. M. Adams: Some Recent Applications of Remote Sensing to Water Resources in Hawaii, in K. P. B. Thomson, R. L. Lane, and S. C. Csallany (Eds.), Remote Sensing and Water Resources Management, *Proc. American Water Resources Association*, no. 17, Urbana, Ill., pp. 89–103, 1973.

34. Scherz, J. P., D. R. Graff, and W. C. Boyle: Photographic Characteristics of Water Pollution, *Photogrammetric Engineering*, vol. XXXV, no. 1, pp. 38–43, 1969.

35. Reeves, R. G. (Ed.): "Manual of Remote Sensing," 2 vols., American Society of Photogrammetry, Falls Church, Va., 1975.

36. Munday, J. C., W. G. MacIntyre, M. E. Penney, and J. D. Oberholtzer: Oil Slick Studies Using Photographic and Multispectral Scanner Data, in *Proc. Seventh International Symposium on Remote Sensing of Environment*, vol. II, Environmental Research Institute of Michigan, Ann Arbor, Mich., pp. 1027–1043, 1971.

37. Scherz, J. P.: Remote Sensing Considerations for Water Quality Monitoring, in *Proc. Seventh International Symposium on Remote Sensing of Environment*, vol. II, Environmental Research Institute of Michigan, Ann Arbor, Mich., pp. 1071–1087, 1971.

38. Hoffer, R. M.: Unpublished File Documents for Research Study on Strip Mine Waters of Southern Indiana, 1974.

39. Rogers, R. H., L. E. Reed, N. J. Shah, and V. E. Smith: Automatic Classification of Eutrophication of Inland Lakes from Spacecraft Data, in *Proc. Ninth International Symposium on Remote Sensing of Environment*, vol. II, Environmental Research Institute of Michigan, Ann Arbor, Mich., pp. 981–987, 1974.

40. Wezernak, C. T., and F. C. Polcyn: Eutrophication Assessment Using Remote Sensing Techniques, in *Proc. Eighth International Symposium on Remote Sensing of Environment*, vol. I, Environmental Research Institute of Michigan, Ann Arbor, Mich., pp. 541–551, 1972.

41. Hoffer, R. M., and Staff: "Natural Resource Mapping in Mountainous Terrain by Computer Analysis of ERTS-1 Satellite Data," Research Bulletin no. 919, Agricultural Experiment Station, Purdue University, West Lafayette, Ind., 1975.

42. Hoffer, R. M., and Staff: Computer-Aided Analysis of SKYLAB Multispectral Scanner Data in Mountainous Terrain for Land Use, Forestry, Water Resource and Geologic Applications, Final report, NASA Contract no. NAS9-13380, SKYLAB EREP Project 398, *LARS Information*

Note 121275, Laboratory for Applications of Remote Sensing, Purdue University, West Lafayette, Ind., 1975.

43. O'Brien, H. W., and R. H. Munis: Red and Near Infrared Spectral Reflectance of Snow, in A. Rango (Ed.), "Operational Applications of Satellite Snowcover Operations," Workshop proceedings, NASA Special Publication SP-391, Washington, D.C., 1975.

44. Hoffer, R. M.: "Interpretation of Remote Multispectral Imagery of Agricultural Crops," vol. 1, Research Bulletin no. 831, Agricultural Experiment Station, Purdue University, West Lafayette, Ind., 1967.

45. Laboratory for Agricultural Remote Sensing: "Remote Multispectral Sensing in Agriculture," vol. 2, Research Bulletin no. 832, Agricultural Experiment Station, Purdue University, West Lafayette, Ind., 1967.

46. Hoffer, R. M.: Interpretation of Color Infrared Photography, in "Fundamentals of Remote Sensing," Minicourse Series, Purdue University, West Lafayette, Ind., 1976.

47. Manzer, F. E., and G. R. Cooper: "Aerial Photographic Methods of Potato Disease Detection," Research Bulletin no. 646, Maine Agricultural Experiment Station, Orono, Maine, 1967.

48. Hope, J. R.: Pathway of Heavy Rainfall Photographed from Space, *Bulletin of the American Meteorological Society*, vol. 47, no. 5, pp. 371–373, 1966.

49. Hoffer, R. M.: Remote Sensing Potentials in Resource Management, in E. J. Monke (Ed.), *Biological Effects in the Hydrological Cycle—Terrestrial Phase*, Proc. Third International Seminar for Hydrology Professors, Agricultural Experiment Station, Purdue University, West Lafayette, Ind., 1971.

50. Hoffer, R. M., and J. C. Lindenlaub: Interpretation of Multispectral Scanner Images, in "Fundamentals of Remote Sensing," Minicourse Series, Purdue University, West Lafayette, Ind., 1976.

51. Bartolucci, L. A., R. M. Hoffer, and T. R. West: Computer-Aided Processing of Remotely Sensed Data for Temperature Mapping of Surface Water from Aircraft Altitudes, *LARS Information Note 042373*, Laboratory for Applications of Remote Sensing, Purdue University, West Lafayette, Ind., 1973.

52. National Aeronautics and Space Administration: "Corn Blight Watch Experiment Final Report," vol. III (Experiment Results), NASA Johnson Space Center, Houston, Texas, 1973.

53. Coggeshall, M. E., and R. M. Hoffer: Basic Forest Cover Mapping Using Digitized Remote Sensor Data and ADP Techniques, *LARS Information Note 030573*, Laboratory for Applications of Remote Sensing, Purdue University, West Lafayette, Ind., 1973.

54. Silva, L. F., and L. L. Biehl: A Study of the Utilization of EREP Data from the Wabash River Basin, *LARS Information Note 012276*, Laboratory for Applications of Remote Sensing, Purdue University, West Lafayette, Ind., 1976.

SIX

APPLYING THE QUANTITATIVE APPROACH

John C. Lindenlaub and Shirley M. Davis

This chapter considers the application of quantitative remote sensing to specific earth resources management problems. In the first section the key steps in carrying out a remote sensing applications project are enumerated. In a sense, this section presents the viewpoint of "middle management" (operations manager), describing the various considerations which must be taken into account when planning and carrying out the application of remote sensing techniques to an earth resources management problem.

Next, a specific application is discussed in detail. The presentation is in the format of a case study and is designed to give the reader insight into the operational aspects of a remote sensing project, considering such things as reference data requirements, analysis procedures, and output products.

The chapter concludes with brief descriptions of five additional applications of quantitative remote sensing. These examples were selected from research and applications reports on the basis of proven feasibility. Together they suggest the range of disciplines in which remote sensing can be effectively applied and the differing means of data collection and analysis which can be implemented as part of the quantitative approach.

Study objectives

After studying Chap. 6, you should be able to:

1. List the key steps in applying remote sensing to earth resources management problems

2. Analyze a description of the application of remote sensing to an earth resources problem, identify the key steps, and evaluate decisions stated or implied concerning the data-collection system used, frequency of data collection, reference data used, and preprocessing and analysis procedures.
3. Given a description of an earth resources management problem, evaluate the situation as to the feasibility of using remote sensing to aid in the solution of the problem. If the use of remote sensing is warranted, propose a scheme for supplying the information necessary to solve the problem. Your proposal should include a statement of the objectives of the project, a mechanism for verifying the feasibility, and a project plan.

6-1 KEY STEPS IN CARRYING OUT A REMOTE SENSING APPLICATIONS PROJECT

Remote sensing projects are carried out to furnish information that is useful to the earth resources manager. Such projects using remote sensing techniques can be conveniently subdivided into five key steps:

1. Stating user requirements or objectives
2. Establishing feasibility
3. Planning the project
4. Implementing the project
5. Assessing the results

The essential first step is the *development of a statement of the information required by the user.* Exactly what information will the remote sensing project need to produce? Since the user is often not familiar with the capabilities of remote sensing and the remote sensing specialist is usually not responsible for making earth resources management decisions, it is important that these information requirements be jointly established. This approach is more likely to yield a set of specifications which are more realistic from the viewpoint of both the earth resources manager and the remote sensing specialist and avoids oversimplified objective statements like, "Give me all the information I need to solve my problem."

The following list of questions provides a useful starting point for establishing the requirements of the user:

1. What are the earth surface features or cover types of interest?
2. Exactly what does the user want to know about these earth surface features (e.g., their locations, areal extent, condition)?
3. What size area is involved?
4. In what format are the results needed (maps, tables, or both)?
5. How accurate must the results be?

6. Is complete coverage of the area required or will a sample of the entire area be sufficient to yield the necessary information?
7. Are there any special temporal considerations (time of day or season of the year)?

Once the user has answered questions like these, it is possible to establish the feasibility of using remote sensing techniques to satisfy the user's requirements.

Establishing the feasibility of the project should be the next step undertaken. It is at this point that a decision needs to be made as to whether or not quantitative remote sensing is a reasonable approach to use. If, for instance, the user requires very detailed and highly accurate information over a very small area, say 100 km², then a set of medium-scale aerial photos would probably be the most reasonable method of data acquisition, and manual air-photo interpretation would be the most reasonable analysis procedure. On the other hand, if information is required over several hundred thousand square kilometers, it would seem reasonable that information could best be gathered from a satellite system, and computer-aided analysis of the data would probably be the most appropriate method for obtaining the information required. A feasibility study should include consideration of whether:

1. The cover types of interest have distinguishable spectral, temporal, or spatial characteristics.
2. Suitable data-collection instruments are available.
3. A suitable data-handling and analysis software system is available.
4. Multivariate analysis procedures will provide the kind of information required.

If the answers to these questions are in the affirmative, one can move on to planning the project.

The third step, *project planning*, is more specific than determining the feasibility of the project. A product of the feasibility study is often a list of optional approaches, all of which could lead to the desired results. The project-planning step narrows these choices to a specific approach. Decisions must be made with respect to:

1. Data-collection system used and frequency of data collection.
2. Reference data and ground observation requirements.
3. Preprocessing requirements.
4. Data analysis procedures.

These factors are usually constrained by considerations of both cost and project schedule.

Implementation of the project plan is the next step in carrying out the application. It is here that the previous decision-making steps are put into action. The operational and management aspects of the project include:

1. Collecting the data.
2. Preprocessing the data.
3. Analyzing the data.
4. Producing output products.

Management of this phase of the project includes ensuring that there are sufficient personnel and computer resources to carry out the project without creating backlogs.

Specifying means of *assessing the results* is just as important a step as stating the user requirements. In fact, one of the criteria used to assess the results is whether the information derived from the remotely sensed data does indeed satisfy the requirements of the user. The ultimate success of the project is determined by whether or not the user is better able to solve the problem or make better earth resources management decisions as a result of the information supplied through the analysis of remote sensing data.

The five steps in a remote sensing project are summarized in Table 6-1. These five steps serve as a convenient way to describe a remote sensing applications project, but one should not assume that these are distinct nonoverlapping steps. As pointed out in the discussion of the first step, stating user requirements is best carried out jointly by the user and the remote sensing specialist. Similarly, the project-planning step needs to be undertaken jointly so that the circumstances under which the project will be implemented can be considered from all sides. The steps overlap in time also. Project planning does not cease as soon as implementation begins. At a given point in time, the level of effort devoted to some step in the project will be higher than for the other steps, but at least some effort will be devoted to all the steps all the time. Although the degree of interaction varies, there is some interaction among all steps of the project. For example, especially important interaction exists between the assessment of results and the user's requirements; if the output products do not meet the user's requirements, corrective action needs to be taken.

In the section that follows, the five steps outlined above will be expanded upon as they relate to a specific remote sensing applications project.

Table 6-1 Key steps in carrying out a remote sensing applications project

1. State user requirements or objectives	What needs to be done?
2. Establish feasibility	Can it be done?
3. Plan the project	What procedures should be used?
4. Implement the project	Do it!
5. Assess the results	Did it work?

6-2 CROP-IDENTIFICATION AND ACREAGE-ESTIMATION CASE STUDY

This case study is based on the work of M. E. Bauer.[1] While Bauer's work was a research project, for tutorial purposes it is presented here as a case study of an operational program. Results quoted are factual, but in some cases the descriptions of the manner and sequence in which they were arrived at have been idealized, drawing upon lessons learned in the research.

The value of having information about expected crop production is substantial. For example, Eisgruber[2] estimated that the overall social benefits from reducing the error in estimates of United States production of corn, soybeans, and wheat from 3 to 2 percent would amount to $14.1 million annually. He further showed that additional social benefits of $8.5 million could be obtained by reducing the error from 2 to 1 percent. The gains would result from improvement in decisions regarding inventory management and shipping of crops based on improved information. These estimates are based on 1964–1970 prices and reflect neither the increase in the value of grain since 1970 nor the benefits that would result from improved crop-production estimates in those parts of the world that do not have the kind of crop-reporting service found in the United States.

In addition, Ewart[3] has shown that more frequent information, as might be provided with remote sensing techniques, would decrease social loss even without improvement in the crop-estimate error. Thus there is considerable social and economic motivation for applying remote sensing techniques to crop identification and acreage estimation, both of which are essential for crop-production estimates.

Stating User Requirements or Objectives

Crop identification and acreage estimation is of interest to both governmental and private-sector organizations. The Statistical Reporting Service of the U.S. Department of Agriculture (USDA) collects and disseminates historical information concerning acreage yields, reports on current crop conditions, and provides estimates of expected yields. The publicly disseminated information is augmented by the efforts of private-sector organizations, especially those in the agribusiness community which have established their own data-collection and estimation procedures. While the application of remote sensing techniques to crop identification and acreage estimation over a very large area would undoubtedly require monetary resources available only to governmental organizations such as USDA or the United Nations Food and Agriculture Organization (FAO), applications involving a smaller geographic region might be undertaken by either governmental or private organizations. This case study falls in the latter category and deals with the agricultural regions of the states of Illinois, Indiana, and Kansas.

The project was initiated by answering a list of questions like those given in Sec. 6-1 to help establish the user requirements. The earth surface features of interest were agricultural crops, specifically corn, soybeans, and wheat. Initially the user, who typically might have been a large grain-exporting company, requested estimates of the acreage and expected yield of those crops, but after discussing this goal with personnel experienced in remote sensing applications, it was decided that a more realistic goal would be to obtain acreage estimates of the various crops several times during the growing season. These figures would then be used as one of the input variables to a crop-production model. The primary format requested for the results was tabular; tables of acreage estimates on a county, crop-reporting district and statewide basis. Complete coverage of the three-state area was specified, and the user asked that acreage estimates be supplied as soon as possible after data collection. After discussions between the user and remote sensing specialists, the following objective statement was agreed upon:

> Provide acreage estimates of corn, soybeans, and wheat present in the states of Indiana, Illinois, and Kansas. Acreage estimates are to be made for state, crop reporting district and county-sized areas. Estimates are to be provided at four different times distributed throughout the growing season and are to be completed within three weeks of the time data are collected.

Feasibility Study

Determining whether remote sensing techniques can meet the objectives of the user is the next step in an applications project. A number of data-collection systems could conceivably have been used for the project: Landsat multispectral scanners, aircraft-mounted multispectral scanners, or airborne camera systems.

The basic capability of remote sensing techniques for crop estimates had been demonstrated during the 1971 Corn Blight Watch Experiment.[4] Both airphoto interpretation and computer-aided analysis of aircraft multispectral scanner data were used in this experiment. The experiment covered the seven-state area of the central U.S. Corn Belt. Furthermore, the feasibility of using a pattern-recognition approach to analyze Landsat multispectral scanner data for crop identification had been demonstrated by Bauer and Cipra.[5] Although carried out over only a three-county area, the acreage estimate results compared well with U.S. Department of Agriculture estimates (see Table 6-2). In both of these studies, data analysis was carried out using the LARSYS software system (Chap. 4) and a maximum-likelihood classification algorithm (Chap. 3).

An important part of establishing the feasibility of a project is determining whether adequate data-handling and analysis facilities are available. In the Corn Blight Watch Experiment, multispectral scanner data had been collected and analyzed only over the western half of Indiana; the number of data vectors classified during each two-week period was approximately 37.2 million. For the present study, the number of Landsat resolution elements contained in the study area was computed to be 98.3 million. In the Corn Blight Watch Experiment, the test segments were analyzed 10 times during the growing season for a total of 372 million data points analyzed. In this study results were required four times during the season, for a total of 393 million data vectors to be classified. Based

Table 6-2 Comparison of USDA area estimates (ground based) with estimates derived from computer analysis of Landsat 1 data for Dekalb, Ogle, and Lee Counties, Illinois[5]

| | Percentage of total area ||
Crop type	USDA	Landsat 1
Corn	40.2	39.6
Soybeans	18.0	17.8
Other	41.8	42.0

on these calculations it was concluded that the computational requirements of the project were within existing capabilities.

When large geographic areas are classified, the computer costs associated with running a maximum-likelihood classification program can become intolerable. But when the features to be classified (agricultural fields in this case) are large compared to the resolution of the remote sensing data-collection system, it is sometimes reasonable to reduce the amount of computer time required for classification by classifying only a sample of the available data vectors. For instance, classification time would be reduced by a factor of four if, instead of classifying every data vector, only the vectors in every other row and column were classified. This possibility was considered,[6] and it was determined that standard sampling techniques[7] could be used to estimate the errors introduced by such sampling. For instance, in wheat versus nonwheat classification of Rice County, Kansas, comprising 255,012 Landsat resolution elements, the results of using an 11.1 percent sample (obtained by classifying every third line and column) were compared to the results obtained by classifying every line and column. Nine different 11.1 percent samples were taken from the same area by selecting a different starting point for each sample. The variance of the nine sample estimates was compared to that determined by theoretical considerations. A similar procedure was used for sampling rates of 50, 33.3, 25, 10, 6.25, 4, and 2.8 percent. In all cases the variances among repeated estimates of the same sample size were not significantly different from the theoretical prediction. It was therefore concluded that the theoretical estimate of the sampling error can be used as the basis for a decision on the appropriate sampling rate to use in situations where the resolution element size is small compared to average "field" size. Obviously, the smaller the sampling rate, the greater the saving in computational costs.

Thus, based on the current state of the technology, previous experience, and the results of related research projects, it was decided that, indeed, suitable data-collection systems were available, that the cover types of interest could in fact be distinguished based on spectral and/or temporal variations, that suitable data-handling and analysis facilities were available, and that multivariate analysis procedures would provide the kind of information required by the user. Furthermore, since data for the Corn Blight Watch Experiment had been collected and analyzed at two-week intervals, it was concluded that it would be feasible to meet both the information and timeliness requirements of the user, provided data could be made available within one week after collection.

Project Planning

Having established the feasibility of the project, specific details had to be worked out. The data-collection system to be used and the frequency with which data would be collected had to be decided upon. These questions are never completely independent of the data-analysis procedures to be used. Because of the large geographic area involved, computer-aided analysis was indicated as opposed to manual image interpretation. This in turn suggested a multispectral scanner as the data-collection system best able to provide high-quality data. Although the use of an aircraft-mounted multispectral scanner would have allowed for scheduling flights to avoid bad weather conditions, the frequency of Landsat coverage (two satellites were operational at the time so that data were available every 9 days) largely negated any advantage an aircraft system might have had in this application. The use of Landsat data was further supported by the relatively low cost of the data, the frequency of coverage required, and the fact that the feasibility study indicated that the spectral and spatial resolution of the Landsat multispectral scanner were sufficient to meet the information needs of the user.

Three types of reference data were chosen to support the analysis of the Landsat data: large-scale aerial photography, topographic maps, and historical information. Color and color infrared photography were to be collected several times during the year for aiding in the selection of training samples, evaluating the accuracy of the classifications, and making acreage estimates for those areas which were covered by clouds during the satellite pass. United States Geological Survey (USGS) topographic maps at a scale of 1:250,000 were chosen to assist in locating county boundaries and transforming county boundary coordinates to the row and column coordinates of the Landsat data so that the data could be analyzed on a county-by-county basis. This scale was judged to be large enough to permit early and accurate location of political boundaries and yet not so large as to render handling of the maps unreasonable. Plans were made to use larger scale $7\frac{1}{2}$-minute (1:24,000 scale) USGS quadrangle maps for those areas in which aerial photography was to be obtained. The larger scale maps were judged to be more useful in aiding the analyst in the selection of training and test fields. The *Statistical Abstract of the United States, U.S. Agricultural Statistics*, and the *Agricultural Census* are examples of references selected to provide historical data to aid in the establishment of homogeneous subregions and to provide estimates of the a priori probabilities of the ground cover types of interest.

Since the decision had been made to work with Landsat multispectral scanner data, one of the preprocessing requirements was reformatting the Landsat data to be compatible with the analysis software. Multitemporal analysis had not been considered in the feasibility study, but, in case it might be needed, plans were made to take data for the same geographic area but gathered during different phases of crop maturity and overlay or register them. As noted in Chap. 4, the registration process merges data from different Landsat passes onto a single tape with the image points stored in such a manner that the data from any given point in the scene are in geometric coincidence for all passes and can therefore be addressed by a single set of row and column coordinates. This concept is illustrated in Fig. 6-1.

It was also planned to geometrically correct those portions of the area over which aerial photography was to be gathered. In this preprocessing procedure, the data are rotated, deskewed, and rescaled such that image products obtained from the data are north oriented and scaled to conform to available reference data.[8] Use of a common scale for both multispectral scanner images and reference data makes it easier for the analyst to locate corresponding points on the two images.

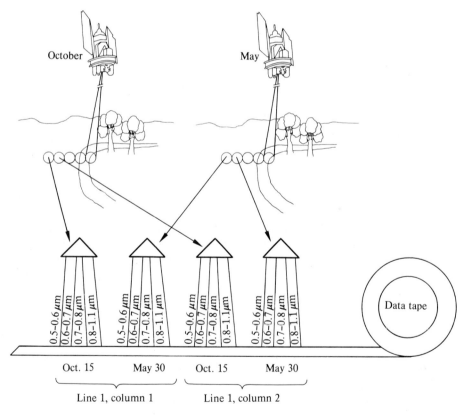

Figure 6-1 Merging Landsat data collected over the same scene at two different times to form a multispectral-multitemporal data set.

Data-analysis plans were formulated next. Aerial photography would be used to aid in the selection of training samples for each class of interest. Training fields would be delineated on computer line-printer maps of the Landsat data. Cluster analysis would determine whether the crop types of interest were spectrally homogeneous or if subclasses had to be defined. Next, an evaluation of the spectral/temporal channels was planned using a feature-selection algorithm based on transformed divergence. In order to reduce the computer time required for classification, plans were made to classify only a fraction of the total number of picture elements in the three-state area. The results of the feasibility study showed that the errors introduced by sampling, i.e., by classifying, say, every second or every third line and column, could be predicted and hence maintained within tolerable limits. Classification would be done using a maximum-likelihood classifier with a priori probability estimates obtained from the historical reference data. The classification results were to be summarized in tabular format and converted to area estimates.

Summarizing, project planning resulted in decisions to use Landsat multispectral scanner data with various maps, aerial photography, and historical documents serving as reference data. Preprocessing requirements and data-analysis procedures were also detailed.

Project Implementation

Step 4 is the phase of the project in which the work actually gets done. A project of this size involves many people, consumes a lot of computer resources, and requires careful management to ensure a smooth operation. While the management aspects of a remote sensing applications project must not be neglected, it seems appropriate, in the context of this book, to emphasize the technical aspects of the project. Therefore, we will make no attempt to address project management in detail.

Collecting the data Figure 6-2 shows the study area and the nominal centers of the 33 Landsat frames required to cover the area. Data were required for the period September through December for winter wheat identification and for April

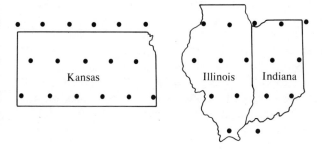

Figure 6-2 Approximate centers of Landsat frames covering the study area.

through August for identification of wheat, corn, and soybeans. Since an important requirement of the project was that the crop acreage estimates be produced within three weeks of the date of collection, standing orders covering the periods of interest were placed with the data-distribution agency† for all frames with centers falling in the study area and having less than 70 percent cloud cover. Under this arrangement, both imagery and computer-compatible tapes (CCTs) for all frames meeting specifications would be shipped as soon as available.

The decision to place a standing order for data with as much as 70 percent cloud cover might seem questionable. But even with such a high percentage of the frame being cloud covered, it was still quite possible that the areas to be analyzed could be cloud free. An alternative approach would have been to request on standing order only enough imagery for screening purposes and, after arrival of this imagery, to order the balance of the imagery and the CCTs for those frames observed to be suitable. The receipt and analysis of the data would have been delayed significantly by such a scheme. Weighing the timeliness objectives of the project against the cost of receiving possibly useless data products, it was clear that the standing order of all required data products was the appropriate strategy.

The standing order also included 70-mm negative transparencies of Landsat channels 5 (0.6 to 0.7 μm) and 7 (0.8 to 1.1 μm) of *all* multispectral scanner frames covering *any part* of the three-state area, regardless of whether their centers fell within the study area. These channels were chosen to provide one visible and one near-infrared film product for each frame. The 70-mm transparencies were used to visually screen the Landsat data for cloud cover and other data quality factors. This screening operation accomplished two things: first, it provided a quick and inexpensive way to determine whether it was worth while to preprocess the Landsat CCTs into the format compatible with the analysis software system and, second, it provided a rapid and efficient means of identifying frames containing useful information but which had not been included in the standing order for CCTs. For example, a frame center might fall outside the study area with only a very small portion of the data falling within the study area. But if the *cloud-free* portion of, say, an 80 percent cloudy frame fell within the study area, it might be desirable to have those data available for analysis. A special order would then be placed for the CCTs since, under the standing order, data tapes for a frame having 80 percent cloud cover would not be received.

In addition to the Landsat data, large-scale color and color infrared photography was collected over 40 percent of the counties in the study area to serve as reference data. Two Model 103 Hulcher 70-mm aerial cameras aboard a Cessna 310 aircraft were used to collect the photography six times during the year: September, November, April, June, July, and August. The flightlines followed state highways in order to minimize the chances of navigation errors and were flown at an altitude of approximately 1500 m (5000 ft). The five Indiana flightlines

† EROS Data Center, U.S. Geological Survey, Sioux Falls, S. D.

are shown in Fig. 6-3. Data-collection requirements stated that the photography must have less than 15 percent cloud cover, that the sun angle must be greater than 35° and, when possible, that acquisition be made within 7 days of a cloud-free Landsat pass. The sun-angle requirement was intended to keep shadow effects on the imagery within tolerable limits. Similarly, the look angle for the

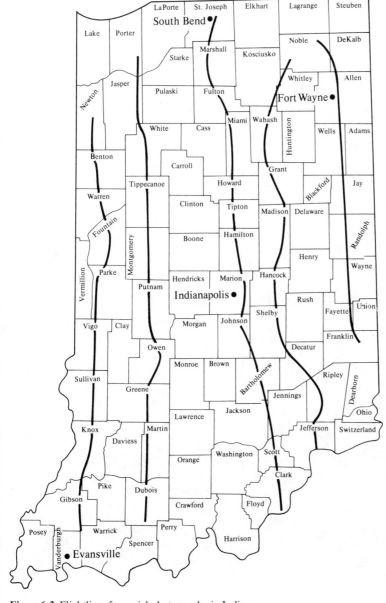

Figure 6-3 Flightlines for aerial photography in Indiana.

photographic missions was specified as less than 40° in order not to complicate interpretation of the photos. Forward overlap of 60 percent was specified to allow stereoscopic viewing. The scale of the photos was to be approximately 1:30,000.

Based on historical meteorological data, it was expected that an average of 20 of the 33 Landsat frames of digital data and related photographic images and reports would be received every 18 days. To handle this volume of data, the system shown in Fig. 6-4 was used. Once the incoming images, reports, and data tapes were logged into the data library, they were ready for the analyst.

Preprocessing As mentioned previously, 70-mm Landsat film products were used to screen the Landsat data for cloud cover and other data quality factors. The data tapes for the frames which passed this inspection were then reformatted so as to be compatible with the analysis software. To assist the analysts in locating training areas, the multispectral scanner data from those portions of the area over which aerial photography had been gathered were geometrically corrected. Scale factors were chosen such that when the data were displayed by means of a standard computer line printer, they would have a scale of 1:24,000, the scale of USGS $7\frac{1}{2}$-minute quadrangle maps.

Analyzing the data The procedures used to analyze the multispectral scanner data are summarized in Fig. 6-5. The first step in the procedure is to assess data quality. A general impression of data quality had already been obtained by visual

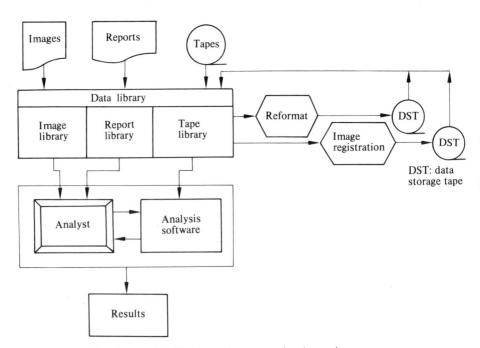

Figure 6-4 Data flow for crop-identification and acreage-estimation project.

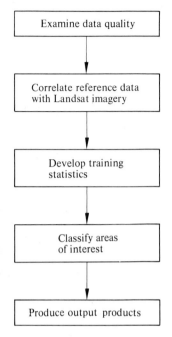

Figure 6-5 Analysis procedure for multispectral scanner data.

inspection of the Landsat 70-mm film products. Assuming the data have passed the film-screening step and have been reformatted, the analyst examines the data on a digital display device. It is important that the analyst be able to look at the entire frame, enlarge selected portions, and switch from band to band. During this step in the analysis, the analyst checks for cloud cover, bad data lines, and other anomalies introduced by the data system, and generally familiarizes himself with the geographic area.

The next step in the analysis is to correlate the Landsat multispectral scanner data with reference data. Recall that aerial photographs were obtained over selected portions of the three-state study area. These photographs, along with USGS maps, constituted the primary reference data available to the analyst. As mentioned in the discussion of data preparation, those portions of the Landsat frames for which aerial photography was available had been geometrically corrected to aid the analyst in identifying corresponding features on the Landsat images and aerial photography.

Correlating reference data with Landsat imagery leads directly into the next analysis step—developing training statistics. The objective of this step is to produce estimates of the statistical characterization of each ground cover class of interest. At this stage in the analysis, the analyst's talent plays an important role. Knowledge of the principles of remote sensor instrumentation systems (Chap. 2), the interaction between the sun's radiation and earth surface materials (Chap. 5), and the basics of pattern recognition (Chap. 3) all contribute to the analyst's ability to do a good job of "training the classifier." A well-designed interactive subsystem (Chap. 4) is also important.

The task of developing the training statistics can best be described in terms of a number of substeps, shown in Fig. 6-6. The procedure is very typical of that used in a wide range of similar applications. Training areas of 100 lines by 100 columns of Landsat data were selected in areas corresponding to aerial photography. In order to adequately represent the variation present in the county, four to six areas were selected in each county, with at least one each in the northern and southern portions of the county.

To facilitate locating agricultural fields in the Landsat data, clustering was used to produce a spectral class (cluster) map (refer to the discussion of unsupervised analysis in Sec. 3-10). Generally, six to eight classes were sufficient to provide a spectral class map on which agricultural fields were easily identifiable. This approach was found to be more satisfactory than working with grayscale maps of single spectral bands. An example of a cluster map is shown in Fig. 6-7. An example of color infrared photography of a portion of the same area is shown in Fig. 6-8 (between pages 36 and 37) in the color illustration section.

As part of the standard analysis procedure, the analyst separated the fields that had been identified as belonging to specific cover types into two groups. One group, known as *training fields*, was used to "train the classifier," i.e., to determine the statistical characteristics of each training class. The other group of fields, known as *test fields*, was reserved for later use in the analysis to aid in the evaluation of the classification results.

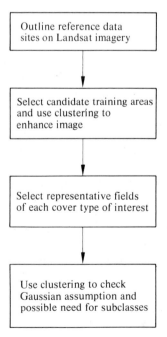

Figure 6-6 Steps in developing training statistics.

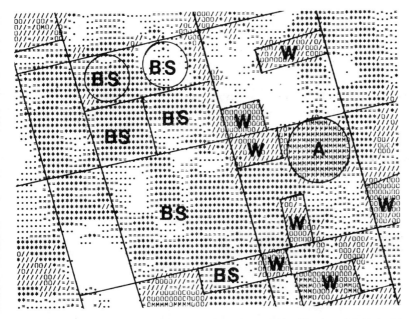

Figure 6-7 Example of cluster map used for location and identification of fields in Finney County, Kansas (W = wheat, A = alfalfa, BS = bare soil). Outlined area corresponds to Fig. 6-8.

The fields designated as training fields were then clustered. As noted in Sec. 3-10, clustering is an unsupervised classification technique that divides the data into spectrally distinct classes or subclasses. It is used at this stage in the analysis sequence to determine whether or not the training fields for each cover type of interest are spectrally homogeneous. If they are not, each of the non-homogeneous cover types would have to be subdivided into a number of homogeneous subclasses in order to satisfy the Gaussian assumption of the maximum-likelihood classification algorithm (Sec. 3-6). Experience gained in the early stages of the project suggested that the analyst initiate the cluster analysis by requesting four clusters for wheat training fields and five clusters for the nonwheat classes. Working with the cluster analysis output, the analyst refines the candidate training areas and chooses the final training areas and number of subclasses.

Since aerial photography was collected for only 40 percent of the counties in the study area, in order to classify each of the remaining counties the analyst used training statistics from an adjacent county. When there was a choice of adjacent counties, the analyst tended to favor a neighboring county which had similar climatic conditions (rainfall was the chief indicator here) and similar soil types and agricultural practices. Historical reference data were especially helpful in these situations.

The next step in the analysis procedure is classification of the areas of interest. The project objective called for acreage estimates on state, crop-reporting district, and county-sized areas. To accomplish this, a table digitizer was used to digitize county boundaries shown on a 1:250,000-scale map. Coordinates of easily identifiable checkpoints, such as major highway intersections and airports, were also digitized. These checkpoints were then located precisely in the Landsat imagery in order to determine the geometric transformation needed to convert the county boundaries from the map to Landsat coordinates. The boundary locations were then used in conjunction with a maximum-likelihood classification algorithm to classify and tabulate results separately for each county in the study area.

To economize on computer time the sampling scheme described earlier was used. Instead of classifying every Landsat data vector, 25 percent of the data vectors were classified—those in every other line and every other column.

Producing output products The classification program which was used in this study creates a results file containing the class assignment for each data vector classified. The results file is then used as input to other processing functions which produce various types of output products. The chief output product in this project was a set of tables showing the percentage of area and number of hectares (acres) for each agricultural crop of interest. Results were obtained for each county and also aggregated into crop-reporting districts. An example of the tabular results for wheat in the South Central crop-reporting district of Kansas is shown in Table 6-3. These results are compared to USDA Statistical Reporting Service (SRS) estimates as part of the results assessment in the next section.

Similar products were produced four times during the calendar year, as required by the project objectives. Cloud-cover conditions prohibited the collection of data over the entire study area on any single Landsat pass, so that none of the four reports actually reflects a "snapshot" of crop-acreage conditions at one point in time, but rather each is a compilation of classification results which show the best estimates available at each of the four different times selected.

Table 6-3 Landsat estimates of wheat in the South Central crop-reporting district of Kansas from data acquired during April 1975

County	Hectares (000)	Proportion, %
Barber	23.1	7.8
Comanche	31.1	15.0
Edwards	58.0	36.4
Harper	110.0	53.4
Harvey	55.3	39.5
Kingman	113.7	50.8
Kiowa	43.2	23.2
Pratt	91.3	48.3
Sedgwick	71.0	27.5
Sumner	217.0	70.9

Assessment of Results

The last key step in carrying out a remote sensing project is assessing the results. As discussed in Chap. 3, classification accuracy may be evaluated by examining training field results and test field results and by comparing the classification results with independent estimates. All three schemes were used in this project.

Training and test field classification accuracies for 14 counties in Kansas are shown in Table 6-4. The accuracy of the classification as determined from training field results is not significantly different from that obtained from test field performance. The overall classification accuracies were generally 85 percent or higher, and it was concluded that accurate area estimates could indeed be obtained from these results.

The classification performance of test (or training) fields can be used to correct for classification bias in the area estimates.[1] In the research project upon which this case study is based, the bias-corrected results were compared to USDA Statistical Reporting Service (SRS) estimates of wheat harvested. Comparisons were made at three levels: county, crop-reporting district, and state. A summary of these comparisons is shown in Table 6-5. It should be noted that in comparing analysis results to SRS figures, the latter are also estimates (and thus subject to sampling error). The accuracy of the SRS estimates is greatest at the state level and least at the county level.

Statistical testing procedures[7] were used in comparing the estimates obtained by remote sensing techniques to the SRS estimates. The correlation coefficient

Table 6-4 Training and test field classification accuracies for selected counties in Kansas

County	Training field classification accuracy			Test field classification accuracy		
	Wheat	Other	Overall	Wheat	Other	Overall
Sherman	70.3	97.5	89.5	75.4	89.0	85.0
Greeley	82.7	93.8	90.0	84.8	93.0	89.9
Trego	76.8	77.1	77.1	86.7	81.1	82.4
Saline	72.3	92.7	82.5	83.5	94.5	87.5
Ford	94.9	98.8	97.4	93.7	97.0	95.7
Hamilton	75.3	55.5	61.9	94.2	78.4	82.5
Hodgeman	86.3	79.3	81.3	89.4	77.7	80.9
Stanton	66.8	62.9	63.6	62.5	79.1	75.5
Barber	96.3	99.7	98.1	92.7	88.8	90.4
Harvey	98.1	93.7	95.5	93.6	98.2	95.6
Pratt	99.8	94.8	97.0	92.7	95.6	93.8
Stafford	94.8	98.5	96.4	99.5	93.4	96.0
Sumner	93.4	95.3	94.3	92.6	89.2	91.2
Allen	94.2	94.5	94.4	95.3	89.7	90.7

Table 6-5 Summary of USDA/SRS and Landsat estimates of area and proportion of wheat in Kansas

Region	Area, hectares (000)			Proportion, %		
	USDA/SRS (1)	Landsat (2)	Difference (2)–(1)	USDA/SRS (3)	Landsat (4)	Difference (4)–(3)
State	4555	4613	58	26.2	26.6	0.4
District						
Northwest	470	387	−83	23.3	19.2	−4.1
North Central	578	575	−3	25.1	25.0	−0.1
West Central	522	579	57	25.2	28.0	2.8
Central	770	956	186	33.1	41.2	8.1
Southwest	784	715	−69	25.6	23.3	−2.3
South Central	1164	1158	−6	40.2	40.0	−0.2
Southeast	267	242	−25	10.0	9.1	−0.9
Counties (median and median difference)	55.0	53.4	0.6	24.85	26.25	0.4

of the USDA/SRS county estimates of wheat area and the Landsat estimates was 0.80. The Landsat wheat-proportion estimates for 49 percent of the counties were within ±5 percent of the SRS estimates; 81 percent were within ±10 percent. At the crop-reporting district level there was only one significant difference between the Landsat and SRS estimates out of the seven districts analyzed in Kansas. In that one district, the differences, although small, were all in one direction. For the state, the SRS estimate was 4,555,000 hectares compared to the Landsat estimate of 4,613,000 hectares, a relative difference of only 1.27 percent.

Based on the training and test field accuracies, together with the comparisons of estimates based on remote sensing data against those obtained by SRS sampling procedures, it was concluded that the objectives of the crop-identification and area-estimation project were satisfied.

6-3 SAMPLE APPLICATIONS

The remainder of Chap. 6 is devoted to the presentation of five application examples which show in some detail how remote sensing can be used to meet a variety of information needs. Each of the examples was selected because it highlights at least one important aspect of remote sensing, such as project planning, sensor characteristics, a special analysis technique, or special requirements for results presentation. Together these examples suggest the broad range of disciplines in which quantitative remote sensing can be used.

All examples were abstracted from remote sensing research reports and thus do not necessarily represent techniques that have reached the operational level. They are presented here because the decisions made in these applications can serve as models for the design of other remote sensing projects. Of course, many alternative examples might have been chosen; the reader interested in a still broader view of possible applications should consult the open literature, especially Refs. 9 and 10.

The descriptions that follow adhere to the same basic format as the preceding case study, but while the intention in the case study was to define and describe the key steps of a single application, the emphasis here is on highlighting the especially significant or interesting aspects of several diverse projects. It is assumed that the reader will already be familiar with the five basic steps previously described.

Example 1: Large Area Land-Use Inventory[11]

Objectives, feasibility, and planning In 1972 the United States and Canada signed the Great Lakes Water Quality Agreement, which called for assessing the pollution of the Great Lakes system. Implementation of this agreement required a series of studies to determine the various sources of pollution in the Great Lakes. The job of one of the task groups formed was to study pollution sources related to land use.

As their work began, this group was faced with the problem of obtaining a current land-use inventory of the 65,000,000 hectares (160,000,000 acres) of land in the Great Lakes drainage basin. After examining the results of a six-county pilot study which used numerical analysis of Landsat 1 data, the international group decided to use this approach to inventory the entire 191 counties which make up the United States portion of the Great Lakes basin, an area approximately 1450 × 1000 km (900 × 620 mi) covering portions of eight states (Fig. 6-9). The area contains forests, lakes, and wetlands in the north, agricultural land in the south, and several large, metropolitan regions encompassing residential, commercial, and industrial areas.

The study group needed the results in two forms: (1) a geometrically corrected, color-coded map of level I land-use categories for each county, and (2) statistical tables containing level I and level II land-use information on a county-by-county basis. The level I and II categories are those defined by USGS (Table 6-6)[12] but for this use they were modified as shown in Table 6-7. A further requirement was that the analysis be completed in less than one year and at the least expenditure possible. This meant that effective interagency communication was essential as well as effective project planning to accomplish the task with a minimum of both primary data and reference data and a minimum of computer time.

Implementation The primary data used for this classification were multispectral scanner data obtained by Landsat 1 over the Great Lakes basin between

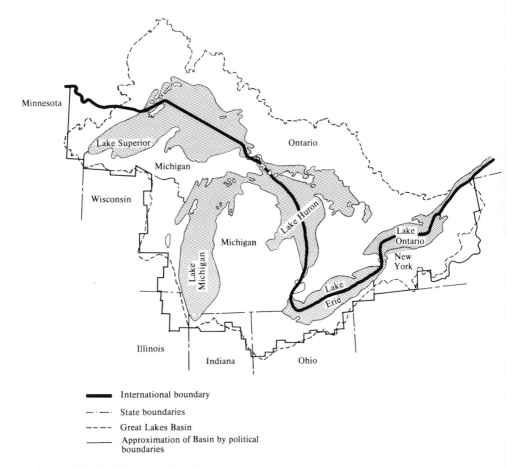

Figure 6-9 United States portion of the Great Lakes basin.

July 1972 and October 1973. For budgetary reasons, it was decided that multi-temporal analysis would not be used, and therefore it was necessary to select, from the available Landsat data, single images which (1) were cloud free, (2) were not seriously affected by problems in the data-collection system, and (3) were obtained during the growing season, after the spring thaw and before the first killing frost in the fall. At this time of year the maximum number of desired land-use classes would be spectrally separable. It would have been ideal to have had data obtained from the entire study area during consecutive satellite passes and at times when both underflight and ground reference data could have been collected, but cloud cover precluded this; in fact, there were some portions of the watershed for which good quality, cloud-free Landsat data were never available during the allotted time.

In addition to Landsat data, 3000 m (9600 ft) aerial photography covering approximately 4 percent of the total area was collected as one source of reference

Table 6-6 Level I and level II land-use categories as defined by U.S. Geological Survey Circular no. 671[12]

Level I	Level II
Urban	Residential Commercial/industrial Extractive Transportation
Agriculture	Row crops Close-grown crops Pasture/meadow Orchard/vineyard
Forest	Forest
No major usage	Water Wetlands Barren lands

data. This underflight photography consisted of 70-mm color and color infrared photography in flightlines laid out to sample all important land-use classes in the watershed (Fig. 6-10). Analysis of this photography provided the most useful source of reference data. Additional reference data, used in the analysis and for evaluation of the results, included city and county maps, U.S. Geological Survey topographic maps and USDA reports (*1967 Conservation Needs Inventory* and 1972 and 1973 reports by the Statistical Reporting Service/USDA).

Since the results of this project were to be presented on a county-by-county basis, the decision was made to carry out a separate analysis for each county. In order to locate the counties in the data, the reformatted and geometrically corrected data were displayed on a digital image-display unit, and, by correlating

Table 6-7 Level I and level II land-use categories used in the Great Lakes basin project

Level I	Level II
Urban	Residential Commercial/industrial
Agricultural	Row crops Close-grown crops Pasture
Forest	Forest
No major use	Water Wetlands

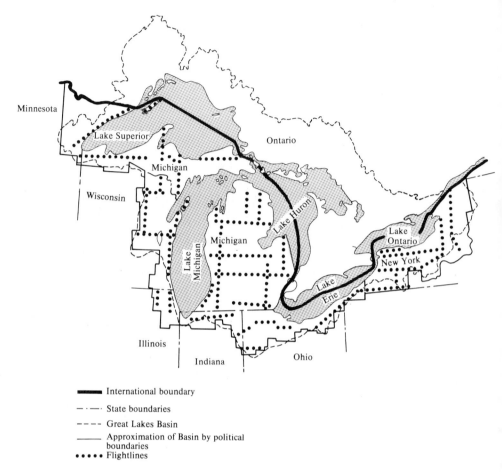

Figure 6-10 Flightlines designed to sample all important land-use classes in the study area.

visible features in the data (roads, rivers, etc.) with topographic maps, analysts were able to delineate individual counties with a light pen. This procedure established the line and column coordinates needed to approximate the county boundaries by a series of straight lines. Data tapes could then be created which contained data values for individual counties.†

† Special treatment was required for those cases in which portions of two Landsat frames were needed to cover an entire county. Because of the generally north-south orbit of the satellite, north-south adjoining frames were often collected on the same day and under nearly identical calibration and atmospheric conditions. Therefore these frames were simply joined together appropriately in the computer and handled in the same way as the "single-frame counties." But east-west adjoining frames would have been collected at least 24 hours apart due to the orbit parameters. In this case, since the characteristics of the data in the two frames could differ considerably, it was decided to analyze the two county sections separately.

The analysis approach used was unsupervised classification of a systematic random sample of 25 percent of the data vectors, those falling in every other line and every other column of the data. Using grayscale printouts of two channels of data, analysts selected homogeneous areas from the counties for which there was aerial photography and calculated statistical information concerning the homogeneity and separability of the spectral classes. Homogeneous clusters were then related to identifiable land-use classes, with some spectral classes combined or deleted as needed to match the information classes. Data from three of the four data channels were used to classify the county using a priori probabilities derived from the *1967 Conservation Needs Inventory*. The decision to classify 25 percent of the data vectors and to use only three channels of data was made to conserve computer time. The pilot studies had indicated that the accuracy of the results would not be significantly lowered by taking these measures.

Counties over which aerial photography had been obtained were classified using training statistics generated within that county. Counties for which there were no aerial photographic data were classified using training statistics from an adjoining or nearby county, provided the statistics did not have to be extended beyond about 100 km or to a different frame of data.

In working with the spectral classes, as determined by clustering, analysts found there was not a one-to-one relationship between the spectral classes and the USGS land-use classes. The accuracy of the results depended on the analysts' skill in augmenting the information available from the spectral data with various kinds of reference information. For example, although most of the level I and level II classes were spectrally distinct, it was difficult to distinguish between the level II categories of "forest" and "residential." Therefore "forest" and "residential" were treated as one "forest" class throughout the classification, but in the final preparation of results, percentages for residential areas, determined from reference data, were subtracted from "forest."

Problems due to clouds were inescapable. Cloud-free Landsat data were never available for six of the counties, and for these the *1967 Conservation Needs Inventory* data had to be substituted for the classification analyses. In other counties where "cloud" and "cloud shadow" classes were produced, the tabular results were adjusted by assuming that the land use in the cloud-affected areas was in the same proportion as in the cloud-free areas.

Results presentation and assessment To prepare the county maps containing color-coded level I land-use information, a digital laser printer was used to create the color separations needed for printing at a scale of 1:215,000. Figure 6-11 (between pages 36 and 37) shows the resulting map for Winnebago County, Wisconsin. The tabular results (Table 6-8) give the number of acres, hectares, and percentage of each of the level I and level II land-use categories for the county. The area statistics have been rounded off to the nearest 4-hectare (10-acre) unit.

When preparing a land-use inventory from satellite data, the detail of the inventory is limited by the resolution of the scanner. In the case of Landsat 1,

Table 6-8 Level I and level II classification results for Winnebago County, Wisconsin

Winnebago County, Wis.	Acres		Hectares		Percentage	
	Level I	Level II	Level I	Level II	Level I	Level II
Urban-commercial-industrial	26,930		10,900		7.3	
Residential		26,930		10,900		7.3
Commercial		—		—		—
Agriculture	203,360		82,320		55.0	
Row crop		134,200		54,330		36.3
Close-grown crop		23,200		9,390		6.3
Pasture		45,960		18,600		12.4
Forest	56,310	56,310	22,790	22,790	15.2	15.2
No major use	83,330		33,730		22.5	
Water		83,330		33,730		22.5
Wetland		—		—		—
Totals	369,930	369,930	149,740	149,740	100.0	100.0

the resolution was approximately 0.6 hectares. Since the sampling scheme selected called for only one of four data points to be classified, the *effective area* represented by each data point was 2.4 hectares. Higher accuracy depends on the collection of data of good quality at high resolution and concurrent collection of adequate ground data. Even under less than ideal conditions and with the need to complete the project at the least possible expense and within the stated time frame, this project demonstrated the utility of numerical analysis of multispectral scanner data for producing large-area land-use inventories suitable for meeting specific information needs.

Example 2: Forest-Cover Mapping in Mountainous Terrain[13]

Requirements and objectives Mapping land use in mountainous areas presents problems which are not encountered in mapping flatland areas having little topographic variability. The natural vegetative cover of the earth's surface is determined by complex interactions of soils, topography, and climate; and the type and density of the vegetation are affected by elevation-related variations in precipitation and solar radiation and by the slope and aspect of the surface. Variations in density and species composition within single forest cover types often cause changes in spectral reflectance, making it more difficult to define areas of homogeneous characteristics.

Feasibility and planning The primary data used for the mapping task were multi-spectral scanner data collected by Skylab on June 5, 1973, over a portion of the San Juan Mountains of southwestern Colorado (U.S.A.). In this region of commercially important timber resources, occurrence of a single forest cover type is generally restricted to a particular elevation range, and within that elevation range the occurrence is further affected by the aspect of the terrain (Fig. 6-12). Topographic data (elevation, slope, and aspect) were therefore digitized and registered with the Skylab multispectral scanner data to aid in the analysis. The wide spectral range of the Skylab multispectral scanner offered for the first time the kind of data believed to be needed for effective mapping of forest cover types in a topographically complex area using computer-aided analysis techniques. Reference data required for training the computer and testing the

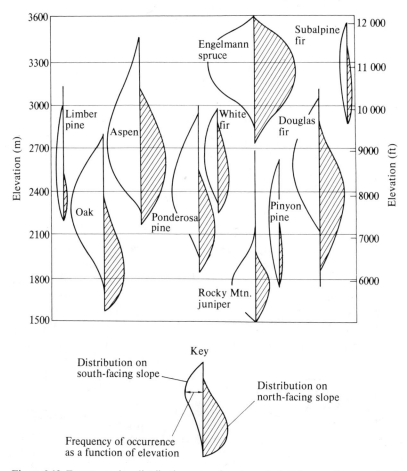

Figure 6-12 Forest species distribution as a function of elevation and aspect in the San Juan Mountains.

results included aircraft-collected infrared photography and detailed data collected on the ground.

Implementation The analysis method used was a hybrid of the supervised and unsupervised approaches (see Table 3-1), a method which is especially well suited to the analysis of spectrally complex areas. It does not require the analyst to select and identify homogeneous samples of all possible variations in spectral response in each cover type (as the supervised approach does), a task which is extremely difficult in complex areas like these. This "modified clustering" technique consists of four basic steps which allow the analyst more effective interaction with the data than other methods:

Step 1. Locate and define, by line and column, several relatively small blocks of data dispersed over the entire study site, each containing three to five cover types: these areas are called the "candidate training areas."

Step 2. Cluster each candidate training area separately; compare cluster maps with reference data and recluster if the spectral cluster classes do not correlate well with the informational classes of interest.

Step 3. With the help of a separability algorithm which computes pairwise statistical distances between clusters, develop a single set of training statistics by combining the clustering results of all training areas.

Step 4. As a preliminary test of the adequacy of the training statistics, classify the training areas using the statistics and a maximum-likelihood classifier; compare the results with reference data to ensure that no labeling errors were made and that no desirable classes were omitted; then classify the entire area of interest.

In this study involving 13 wavelength bands of Skylab data, the clustering sequence in step 2 produced 135 spectrally separate cluster classes within the nine training areas selected in step 1 (Fig. 6-13). To reduce the number of spectral classes, transformed divergence was used as a separability measure (Sec. 3-8) to combine similar cluster classes; specifically, classes with a transformed divergence of less than 1.50 were considered to be spectrally similar and were combined. To further reduce the number of classes, those with a transformed divergence between 1.50 and 1.70 were also candidates for combining or even for deleting in the case of classes which were relatively insignificant in the test area. Through the judicious use of the separability measure and reference data, the 135 cluster classes were reduced to 31 spectral classes which allowed for a reasonably accurate but not overly complex classification. Attention to the specific objectives of the analysis was important in determining the extent to which this reduction could be carried out. Through comparison with the underflight photography, these spectral classes were associated with earth surface features and could be considered henceforth "spectral/informational classes." As a final check of the training statistics, a classification of the training area was carefully evaluated before classification of the entire area was begun.

Figure 6-13 Image derived from multispectral data showing nine training areas used in the analysis.

The analysis approach discussed so far in this study has been a point-by-point classification technique in which each data vector is treated individually by the maximum-likelihood classifier. This approach produces a very detailed results map which may contain even more detail than is actually needed by the resource manager. For example, although a single hectare identified as pasture in the midst of a large forest might interest the wildlife manager as a potential site for a wildlife habitat, this degree of detail would be distracting to the timber manager who probably would prefer that 5-hectare areas be the minimum mapping unit.

To achieve more generalized maps, a second classification of this Skylab data was carried out using an algorithm that first locates areas that are spectrally homogeneous and then classifies these areas as whole units (see Kettig and Landgrebe[14]). The maps resulting from these two approaches, shown in Fig. 6-14 (between pages 36 and 37), appear somewhat different; the point classifier produced a map with a salt-and-pepper effect, whereas the sample classifier produced a map in which spectrally similar data points were assembled into larger units and classified as a single cover type. Evaluation of these results by resource managers indicated that the second map was both more compatible with existing map systems and, in some cases, potentially more useful for operational resource management.

Assessment An important aspect of this study was the variety of methods used for evaluating results and verifying that the classification maps and tables would meet the user's needs. Three techniques were used for evaluation, one of them qualitative in approach and the other two quantitative; together they provide insight into the range of evaluation techniques available to the data analyst.

For the qualitative evaluation, the computer classification map was visually compared to forest-cover maps made by manual interpretation of aerial photography. The degree of correspondence gave a general indication of the computer performance, but precise (quantitative) comparisons were not obtained in this way. Qualitatively the classification appeared to be reasonably accurate.

To obtain a quantitative evaluation of the classification, a grid was placed over the analysis area and potential test areas, of 7.36 hectares (18.4 acres) each, were defined at intervals throughout the area. Each test area consisted of a square segment with 16 resolution elements on a side, plus a surrounding buffer strip one resolution element wide; the latter was added to allow for small registration errors (Fig. 6-15). This set of potential test areas was superimposed on a cover-type map of the area which had previously been developed by photointerpretation, and test areas containing more than one cover type were eliminated. This procedure resulted in 540 test areas, representing in total about 5 percent (4,000 hectares; 10,000 acres) of the entire area. The results of this evaluation,

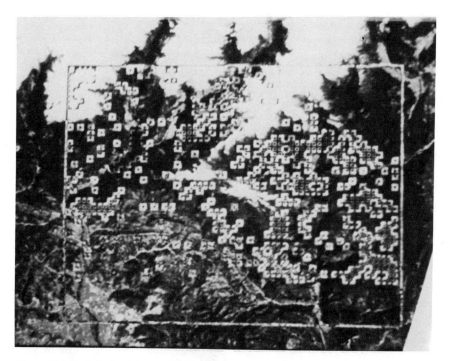

Figure 6-15 Statistically defined grid of test areas.

Table 6-9 Classification performance in test areas for forest cover types from Skylab multispectral scanner data

Cover type	Per point classifier, % correct	Per sample classifier, % correct
Water	94.8	95.8
Snow	100.0	100.0
Pasture	52.3	67.2
Deciduous	61.7	70.7
Oak	63.7	67.5
Aspen	54.8	65.4
Coniferous	90.9	91.0
Ponderosa pine	72.9	82.9
Douglas/white fir	76.8	79.1
Spruce/fir	50.0	52.7
Overall: major cover types	85.0	87.2
Overall: forest cover types	71.0	76.0

summarized in Table 6-9, show an overall classification accuracy of 71 percent, which was judged to be relatively good for an area with the vegetative and topographic complexity of this region. It was felt, however, that an effective method for incorporating the digitized topographic information into the actual classification process (which had not been done in this case) could result in significant improvement in identification of individual forest cover types.

To evaluate the forest cover-type map in another quantitative way, estimates of the total area of various forest cover types obtained from the computer-aided classification of the Skylab data were compared to estimates made from manual interpretation of the forest cover maps previously developed from aerial photographs. The computer tabulated the number of resolution elements in the test area that were classified into each forest type of interest and then multiplied that number by the area on the ground represented by each resolution element. These results were compared with area estimates obtained from aerial photographs through standard photointerpretation dot-grid and planimetric methods. The results of this comparison, on a quadrangle-by-quadrangle basis, are shown in Table 6-10. In many cases the area estimates from the two sources are quite similar, but in others the differences appear significant. For large geographic areas, however, overestimation of a forest type in one quadrangle is often balanced by underestimation of the same cover type in another quadrangle. The effect of this is to make the total area estimates shown at the bottom of the table more nearly equal, which supports the conclusion that accurate area estimates of forest cover types can be made from computer-aided analysis of multispectral scanner data from satellite altitudes, provided relatively large areas (e.g., several thousand acres) are involved.

Table 6-10 Comparison of area estimates (hectares) obtained from cover-type maps (CTM) and computer-aided analysis techniques (CAAT)

Cover type	Vallecito		Granite Peak		Ludwig Mtn.		Baldy Mtn.		Devil Mtn.		Totals	
	CTM	CAAT	CTM	CAAT	CTM	CAAT	CTM	CAAT	CTM	CAAT	CTM	CAAT
Exposed	449	155	495	217	170	665	0	186	15	108	1,129	1,311
Water	1,160	928	0	31	139	0	16	0	0	0	1,315	959
Grassland	1,021	232	124	46	2,290	1,114	542	93	263	93	4,240	1,578
Oak	340	108	15	31	3,945	4,409	2,058	2,150	402	317	6,760	7,015
Aspen	2,862	1,609	990	1,129	1,114	1,067	2,243	2,259	1,887	1,439	9,096	7,503
Ponderosa	774	2,553	0	1,006	5,616	6,159	3,264	6,312	2,831	4,084	12,485	20,114
Douglas and white fir	5,368	5,740	4,131	5,198	2,181	1,887	5,136	2,924	8,772	8,354	25,588	24,103
Spruce/fir and snow	3,496	4,146	9,716	7,813	15	170	2,212	1,547	1,300	1,021	16,739	14,697

Example 3 : Snow-Cover Mapping[15]

Requirements and objectives The water supply for much of the Southwestern United States is provided by the snow that falls in the Rocky Mountains, then melts, and runs through the valleys below. In order to make good use of this important resource, hydrologists must be able to make accurate predictions of the rate of run-off, and to do this they have sought ways to measure the amount of snow that has accumulated during the winter months. One strategy has been to use aerial photography to estimate the extent of the snow cover; another has been to make depth measurements of the snow at specific points on the ground. The synoptic view that can be obtained from orbiting satellites now offers hydrologists a way to view entire watersheds over a short period of time and, with appropriate sensors, to collect data from which they can estimate both the extent and the condition of the snow cover over wide areas.

The analysis objectives of the study summarized here were to map the extent of snow cover in a single watershed, to distinguish among different spectral classes within the snow pack, and to correlate these classes with elevation information. By meeting these objectives, hydrologists would have information about the location and extent of the snow, as well as information that would help estimate the amount of water present in the snow.

Feasibility The feasibility of mapping the snow cover with Skylab data was based on earlier studies which singled out the middle-infrared channels as best for this task (see Sec. 5-4). Prior to Skylab, only visible and near-infrared data had been available from satellite altitudes, and the problem of differentiating snow-covered areas from cloud formations hampered early snow mapping efforts; as recently as 1973, mapping snow cover from Landsat imagery was limited to cloud-free dates.[16] Skylab data provided the first opportunity to test snow mapping from satellite altitudes using data in a full range of reflective wavelengths.

Project planning and implementation The test site chosen for the snow-cover mapping study was the 18,000-hectare Lemon Reservoir watershed in the Rocky Mountains of southwestern Colorado (U.S.A.). The topography of the test site area is rugged, ranging in elevation from less than 2000 to over 4200 m. Within this range of elevation is a complex mixture of forest types, rangeland, alpine tundra, agricultural areas, water bodies, geological features, and various artificial features. The timberline occurs at approximately 3600 m, and below this elevation there are many variations in stand density between and within different forest cover types.

In order to map the snow cover of the Lemon Reservoir watershed, Skylab multispectral scanner data from June 5, 1973, were obtained. At this time the snow pack was in the process of melting, particularly at the lower elevations, and the elevation-related variations in the condition of the snow could be observed. The Skylab scanner collected reflectance data in 13 wavelength bands, five in the visible wavelengths, three in the near infrared, four in the middle infrared, and one thermal infrared channel. Pictorial representations of the 13 channels of data are shown in Fig. 6-16. In addition to these primary Skylab data, underflight photography was collected at 18,000 m (60,000 ft), and elevation information from the U.S. Geological Survey was overlaid on the scanner data.

There were several steps in the analysis of the multispectral scanner data that merit comment here. First, the subclasses of snow were statistically defined

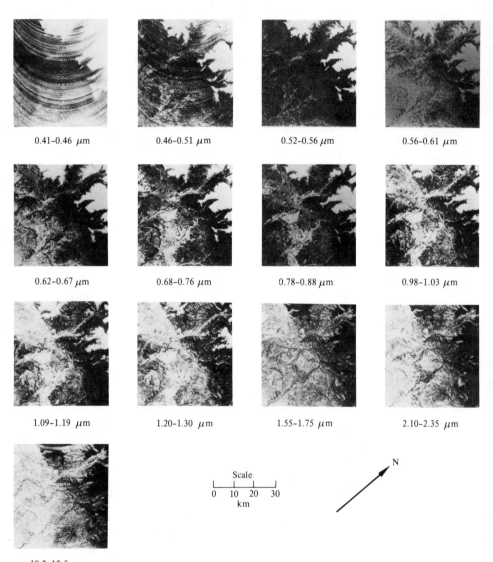

0.41–0.46 μm   0.46–0.51 μm   0.52–0.56 μm   0.56–0.61 μm

0.62–0.67 μm   0.68–0.76 μm   0.78–0.88 μm   0.98–1.03 μm

1.09–1.19 μm   1.20–1.30 μm   1.55–1.75 μm   2.10–2.35 μm

Scale

0   10   20   30

km

N

10.2–12.5 μm

Figure 6-16 Thirteen channels of Skylab multispectral scanner data, San Juan Mountains (Colorado), June 5, 1973.

using the clustering procedure, without the use of ground information (Sec. 3-10). This use of the purely unsupervised classification approach resulted in five distinct spectral classes of snow which could be distinguished in the data. The supervised approach was used to train the computer to recognize other major ground cover types in the area, i.e., water, forest, agriculture, and a specular reflection class for some surface water bodies. In total, therefore, there were nine spectral classes defined for the final classification.

The second analysis technique worth special mention is the use of layered classification.[17,18] In essence, the layered classifier uses a maximum-likelihood classification algorithm in a sequential or layer-by-layer mode. The major advantage of this approach is that it allows the multispectral discrimination between subsets of classes to be accomplished using the optimal combination of spectral channels for each discrimination. Figure 6-17 illustrates the decision tree developed for this snow cover mapping task. In order to separate the agriculture, forest, and water classes from the five classes of snow and the specular reflection class, only channel 5 (0.62 to 0.67 μm) was needed. If a further discrimination among crops, trees, and water had been requested, this could have been accom-

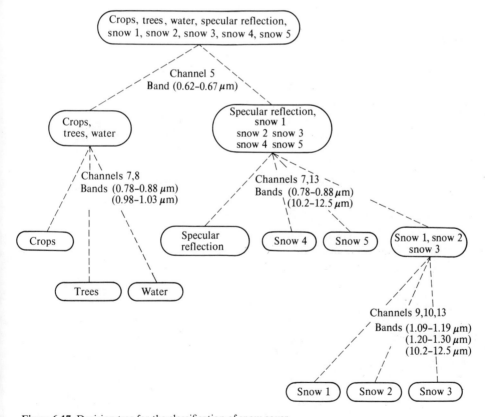

Figure 6-17 Decision tree for the classification of snow cover.

plished by using channels 7 and 8. A convenient way for the analyst to see where the training classes fall in the range of responses for each channel is through use of the coincident spectral plot (Fig. 6-18). In this case, the range for each of the nine spectral classes (identified as A through I) is shown for each of the 12 wavelength bands. By looking at this plot, the analyst could determine that in channel 5 there is a clear distinction between the group of classes A, B, and C (crops, trees, and water) and the remaining six classes. Similarly, the plots for channels 7 and 8 show the spectral separability possible among classes A, B, and C.

Differentiation among the remaining six classes was more complex and required two sequential classification steps. The coincident spectral plot shows that channel 7 could be used to separate out three of the remaining spectral classes (D, I, and H) and that the remaining three could be subsequently separated by either channel 9 or 10. Thus, using the layered classification approach required three classification steps, but because the process was carefully controlled by the analyst, the results that were obtained were 2.2 percent more accurate than those obtained by the standard supervised classification procedures (99.7 instead of 97.5 percent for snow-clouds-other) and the computation time was reduced by 15.75 percent (198 instead of 235 s).

The output products resulting from the classification of the Lemon Reservoir watershed were (1) a snow cover map showing five classes of snow (Fig. 6-19, between pages 36 and 37) and (2) a table listing the areal extent in hectares of the total snow cover as well as of the five individual snow classes (Table 6-11). The boundaries of the watershed had been digitally defined in the data, and the map and table prepared could therefore be limited to the points lying within the boundaries. It is interesting to note that it is often the tabular output results that are more likely to be of use to the hydrologist than the maps, because they give a *quantitative* determination of the area covered by snow.

Assessment The snow-cloud differentiation study clearly indicated that spectral differentiation between snow and clouds could be achieved using data in the

Table 6-11 Areal extent of five classes of snow in the Lemon Reservoir watershed (June 5, 1973)

Watershed	Area, hectares	Cover, %
Entire area	17,866	100.0
Snow 1	1,021	5.7
Snow 2	3,717	20.7
Snow 3	3,395	19.0
Snow 4	2,221	12.3
Snow 5	1,322	7.3
Total snow cover	11,676	65.0

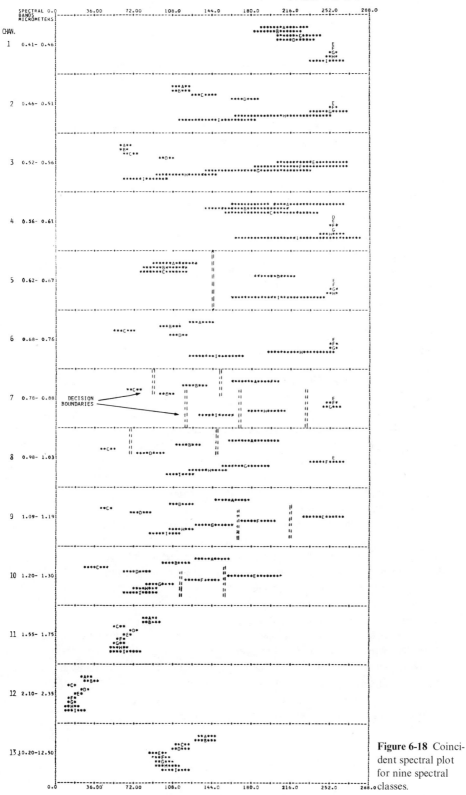

Figure 6-18 Coincident spectral plot for nine spectral classes.

325

middle-infrared portions of the spectrum. Furthermore, it was found that five spectral classes of snow could be defined and mapped and that these classes were closely related to elevation; overlaying digital topographic data with the Skylab data revealed that the five spectral classes of snow tended to follow the elevation contours. Air-temperature data obtained within 15 min of the Skylab overpass indicated a temperature range in this region from 7 to 15°C, with the higher temperatures at the lower elevations. At these temperatures, the snow pack was melting, with the snow at the lower elevations melting at a higher rate. The more rapidly melting snow had a higher water content, which resulted in lower reflectance measurements in the near-infrared wavelengths, making the classes spectrally separable. The ability to relate spectral snow-cover classes to topographic position is of benefit to the hydrologist in another way: with this information the snow accumulations can be calculated at various elevations based on snowfall prediction models. Thus the combination of computer-obtained estimates of the areal extent of the snow cover at different elevations and information about the predictable water equivalents at various elevations can permit accurate and timely predictions of run-off within individual watersheds.

Example 4: Geologic Mapping[19]

Objectives and feasibility Geomorphological mapping through remote sensing presents a problem to the data analyst that is quite different from any we have looked at so far. It is one that is related to the very nature of geologic studies. Multispectral sensors are designed to detect the spectral reflectance characteristics of earth surface materials, but only seldom are the geologic forms of interest exposed at the surface and, thereby, directly detectable. The identity of geologic materials which lie under the earth's surface can therefore only be inferred from the information about the surface materials present.

Research has shown that many biologic cover types have an affinity for soils with particular chemical and physical composition and that these soil properties are in turn greatly dependent on the geologic parent materials from which they have been derived. Therefore, mapping geologic materials by using spectral data is based on the assumption that subsurface materials will manifest themselves as spectrally separable classes at the earth's surface. Geologic mapping by inference is a valid approach, particularly in areas where the natural vegetation has not been disturbed by human activity.

The objectives of the experiment described here included delineation of surface landforms in an alpine border area through analysis of spectral reflectance measurements. This procedure included inferring the botanical, hydrologic, and cultural patterns and, from this information, inferring the landforms. It was required that the final results show both the physiographic and the geologic features of the area, and at appropriate scales.

Project planning The study area chosen was a 520 km² (200 mi²) area located 2.3 to 8 km (2 to 5 mi) east of Durango, in the San Juan Mountains of southwestern Colorado (U.S.A.). This area was selected because, owing to its relatively arid nature, it had a less complex ecosystem than more humid areas. General accounts of the topography and geology of the region were available in studies by Atwood and Mather[20] and Mather.[21]

When delineating subsurface geologic materials, a great deal of human judgment and manual interpretation may be called for in order to extract needed information from the data. The need for manual interpretation is heightened because spatial features may play an important role in developing the final classifications, but computer-implemented techniques for handling spatial information have not yet been developed to an adequately sophisticated level for this application. Nevertheless, computer-aided classification is an important part of the study described here and is shown to produce high-quality results when augmented by manual interpretation.

Implementation The primary data used in the analysis were collected by the Landsat 1 multispectral scanner on August 21, 1972. This was during late summer when analysis of this mountainous area was not complicated by the presence of snow. Because Landsat 1 passed over the area at about 10:30 a.m., local time, all west-facing slopes and declivities were in shadow when the data were recorded. Reference data, used to provide basic geographic information about the area, were obtained from 1949 maps developed by Zapp.[22]

Because simple classification of the spectral characteristics of the area would not provide adequate information for the desired final results, a procedure was developed that combined geologic expertise with unsupervised and supervised classification. In order to select representative training samples, analysts used information from the reference maps to isolate three discrete lithologic types for analysis—sandstone, shale, and alluvium—which cumulatively are sufficient to describe all surface geologic materials in the study area. Shadows present in the data were initially treated as a fourth class. Four subareas, each including the rock types and topographic forms of interest, were selected for unsupervised classification. The resulting computer line-printer image provided reliable visual separation for training the computer to recognize two of the four classes: shadow patterns and alluvial areas.

For further refinement of the training classes, it was essential that the geologists have a clear understanding of the general spectral characteristics of each lithology and its general spatial distribution within the training area. The shadow patterns helped locate the highly reflective dip slopes, which have sandstone as their surface rock type, and the strike valleys, known to be underlain with shale. From the unsupervised analysis results, areas in the data were located which contained all four classes of interest, and clustering of these areas produced nine spectral subclasses: four sandstone, three shale, one alluvium, and one shadow. Still further refinement of the training data added two alluvial subclasses, one shale subclass, and two shadow subclasses. The chief advantage of using this approach was that by initially reducing the spectral variance of the clustered regions, the clustering algorithm was able to divide each cover type into spectrally distinct subclasses.

As noted, the objectives of this study called for a physiographic map and a geologic map of the area. Figure 6-20 is a simulated physiographic map that displays the four major classes; the three lightest classes represent areas in sunlight, the darkest class is shadow. This four-class display produces a three-dimensional illusion which gives a sense of the physiography of the study area.

Before the geologic map could be prepared, it was necessary to eliminate the class "shadow." Here again, an understanding of the geology of the area was essential. Because of the relationship that usually exists between bedrock types and their topographic expression, analysts could assume with a high degree of confidence that the areas in shadow were underlain by shale. Making this assumption, they could combine shale and shadow classes and display them in the same way on the results map. Figure 6-21(a) is a classification map of the study area showing the locations of the three geologic features of interest: alluvium, shale, and sandstone.

Assessment Effective geologic mapping from remote sensing data depends on spectral information and on textural and contextual patterns as well. Therefore

Figure 6-20 Computer-generated physiographic map of the Durango, Colorado area.

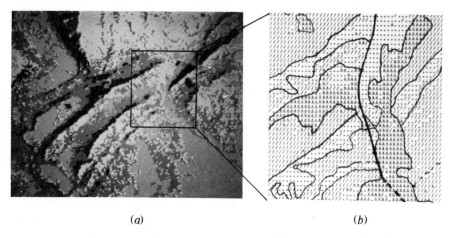

(a) (b)

Figure 6-21 Geologic map of an area around Durango, Colorado. (a) Grayscale classification map (white = alluvium, gray = shale, black = sandstone). (b) Alphanumerically coded display of the area outlined on (a) (A = alluvium, / = shale, • = sandstone). The heavier black line is a fault inferred by interpretation of the classification results.

the analysis process based on spectral information is greatly enhanced by the participation of a human interpreter having an understanding of the local geology sufficient for him or her to recognize these geologic features and to draw inferences from them. The procedures used in this study yielded encouraging results. Figure 6-21(b) is a larger scale alphanumeric display of the outlined subarea of Fig. 6-21(a) with the superimposed lines drawn according to map units derived from Zapp.[22] Visual inspection of this display reveals that there was general agreement between the two maps; where differences exist, there is evidence from ground observation that the machine-produced map often is more reliable than the existing geologic map and could thus contribute to its revision.

Another important technique for presenting results is shown in this example: the use of one set of classification results for producing different map types (e.g., physiographic and geologic) and multiple map scales ranging from 1:24,000 to 1:1,000,000. This procedure provides a potential means for refining existing geologic maps, an important step toward good management of the earth's resources. Future refinements in machine-aided analysis of spatial features should lead to procedures for geologic mapping that require less human intervention than is now needed.

Example 5: Water-Temperature Mapping[23]

Requirements and feasibility The study described in this section had as one of its objectives mapping the temperature changes in the Wabash River (Indiana, U.S.A.), downstream from a fossil-fuel power plant. Prior studies had indicated that the heated effluent from the plant, which significantly affected the fish populations, was not dissipated for many miles downstream, and a more quantitative map of the temperature changes was needed in order to assess the effects.

Remote measurement of surface temperatures is possible because of the existence of two atmospheric windows at 4.5 to 5.5 and 8.0 to 14.0 μm, which allow the radiant energy emitted by earth surface features to pass relatively unattenuated from the surface to the sensor (see Sec. 2-2).

There is a definite relationship between the radiant energy emitted by an object and its radiant temperature. Furthermore, this relationship is almost linear over small temperature ranges (for example, between 20 and 30°C), thus allowing the use of a simple linear calibration function for the conversion of radiant energy measurements into temperature values.

Project planning and implementation The primary data used in this analysis were collected by an airborne multispectral scanner over a 24-km (15-mi) segment of the Wabash River. In addition to recording energy from the ground in the visible, near-infrared, middle-infrared (0.4 to 2.6 μm), and two thermal infrared regions (4.5 to 5.5 and 8.0 to 13.5 μm), the scanner had two thermal reference blackbodies, used to calibrate the measurements of thermal energy. One was maintained below the expected scene temperatures; the other was maintained above the expected scene temperatures.

It is interesting to note that this study describes a mapping task for which aircraft instead of satellite-collected data were used in order to meet the stated objectives. At the time when this project was planned, there was no satellite-borne scanner in operation that collected data in the thermal wavelengths and with adequate ground resolution.

Before the scanner data were analyzed, two significant preprocessing steps were carried out in addition to the standard digitization and reformatting steps. One of these steps was the conversion of the recorded emissive data to temperature information, a relatively simple step because of the almost linear relationship between radiation measured by the scanner and the actual temperature of the surface water. The second preprocessing step worth special mention was the use of scan-line averaging, a technique for using redundancy in the data for noise reduction. In this procedure, data redundancy resulting from overscanning (see Sec. 2-6) provides additional information which improves the data quality. Instead of simply being discarded, the redundant scan lines are averaged, effectively increasing the signal-to-noise ratio and thus improving the resolvability of individual classes, i.e., temperatures in this case.[24] This preprocessing step is useful whenever overscanning occurs in the data and there is a need to reduce the noise before analysis.

The technique used for data analysis was layered classification, a procedure used also in the snow-cover mapping example described earlier. With this approach, spectral reflectance measurements were used first to identify those data vectors which represented water. Then the thermal characteristics of only the water areas were examined. This two-step technique was necessary because a number of other surface materials were in the same radiant temperature range as the water and so could not be differentiated from water using only thermal information. Had this similarity in temperature not existed, computer-generated water-temperature maps might have been created using only the thermal data.

The resulting thermal map, Fig. 6-22 (between pages 36 and 37), shows the water areas with different colors representing the various surface temperatures present.

Assessment In order to check the accuracy of the thermal calibration technique, a number of water-temperature measurements made with a conventional contact thermometer were collected at approximately the same time as the scanner data. As anticipated, the radiant temperatures, which really are indicative of only the top 0.02 mm of water, were consistently lower than the kinetic temperatures because of the cooling effect of evaporation at the surface, but allowing for that consistent and correctable difference, water-surface temperatures could be determined within two-tenths of a degree Centigrade from an altitude of 608 m (2000 ft) using the 8.0 to 13.5-μm band. For comparison, the scanner collected data over the same area at two higher altitudes, 3000 and 1500 m. The temperatures assigned to these data using the same analysis procedures were significantly lower, and it was concluded that even in the so-called "atmospheric windows" the atmosphere was not sufficiently transparent to thermal radiation to let it pass without significant attenuation.

The results of this study showed that by using a multispectral scanner with internal thermal calibration, radiant temperature measurements can be made on water bodies without the need for surface measurements.

6-4 CONCLUSION

In this chapter we have seen some specific ways in which quantitative remote sensing can be applied to earth resources management problems. The application of remote sensing to these problems was discussed in terms of five key steps presented early in the chapter and then elaborated on through the step-by-step discussions in the case study on crop identification. The same key steps also provide a framework for the five more briefly treated analysis examples. Taken together, the case study and analysis examples demonstrate the use of a variety of data-collection and handling techniques, describe a variety of analysis techniques, and show numerous ways in which human judgment and understanding contribute significantly to the analysis procedure. Furthermore, they present remote sensing tasks involving very large geographic areas and relatively small ones, analysis results presented in image format and in tabular format, and both qualitative and quantitative ways of evaluating analysis results.

Through the material contained in this chapter, we can glimpse both the scope of applications possible for quantitative analysis of remote sensing data and the versatility that exists in all aspects of the procedure, from project planning to final results presentation. It is, in fact, a versatility that permeates back to the initial statement of objectives, allowing resource managers to formulate ever more demanding and pertinent analysis objectives as they grow in their understanding of the science. It is this versatility throughout the process which will enable as yet unformulated objectives to be successfully met in the years ahead and will foster the evolution of the technology.

This chapter, then, serves on one hand as a discussion of what has been accomplished through quantitative analysis of remote sensing data, and also as a pool of ideas which can be newly reassembled and further developed as wider understanding of the technology is achieved.

PROBLEMS

6-1 Imagine you are lecturing to a group of forestry students on the application of remote sensing to a forestry problem. List the key steps in carrying out the application that you would like the students to have in their notes.

6-2 The applications examples described in Sec. 6-3 were extracted from the results of remote sensing research projects. Thus in some cases the approach taken may have been constrained in a manner not likely to be encountered in a large-scale application. Select three of these application examples and analyze them, identifying the key steps in the project. Evaluate and comment on the choices made with regard to the data-collection system used, frequency of data collection, sources of reference data, and analysis procedures.

6-3 The developing country of Duepur is faced with two difficult problems. One problem concerns the production and distribution of food within the country. The other problem is concerned with the management of forest areas which are being clear-cut and the lumber sold on the world market at a rapid but unknown rate. The government of Duepur seeks an opinion as to whether or not remote sensing can provide the necessary data upon which to base sound management decisions.

Your job is to write a two- or three-paragraph opinion as to whether or not you feel remote sensing technology offers viable solutions to these problems. Your opinion should be based on sound technical judgment. If a pilot study seems appropriate you should recommend specific sensor systems for data collection, state types of reference data which would be useful, and propose a method for evaluating the pilot project. Technical reasons should be given to support the use of a particular type of data-collection system. Similarly, reasons should be given as to why a particular system would not be useful. A map of the country is shown in Fig. P6-3 and a description of the geography and farming practices is given below.

Duepur is located approximately 18°N latitude. Most of the population lives in the western coastal region. There are two large cities: one is located at the mouth of the Smile River and the other further south along the coast. The coastal zone tends to be cloud-free about 80 percent of the time. This is in sharp contrast to the highland region which is covered with clouds almost every day except for an hour or two after sunrise.

Ninety percent of the country's food production takes place in the sunny coastal region. Farms are small by United States standards, ranging in size from 7 to 10 hectares (15 to 20 acres). Field sizes average 1 hectare. Farm families usually devote about one-half of their land to a "cash" crop. The other half is used for grazing animals and the growing of fruits and vegetables for family use. The primary cash crop is wheat, except in the Smile River Valley area where a lot of rice is grown. It is estimated that modern agricultural methods could produce a twofold increase in productivity, thereby producing a food surplus which could be exported. To accomplish this task, however, a method must be developed for monitoring crop production and developing a transportation and storage system.

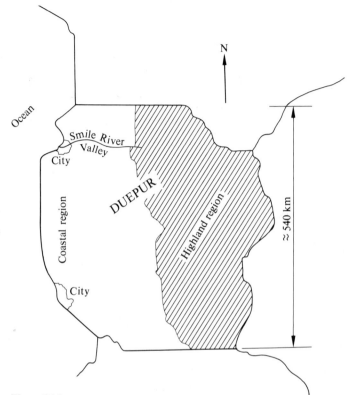

Figure P6-3

The highland region has vast forest areas. This resource is endangered because commercial logging firms have been clear-cutting large areas and shipping the logs across the border. The companies have not taken any steps to replant or to cut in any systematic way. The government is concerned because it has no adequate way to monitor or regulate the loggers.

Reread the second paragraph of this question and proceed to write your answer.

6-4 If agricultural crops are not transported to processing or storage facilities in a timely fashion, a substantial loss may result due to spoilage, rodent infestation, or other hazards. Can you suggest a way in which remote sensing might be used to aid in the allocation of railroad boxcars and trucks for the transportation of these crops?

REFERENCES

1. Bauer, M. E.: Crop Identification and Area Estimation Over Large Geographic Areas Using LANDSAT MSS Data, Final Report on LANDSAT Investigation no. 21330, NASA Contract NAS5-20793, *LARS Information Note 012477*, Laboratory for Applications of Remote Sensing, Purdue University, West Lafayette, Ind., 1977.
2. Eisgruber, L. M.: Potential Benefits of Remote Sensing: Theoretical Framework and Empirical Estimates, *LARS Information Note 030872*, Laboratory for Applications of Remote Sensing, Purdue University, West Lafayette, Ind., 1972.

3. Ewart, R.: "Effect of Information on Market Behavior," Ph.D. Thesis, School of Agriculture, Purdue University, West Lafayette, Ind., 1972.

4. MacDonald, R. B., M. E. Bauer, R. D. Allen, J. W. Clifton, J. D. Ericson, and D. A. Landgrebe: Results of the 1971 Corn Blight Watch Experiment, in *Proc. of the Eighth International Symposium on Remote Sensing of Environment*, vol. I, Environmental Research Institute of Michigan, Ann Arbor, Mich., pp. 157–190, 1972.

5. Bauer, M. E., and J. E. Cipra: Identification of Agricultural Crops by Computer Processing of ERTS MSS Data, in *Proc. of the Symposium on Significant Results from ERTS-1*, NASA Document no. SP-327, Washington, D.C., pp. 205–212, 1973.

6. Bauer, M. E.: Quarterly Progress Report on LANDSAT Investigation no 21330, 1 October to 31 December, 1975, NASA Contract NAS5-20793, Laboratory for Applications of Remote Sensing, Purdue University, West Lafayette, Ind., 1975.

7. Cochran, W. G.: "Sampling Techniques," John Wiley and Sons, New York, 1963.

8. Anuta, P. E.: Geometric Correction of ERTS-1 Digital Multispectral Scanner Data, *LARS Information Note 103073*, Laboratory for Applications of Remote Sensing, Purdue University, West Lafayette, Ind., 1973.

9. *Proc. of the Symposium on Machine Processing of Remotely Sensed Data*, IEEE Cat. No. 76 CH 1103-1 MPRSD, IEEE Single Copy Sales, Piscataway, N.J., 1976.

10. National Aeronautics and Space Administration: *Proceedings of the NASA Earth Resources Survey Symposium*, 3 vols., NASA Document TMX-58168 (JSC-09930), Johnson Space Center, Houston, Tex., 1975.

11. Weismiller, Richard A.: Land Use Inventory of the Great Lakes Basin Using Computer Analysis of Satellite Data, Final Report on U.S. Environmental Protection Agency Contracts No. 68-01-2551 and -3552, *LARS Information Note 011077*, Laboratory for Applications of Remote Sensing, Purdue University, West Lafayette, Ind., 1977.

12. Anderson, J. R., E. E. Hardy, and J. T. Roach: "A Land Use Classification System for Use with Remote Sensor Data," USGS Circular no. 671, U.S. Geological Survey, Washington, D.C., 1972.

13. Hoffer, R. M., and Staff: Computer-Aided Analysis of SKYLAB Multispectral Scanner Data in Mountainous Terrain for Land Use, Forestry, Water Resources, and Geologic Applications, Final Report on NASA Contract no. NAS9-13380, SKYLAB EREP Project 398, *LARS Information Note 121275*, Laboratory for Applications of Remote Sensing, Purdue University, West Lafayette, Ind., 1975.

14. Kettig, R. L., and D. A. Landgrebe: Classification of Multispectral Image Data by Extraction and Classification of Homogeneous Objects, *IEEE Trans. on Geoscience Electronics*, vol. GE-14, pp. 19–26, 1976.

15. Hoffer, R. M., and Staff: Computer-Aided Analysis of SKYLAB Multispectral Scanner Data in Mountainous Terrain for Land Use, Forestry, Water Resource, and Geologic Applications, Final Report on Contract no. NAS9-13380, SKYLAB EREP Project 398, *LARS Information Note 121275*, Laboratory for Applications of Remote Sensing, Purdue University, West Lafayette, Ind., 1975.

16. Hoffer, R. M., and Staff: "Natural Resource Mapping in Mountainous Terrain by Computer Analysis of ERTS-1 Satellite Data," Research Bulletin no. 919, Agricultural Experiment Station, Purdue University, West Lafayette, Ind., 1975.

17. Wu, C. L., D. A. Landgrebe, and P. H. Swain: The Decision Tree Approach to Classification, *LARS Information Note 090174*, Laboratory for Applications of Remote Sensing; also Technical Report TR-EE 75-17, School of Electrical Engineering, Purdue University, West Lafayette, Ind., 1974.

18. Swain, P. H., C. L. Wu, D. A. Landgrebe, and H. Hauska: Layered Classification Techniques for Remote Sensing Applications, in *Proc. of the Earth Resources Survey Symposium*, vol. I-B, NASA Document TMX-58168 (JSC-09930), NASA Johnson Space Center, Houston, Tex., pp. 1087–1097.

19. Melhorn, W. N., and Sinnock, S.: Recognition of Surface Lithologic and Topographic Patterns in Southwest Colorado with ADP Techniques, presented at the Symposium on Significant

Results Obtained from ERTS-1 Data, Goddard Space Flight Center, Greenbelt, Md., March 5–9, 1973. Also, *LARS Information Note 030273*. Laboratory for Applications of Remote Sensing, Purdue University, West Lafayette, Ind., 1973.

20. Atwood, W. W., and K. F. Mather: "Physiography and Quaternary Geology of the San Juan Mountains, Colorado," Prof. Paper no. 166, U.S. Geological Survey, 1932.

21. Mather, J. F.: Geomorphology of the San Juan Mountains, in "New Mexico Geological Society Guidebook," Eighth Field Conference, Southwestern San Juan Mountains, Colorado, 1957.

22. Zapp, A. D.: "Geology and Coal Resources of the Durango Geologic Survey, Oil and Gas Investigations Preliminary Area, La Plata and Montezuma Counties, Colorado," U.S. Map 109, 1949.

23. Barolucci-Castedo, L. A., R. M. Hoffer, and T. R. West: Computer-Aided Processing of Remotely Sensed Data for Temperature Mapping of Surface Water from Aircraft Altitudes, *LARS Information Note 042373*, Laboratory for Applications of Remote Sensing, Purdue University, West Lafayette, Ind., 1973.

24. Lindenlaub, J. C., and J. Keat: Use of Scan Overlap Redundancy to Enhance Multispectral Aircraft Scanner Data, *LARS Information Note 120271*, Laboratory for Applications of Remote Sensing, Purdue University, West Lafayette, Ind., 1971.

SEVEN

USEFUL INFORMATION FROM MULTISPECTRAL IMAGE DATA: ANOTHER LOOK

David A. Landgrebe

In Chap. 1 we examined in generalized overview form a system concept for extracting useful information from remotely sensed multispectral data. In the subsequent chapters the details of each subsystem component for this approach were examined. In this chapter we will return to a broad overview perspective. The purpose will be to present some further fundamentals of the approach and, in the process, to provide the reader with a perspective against which to judge future developments and potential.

Study objectives
After reading Sec. 7-1, you should be able to:
1. Describe a remote sensing system in terms of three large functional components.
2. Describe the relative level of control the system designer and the system operator typically can exercise over each part of the system.

7-1 A SYSTEMS PERSPECTIVE

In connection with Fig. 7-1 (Fig. 1-18, repeated here for convenience), we discussed the organization of a remote sensing system. Let us now build on that discussion. The entire system consists of three distinctly different parts. These are:

1. The scene.
2. The sensor system.
3. The processing system.

The *scene* refers to that part of the system which is in front of the sensor. It includes not only the earth's surface but also the atmosphere through which the energy passes both on the way to the earth's surface from the sun and on the return passage back to the sensor. The distinguishing characteristic of this part of the system is that there is no human control over it, either on the part of the system designer before construction or of the system operator after. Since

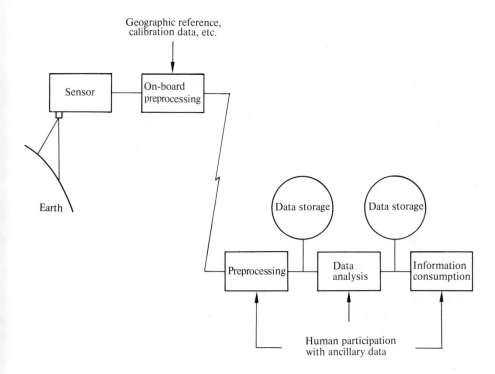

Figure 7-1 Block diagram of a remote sensing system.

there is no such control and we cannot design it nor control its operation, as we can the other system parts, our thrust relative to this part of the system is to learn as much about the scene as possible. Indeed, scene understanding is at least as important to successful overall system operation as understanding the other system parts.

The *sensor system* functions to gather the main body (but not all) of the data about the scene. Its design parameters must be selected so that the scene will be adequately represented by the data for purposes of extracting the needed information. In principle there is no reason why this part of the system cannot be directly under the control of both the designer and the operator of the system. In practice, though, this does not turn out to be the case. Modern sensor systems are large, complex, and expensive. As a result new ones are designed rather infrequently—in the case of satellite sensors, perhaps not much more often than once per decade. Furthermore, they must be designed to serve a broad spectrum of users and uses. As a result the system designer does have design control over the sensor but the system operator (i.e., the analyst) usually does not have operational control over it. Even in the case of the designer, however, there is little opportunity to optimize the sensor system for a given application because of the large number of applications the sensor must serve in order to justify its cost.

All of the remainder of the system, occurring after the sensor system in the data stream, we will refer to as the *processing system*. It is apparent that this is the portion of the system over which both the designer and the operator typically have the greatest number of choices. With due regard to costs and the economies of scale, it is possible to optimize most or all of the processing portion of this system with respect to specific applications both in the design and operation phase.

In the next several sections we shall discuss each of these three parts of a remote sensing information system in more detail and attempt to develop a viewpoint which will provide for the assessment of potential of a total remote sensing system both in terms of meeting a given new need and for further development of the system itself.

PROBLEM

7-1 Complete Table P7-1 by listing the functional parts of a remote sensing system and describing the amount of control exercised over each part by the system designer and the system operator (none, little, some, or considerable).

Table P7-1

System component	Control by designer	Control by operator
1.		
2.		
3.		

Study objectives

After reading the next section you should be able to:

1. Discuss why a stochastic model is the most appropriate way to represent the scene.
2. With respect to the scene, give an example which illustrates the saying that "one man's signal is another man's noise."
3. Illustrate, by drawing an information tree, how a hierarchical structure can aid in scene understanding and lead to a multistage classification procedure.
4. State the requirements for a valid list of classes.

7-2 THE SCENE AND ITS COMPLEXITIES

The scene is the portion of the system which provides us with the greatest challenge. This is true partly because, as previously noted, it is the only portion of the system not under design or operational control; however, and much more significantly, it is by far the most dynamic and complex portion of the system. There are so many different classes of materials which are found on the earth's surface, and they can be found with so many subtle and not-so-subtle variations due to such a large number of factors, that one must strive for a very knowledgeable orderliness and discipline to see them in their proper interrelationship. A very large portion of the errors that we make in design or operation come about because of the underestimation or oversimplification of this scene complexity.

We will suggest two ideas which may be helpful in dealing with scene complexity. First, it is useful to divide the variability factors present in the scene into two categories, those which are related to the information desired and those which are not. The former are then referred to as the *signal* while the latter are the *noise*. Notice first about this viewpoint that a given variability factor (e.g., a set of soil patterns) can for one application be the signal (e.g., for a soil-mapping problem) and in another application be the noise (e.g., in a crop-mapping application). The implication of this viewpoint is that not only will a stochastic model be used for the noise but such a model will also be used for the signal. This is as a direct reaction to the degree of scene complexity; since there are so many causitive factors for the variability present, it is inappropriate and probably impossible to treat them all in a deterministic, cause-effect fashion. This is precisely the situation that stochastic signal models are designed to handle. This does not say that we should not strive to understand the cause-effect relationships present for variabilities in the data. Indeed, such an effort is the key to greater scene understanding. But from a philosophical viewpoint, design of a system using a stochastic approach has proven to be the most effective way to deal with systems which contain many signal factors.

A second idea useful in the face of this high degree of scene complexity is that it will be useful to understand the hierarchy of the classes of materials in the scene. This has frequently proven to be the case when people have faced the problem of understanding a complex situation or set of data. Two examples of this are the construction of the Dewey decimal system for libraries and the taxonomic approach used in classifying plant and animal life. Furthermore, the field of data structures is a very active field of research for information scientists at the present time,[1] and will likely be so for some time in the future. Thus we can look forward to an increasing number of tools in dealing with scene complexity in this way.

To briefly illustrate the concept, an example of the hierarchy present in earth resources classes is displayed in information-tree form in Fig. 7-2. In this figure we see something resembling an inverted tree in which earth surface features have been listed in a taxonomic fashion. At the top are listed the more general classes of surface features. These are then subdivided into appropriate subfeatures and this subdivision can continue through many additional stages.

A set of requirements for a valid list of classes is that:

1. The list of classes must be *exhaustive*.
2. The classes must be of *informational value*.
3. The classes must be spectrally or otherwise *separable* (i.e., distinguishable based on the available data).

The problem of training a pattern classifier resolves itself to devising an exhaustive list of classes which are *simultaneously* of information value and separable. To be valid the information tree must satisfy these same requirements. Even so, in drawing such a tree there are many arbitrary choices; more than one tree configuration may provide valid and useful descriptions of a scene just as more than one list of classes may.

The entries in the tree of Fig. 7-2, for example, are all generic names of classes of informational value. They might just as well have been regions and subregions of *n*-dimensional multispectral feature space, thus providing more direct evidence of spectral separability. The tree concept is intended to be useful in displaying the totality of classes of earth surface features in such a way that the interrelationship of the various classes can be seen. For a particular application one frequently will want to construct a tree, however, so that it has a greater degree of detail in the portion of greatest interest.

Before leaving the subject of the scene and the information tree, let us use it to assess how far we have come in learning to gather earth resources information by remote sensing. We can do this in part by examining how deeply we are able to penetrate into the tree with today's technology. A review of literature especially regarding exploratory studies of Landsat data analysis (see, for example, Ref. 2) would suggest that we have penetrated only to roughly the second layer of classes in this figure. Certainly vegetation can reliably be distinguished from exposed earth; usually one can go beyond this point by at least one layer of categories. Some advanced demonstrational or operational-like programs now extend to forest-type mapping and crop-species identification, for example, but with Landsat 1 and 2 data it has proven difficult to go beyond this point reliably.

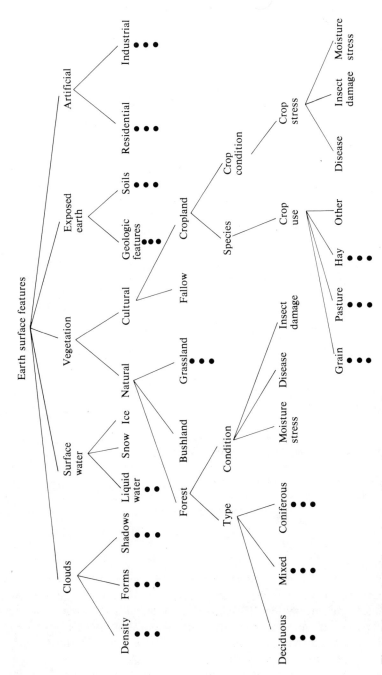

Figure 7-2 A typical information tree for earth resources classes.

As to the future, based on evidence in hand, the potential for functioning successfully well beyond these points can hardly be doubted. The 1971 Corn Blight Watch Experiment[3] demonstrated, for example, that a particular crop stress condition could quite reliably be identified and broken into three levels of stress. This was possible, of course, because, as compared to Landsat 1 and Landsat 2 data, there was a considerably larger number of spectral bands available in the aircraft multispectral scanner data used and there was considerably greater flexibility with regard to temporal control over the sampling. This suggests that the information to distinguish between more detailed classes in the scene is present in multispectral data, and we have but to devise suitable, more advanced sensor and processing systems to be able to extract it.

And, finally, note that we have introduced and used the idea of the information tree for the purpose of perceiving something about the scene that clarifies our understanding of its complexity. This is a valid and useful application of the concept; there are even more direct applications of it. Whole processing systems based upon and taking advantage of such a hierarcical presentation are available. Algorithms with decision logic based upon such a tree have demonstrated improved information-extraction capabilities while at the same time reducing the amount of computation required, because at each decision point in the tree, i.e., a point at which the tree branches, there are fewer alternatives from which to choose. Though potentially quite powerful, such techniques are complex and the details of their use are beyond the scope of this text. Some applications appear in Chap. 6.

PROBLEMS

7-2 Make a list of the key points you would use in arguing that for remote sensing problems a stochastic model is the most appropriate way to model the scene.

7-3 Give an example illustrating that scene variations considered noise in one application might in another application be considered signal.

7-4 Consider a forestry application involving identification of tree species in large natural forest areas. Show how an information tree might aid in this analysis.

Study objectives
After reading Sec. 7-3 you should be able to:
1. Describe in a sentence or two the job or function of the sensor in a remote sensing system. Take into account, at least in a general way, properties of the scene, the processing system, and the needs of the user.
2. List and describe to an imaginary (or real) user the important parameters associated with a sensor system.
3. Demonstrate that the design parameters of the sensor system are not independent of each other.

7-3 THE SENSOR: CHARACTERIZATION OF THE SCENE BY THE DATA

It is the sensor system's job to gather data which will adequately characterize the variations of the scene that are information bearing. To understand this job we must have well in mind the list of parameters which are important in the extraction of information about the scene. Later in this section we will propose that this list consists of the following five entries: the spatial characteristics, the spectral characteristics, the signal-to-noise ratio, the ancillary information available, and the informational classes and their interrelationships. In preparation for this we will first review the results of three research efforts which have appeared in the literature; they will then be used to construct the needed list of parameters significant to the extraction of information from data.

First Research Result: Measurement Complexity

The first result, illustrated in Fig. 7-3, was obtained during a study of predicting, on a theoretical basis, the expected accuracy of pattern-classifier systems.[4] The vertical axis in this figure is the mean recognition accuracy obtainable from a pattern classifier; i.e., the average over *all possible* pattern classifiers. This is plotted as a function of *measurement complexity*. In this case, measurement complexity is a measure of how precise a measurement is taken. In the case of digital multispectral data, it corresponds to the number of brightness-level values recorded, k, raised to the pth power, where p is the number of spectral bands. The more bands one uses and the more brightness levels in each band, the

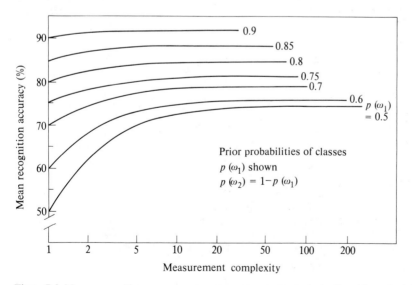

Figure 7-3 Mean recognition accuracy versus measurement complexity (two-class case) for various prior probabilities.[4]

greater the measurement complexity. The assumption was made in the referenced work that there are only two classes to choose between, and the parameter used in the figure is the a priori probability of occurrence of one of the classes.

The results show that as one increases the measurement complexity, one can always expect the accuracy to increase. However, there is a saturating effect which occurs, in that after a certain measurement complexity is reached the increase in accuracy with increasing measurement complexity is very small.

This graph in Fig. 7-3 is the result one obtains if an infinite number of training samples is available by which to train the pattern classifier. During the same study, the author obtained the result for a finite number of training samples; this is shown in Fig. 7-4. In this figure, the two classes are assumed equally likely and the parameter m is the number of training samples used. An unexpected phenomenon is observed here in that the curve has a maximum. This suggests that for a fixed number of training samples there is an optimal measurement complexity; too many spectral bands or too many brightness levels per spectral band are undesirable from the standpoint of expected classification accuracy.

The occurrence of this phenomenon is predictable from the following argument. If one has only a fixed number of samples by which to estimate the statistics that govern the classes, as one increases the measurement complexity by increasing the number of spectral bands, for example, a higher and higher dimensional set of statistics must be estimated with a fixed number of samples. Since we are endeavoring to derive a continually increasing amount of information from a fixed amount of data (the training samples), the accuracy of estimation

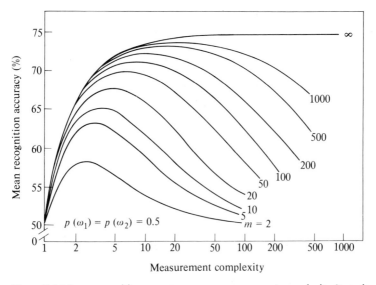

Figure 7-4 Mean recognition accuracy versus measurement complexity (two-class case) for various numbers of training samples, m.[4]

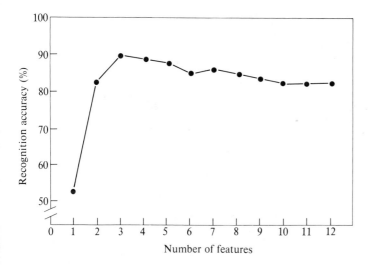

Figure 7-5 Experimental results showing accuracy versus the number of optimally selected features.[5]

must eventually begin to decrease. One could not expect good classifier performance if ten-dimensional statistics were to be estimated from five training samples. For our present purposes, though, the important point is that *there is a maximum*; i.e., there is a best measurement complexity.

Though these results were obtained on a theoretical basis, this phenomenon has been observed in practice. Figure 7-5 shows some results obtained using actual remote sensing data.[5] These results were obtained in a test on classifying 12-channel data. A classification was done using all 12 spectral channels and a given training set. The same classification using the same training set was then used on optimally determined subsets of the 12 features. The figure shows the accuracy obtained versus the number of features. One sees that there is a maximum in this curve just as predicted by the theoretical results. In this case, the greatest accuracy occurred with three features, and this accuracy was significantly greater than when all 12 were used.

Second Research Result: The Effects of Noise

We move now to the second experiment which aids in establishing the sensor system parameter list. In this case a study was carried out to determine the effects of noise on pattern-recognition analysis of multispectral data.[6] The data were gathered from an agricultural area by a multispectral scanner, then simulated noise in varying amounts was added to the data to form modified data sets. Identical analyses were carried out on each and the results compared. Figure 7-6 shows a conventional airphoto of a small portion of the test site used for reference purposes. Figure 7-7 shows a printout of one channel of the data, first with no noise added and then with various levels of noise added. The noise

Figure 7-6 Air photo showing a portion of the test site used in the noise sensitivity experiment.[6]

level is denoted by σ, the standard deviation of the noise in units of brightness levels. In other words, $\sigma = 5$ signifies a noise level in which the standard deviation of the noise was equal to five brightness levels out of the total brightness-level range, which was 256. Figure 7-8 shows the result of classifying the data and allows one to determine on a qualitative basis the manner in which the classification accuracy deteriorated as the noise level increased. In this figure, all points classified into the wheat class are displayed as W's and all those classified into all other classes are displayed as blanks.

Figure 7-9 shows the results of the test in quantitative form. The overall average accuracy and the accuracy for two specific classes are shown. While the general shape of these curves shows the expected downward trend, a significant point to be noted here is with regard to the two individual classes. The data set was gathered in early summer over a test site in the United States Corn Belt. At that time of year, wheat has ripened to a golden brown and is ready for harvest. Soybeans, on the other hand, having been planted only a few weeks before, have not developed a full canopy and the percentage of ground cover is still rather small. As a result, the wheat class is relatively easy to distinguish from the other classes present in the scene, while the soybean class represents a much more difficult discrimination problem. Notice that as noise is added to the data the accuracy of the difficult class, soybeans, deteriorates much more rapidly than that of wheat. Indeed, while the classifier design was good enough to result in the soybean classification being above average in the no-noise case, it quickly fell below average, whereas the wheat class remained well above average at all noise levels. This suggests that a more marginally identifiable class may be more adversely affected by a deteriorating signal-to-noise ratio.

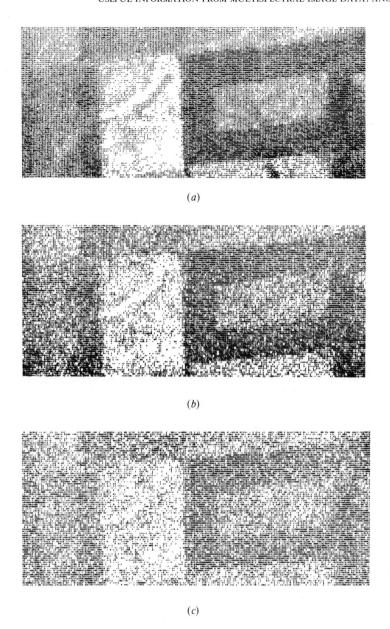

(a)

(b)

(c)

Figure 7-7 Alphanumeric printouts of the reference data and data with noise added.[6] (a) Original data. (b) $\sigma = 5$. (c) $\sigma = 15$.

Third Research Result: Intrinsic Dimensionality

Finally, we move to the third study to be considered before developing the list of sensor system parameters. Data-compression techniques can make possible the

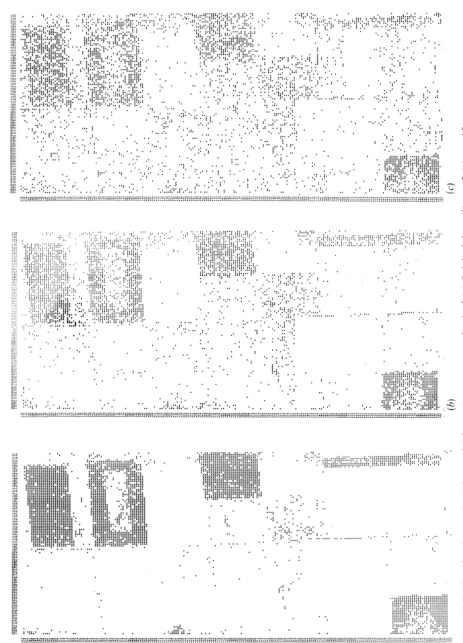

Figure 7-8 Classification results (wheat only) for various amounts of noise.[6] (*a*) Original data. (*b*) $\sigma = 5$. (*c*) $\sigma = 15$.

(*a*)

(*b*)

(*c*)

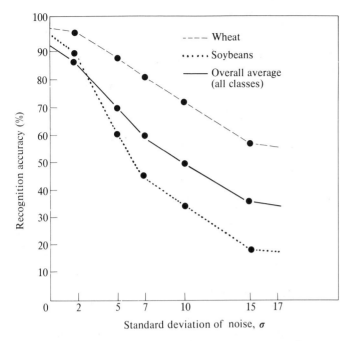

Figure 7-9 Recognition accuracy versus amount of noise added.[6]

transmission or storage of a smaller number of bits of data without the loss of information. Thus, by inference, data-compression studies tend to reveal the amount of real information present in relation to the quantity of raw data. One approach for accomplishing data compression is indicated by Fig. 7-10. Here is shown a hypothetical plot of the distribution of data in one dimension versus that in another. Suppose one has data which are distributed in the region outlined in this two-dimensional space. We may define a new coordinate system with one axis (y_1) oriented in the long direction of the distribution, as shown in the figure, and the second axis (y_2) perpendicular to the first. The mathematical operation required to do this is simple: a linear combination of the data values in the old coordinate system achieves the data values in the new coordinate system, as shown in these equations:

$$y_1 = a_{11}x_1 + a_{12}x_2$$

$$y_2 = a_{21}x_1 + a_{22}x_2$$

where (x_1, x_2) are the data coordinates in the original coordinate system and (y_1, y_2) are the data coordinates in the new coordinate system. By such a transformation it can be seen that the dynamic range for the new coordinate y_1 will be much greater than either of the original coordinates x_1 or x_2, while that of y_2 will be quite small.

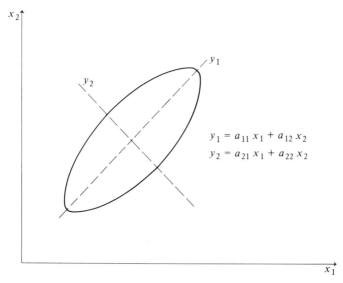

Figure 7-10 Principal component axes in two-dimensional space.

The same operation can be carried out for data in higher dimensional spaces. New coordinate directions are defined one after another, each time taking the direction of orientation of the new coordinate to be perpendicular to all of the previously selected ones and in the direction of the remaining maximum data dynamic range. Such a procedure is called a principal components transformation.[7]

When data are encountered which have some degree of correlation or redundancy between their components, as indicated by an elongated distribution in n-dimensional space (and this is usually the case with multispectral data), this procedure can be used to express the data in an essentially lower dimensional space without significant loss of information. To illustrate this, consider Fig. 7-11. This figure indicates the dynamic range present in each of the new coordinates of the transformed coordinate system for 12-feature multispectral data processed in this way. Given that the data had approximately equal dynamic range in all 12 dimensions before the transformation, we see that they now have nearly all of their dynamic variation expressed in about three of the new dimensions. Thus, the data could be stored or transmitted using a three-dimensional data system as compared to the original 12-dimensional system. The term *intrinsic dimensionality* is sometimes used when referring to the smallest number of dimensions which could be used to accurately represent a data set.

For example, it is known that the intrinsic dimensionality of the Landsat 1 and 2 multispectral scanner is approximately 2. This is illustrated by processing some Landsat data through a principal components transformation and generating imagery from the result. The results of doing this are shown in Fig. 7-12. Notice that though there is a reasonable amount of information-bearing

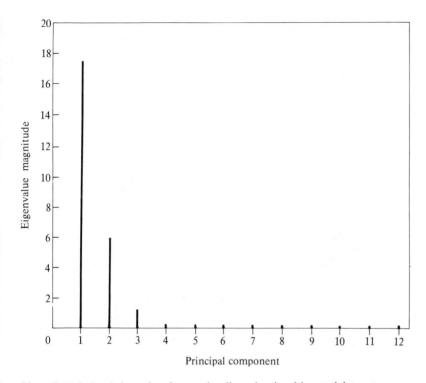

Figure 7-11 Ordered eigenvalues for a twelve-dimensional multispectral data set.

contrast and detail in all four original channels displayed in image form, the principal component features show much detail only in the first two. The amount of detail, as expressed through the image contrast, is much greater in the first principal component—probably greater than can be perceived in an image representation of the data. Considerable contrast is still present in the second principal component, but the third and fourth show mostly noise.

A further demonstration of this concept of intrinsic dimensionality can be obtained by classifying data sets. Returning to the 12-band aircraft data set referred to in connection with Fig. 7-11, standard maximum-likelihood pattern-recognition classifications were carried out on these data. Using the data which had been transformed via a principal components transformation, successive classifications were carried out using first all 12 and then successively fewer components, continuing to only two components.[6] Figure 7-13 shows the results of this test. One sees that essentially constant and very good accuracy is obtained down to the use of three components. Then the accuracy drops off very rapidly when fewer than three of the principal components are used. Apparently, then, the intrinsic dimensionality of this data set is 3.

The purpose of this discussion was to introduce the concept of intrinsic dimensionality and to point out that the number of spectral bands collected in a

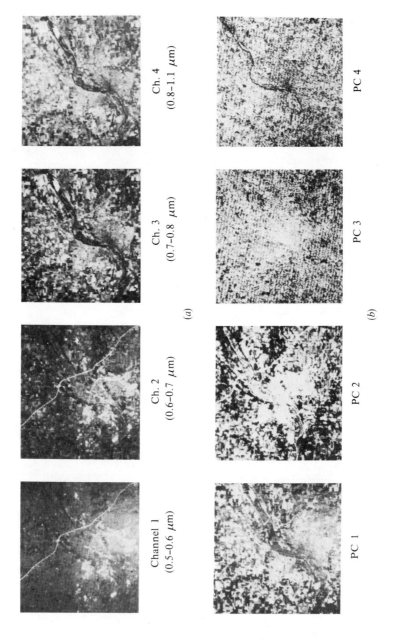

Channel 1
(0.5–0.6 μm)

Ch. 2
(0.6–0.7 μm)

Ch. 3
(0.7–0.8 μm)

Ch. 4
(0.8–1.1 μm)

(a)

PC 1

PC 2

PC 3

PC 4

(b)

Figure 7-12 Black-and-white photos of (a) original Landsat data and (b) principal component images.

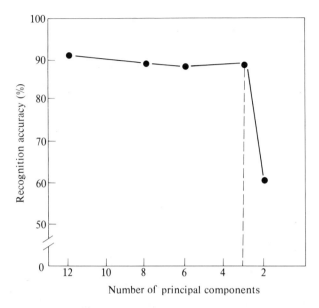

Figure 7-13 Recognition accuracy versus dimensionality for a data set classified in principal components form.

data set does not necessarily indicate directly the intrinsic dimensionality of the data. The actual intrinsic dimensionality of multispectral data from the usable portions of the optical region (0.4 to 15 μm) is not known, but based on current understanding of the phenomena of reflection and emission of energy from earth surface materials, this intrinsic dimensionality is probably somewhere around 6.

The Parameter List

Having now reviewed these three experiments which illustrate various concepts regarding representation of information in data, we will propose the list of parameters which are fundamental to the extraction of information about a scene. The list is as follows:

1. The spatial resolution and detail.
2. The spectral resolution and detail.
3. The signal-to-noise ratio, S/N.
4. The ancillary information available.
5. The informational classes and their interrelationships.

We will discuss each of these in turn, justifying why it should be a member of this important list and providing additional explanatory details about it. The three research results discussed above will prove helpful in this.

Spatial resolution and detail Whenever an imaging sensor system is designed, much attention is focused upon the choice of spatial resolution. The true role of spatial resolution in modern systems, however, is not always accurately perceived. If analysis is to be done by primarily image-oriented techniques, the spatial resolution becomes a prime, almost dominant consideration, for in this case it is the major information-bearing attribute of the data. It conveys to the eye the spatial structure of the scene from which it is possible to deduce much information about the scene.

However, if the analysis is to be done on a multispectral basis, the spatial resolution of the scene has quite a different role. It determines what informational classes can be utilized directly with respect to the data set. For example, data gathered over an urban scene at a 100-m instantaneous field-of-view would permit the direct analysis into such classes as industrial, commercial, residential, etc. At 1-m resolution, on the other hand, the classes would be grass, trees, roof tops, concrete, etc., i.e., the constituent informational classes which go to make up the 100-m informational classes. Thus, with present multispectral processing techniques the question of what resolution is useful rests principally upon what informational classes are desired.

When thinking of machine processing methods, greater resolution will, of course, provide greater detail about the spatial structure of the scene. Current operational machine-based analysis algorithms do not yet significantly rely on this characteristic, however, and we shall return to this point when dealing with new processing algorithms.

Other spatial characteristics are also of interest and importance. For example, the spatial sampling rate chosen is obviously an important consideration. To illustrate, if a sampling rate is chosen so as to provide pixels which significantly overlap one another, it is possible to effect a trade-off between instantaneous field-of-view and sampling rate. In any case, it is clear that *spatial resolution and detail* belongs in the list of fundamental parameters.

Spectral resolution and detail Let us next turn to the spectral sampling scheme. We know that the distribution of energy as a function of wavelength potentially conveys to us much information, and thus spectral resolution and the spectral sampling scheme are surely fundamental system specifications. The most common scheme used in building a scanner is to select specific "spectral windows" through which to view the scene. But how many such windows are needed?

It is common practice to collect data in a larger number of spectral bands but to use only a subset of specifically selected bands for any specific analysis task. At first glance this seems wasteful, as if the "correct" data were not gathered in the first place. We have already seen that the intrinsic dimensionality of typical multispectral data may be significantly less than the actual number of spectral bands.

The reason for this limitation is that current scanner systems do not provide a very complete characterization of the information the scene has to offer in the data they produce. Sampling a smooth curve (the actual spectral response versus wavelength function for a given scene pixel) with nonoverlapping bandpass functions of relatively rectangular shape (the response functions of the spectral bands of the scanner) cannot provide a very complete representation of the information in the smooth curve unless there is a relatively large number of quite narrow such window functions.

Consider the following generalized mathematical model of the situation. The job of the receptor in a pattern recognition system is to measure the spectral response function of the "unknown" and report the results to the classifier in the form of a set of numbers. In the case at hand, the "unknown" is characterized by its spectral response function, denoted by $R(\lambda)$. We can express this as

$$R(\lambda) = c_1\phi_1(\lambda) + c_2\phi_2(\lambda) + \cdots + c_n\phi_n(\lambda) = \sum_{n=1}^{N} c_n\phi_n(\lambda) \qquad (7\text{-}1)$$

The functions $\{\phi_n(\lambda)\}$, $n = 1, \ldots, N$, are referred to as the "basis functions" and, once chosen, the set is used in representing every spectral response function. They are entirely determined by the scanner design, and, as we have said before, for current scanners they are usually a series of nonoverlapping window functions, i.e., functions which are zero everywhere except in a pass band.

The coefficients $\{c_n\}$ contain the results of the measurements on the unknown $R(\lambda)$: that is, given a set of basis functions $\{\phi_n\}$, the coefficients depend only on $R(\lambda)$ and thus carry the information about $R(\lambda)$ to the classifier. What we desire, then, is for the $\{\phi_n\}$ to be designed in such a way that every different $R(\lambda)$ will result in a uniquely different set of coefficients $\{c_n\}$. Indeed, we can only be assured that the set $\{c_n\}$ contains all of the information which $R(\lambda)$ contained if it is possible to exactly reconstruct $R(\lambda)$ from a knowledge of the $\{c_n\}$. This property is referred to as completeness.†

The reason it is necessary to collect data in a larger number of spectral bands and then use various subsets for various specific analyses is that even current airborne scanner systems, which generally have more bands than current spaceborne systems, do not provide this completeness property. And the degree of exactness to which a spectral response function can be reconstructed from data from current spaceborne scanners is very limited.

In the case of the multispectral scanner of Landsat 1 and 2, the basis functions were a series of four windows. Figure 7-14 shows a typical spectral response function for vegetation, the set of four basis functions for the multispectral scanner,‡ and the approximation which it can provide to the vegetative response curve. It is clear that the multispectral scanner on board Landsat did not by any means convey to the processing system all the information which the scene had to offer.

---

† This discussion of the completeness property provides a very simplified overview of the theory of orthogonal functions as applied to signal representation. The mathematically inclined reader may wish to pursue the theory more fully. See Ref. 7.

‡ Actually the basis functions of the Landsat multispectral scanner are not precisely rectangular, but are more rounded, occupying roughly the four regions shown.

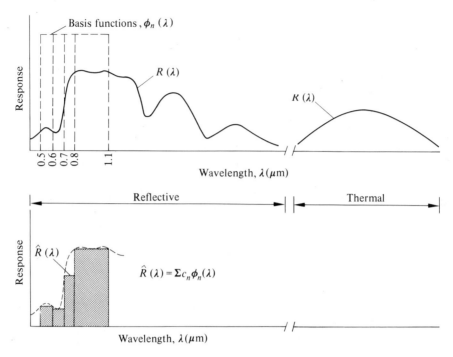

Figure 7-14 The Landsat approximation of a typical green vegetation spectral response function.

Our ability to devise scanners with optimal basis functions rests heavily upon our ability to design and build scanners with (1) adequate numbers of bands and (2) detector sensitivities as a function of wavelength with each band capable of being constructed to mathematically derived specifications. A detailed treatment of this subject is beyond the scope of this text, but we can observe at this point that one measure of the potential for further development of multispectral sensor systems is the degree to which the completeness property is achieved with a number of bands, N, which is close to the intrinsic dimensionality of spectral response functions typical of the earth's surface.

Signal-to-noise ratio Recall from the second experiment described above regarding adding noise to the data that the accuracy of classification continued to improve as long as the signal-to-noise ratio was increased. This makes one tend to request as high a signal-to-noise ratio as possible. However, the matter is not quite that simple because of a strong interrelationship between the spatial resolution, the spectral resolution, and the signal-to-noise ratio. Obviously, there is only a certain amount of energy per unit area rising from the surface of the earth. One may choose to divide up this energy in various ways, trading off among these three variables. For example, if one needs very fine spatial resolution, each pixel, as a result of being smaller, will have a relatively smaller amount of total energy to further divide up spectrally and to achieve the desired signal-to-noise ratio in each spectral band. Because of this trade-off based on the law of conservation of energy, selection of any two of these parameter values causes the third to be fixed.

The important point to note here is the significant impact of the fact that these three fundamental parameters are closely interrelated. The problem of proper scanner system design is more complicated than simply trying to decide independently on the optimal values for each of these three parameters.

Ancillary information The fourth of the fundamental parameters is the ancillary information that is available. Useful ancillary information can arise in both subjective and objective forms, both as quantitative and nonquantitative data. An example of ancillary data in quantitative, objective form would be topographic data registered with multispectral image data on a pixel-by-pixel basis; this will be discussed later in this section. Another use of ancillary data is as reference data for purposes of deriving training statistics for the classifier. This has been discussed in Chaps. 3, 5, and 6; we will show here the interrelationships between these data and the other parameters.

Recall the result from the first experiment described above in which the mean recognition accuracy was displayed as a function of measurement complexity for various numbers of training samples. The number of training samples in this case represents in one sense a quantitative measure of the amount of ancillary information one has about the ground scene. The more one knows about the ground scene the more training samples one can select. Recall that the curves displayed in Fig. 7-4 had a maximum for all finite numbers of training samples, i.e., for all finite amounts of ancillary information. There are two things to note especially about this maximum. First note that the value of the maximum continually increases as the number of training samples increases. This illustrates the fact that the ancillary data parameter is a valid member of our parameter list. The more of it one has, the greater the accuracy one can expect.

The second thing to note about these curves is that the maximum moves constantly to the right. This suggests that the ancillary data parameter is interrelated with the others in the following way. The abscissa, measurement complexity, is directly related to the number of spectral bands and to the number of gray values or brightness levels used in each spectral band. Good engineering design dictates that the number of gray values should be chosen in relation to the available signal-to-noise ratio, since quantizing to a degree of fineness unwarranted by the noise level (or range of uncertainty of the analog signal) would be wasteful. Thus, measurement complexity is related to both the number of spectral bands and the signal-to-noise ratio. The fact that the maximum moves upward and to the right as the number of training samples increases suggests that the more ancillary information one has, the higher the accuracy one can expect, although a greater number of spectral bands and/or greater signal-to-noise ratio will be required to achieve it. In short, the ancillary data parameter is a fundamental one, and is interrelated to the previous three.

Informational classes The last of the five listed fundamental parameters is the informational classes and their interrelationship in the measurement or feature space. In discussing this member of the list it will be helpful to review briefly how a typical feature selection algorithm works. Imagine how multispectral data might appear in n-dimensional space. Data for each of the different classes tend to be clustered in localized regions, usually with slight overlap among neighboring classes. Figure 7-15 illustrates this situation for three classes in a particular n-dimensional space. To increase the clarity of presentation, we have depicted the central region of the clusters but not the overlap.

The feature-selection problem arises because in some subspaces classes may be more separable than in others. Feature-selection algorithms have been devised which determine the degree of separability in each possible subspace (Chap. 3). In Table 7-1 we see a typical type of output from such a feature-selection algorithm. The aircraft-collected data set involved in this case had a total of 12 channels, and the algorithm was used to evaluate all possible subsets of four channels and to determine the best. In the table the second column ("Spectral bands") shows the quadruple indicating the particular channel subset being evaluated in each case. The algorithm first calculates the statistical distance between each class pair in each possible subspace. It then ranks the subsets according to the average of these pairwise separability measures. The third column indicates this average separability, while the columns to the right show the individual class-pair separabilities. The columns are headed by pairs of symbols indicating the class pairs involved, the S for soybeans, C for corn, and so on. Note, for example, that the combination of bands 6, 9, 11, and 12 appears to be more suited for separating soybeans from alfalfa, while the 1, 9, 11, and 12 combination provides greater separability between soybeans and wheat.

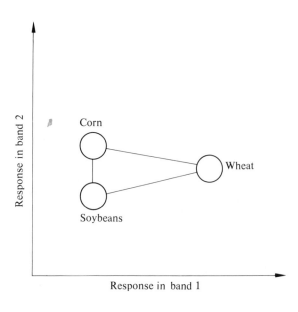

Figure 7-15 Diagram showing relative location and separation (hypothetical) of three classes in two-dimensional space.

Table 7-1 Typical output for a feature-selection algorithm

		Class separability measure (no maximum)					
			Individual class separability				
Rank	Spectral bands	Average separability	SC	SW	SA	WR	WY...
1	1, 9, 11, 12	448	25	190	188	620	58
2†	6, 9, 11, 12	428	26	177	229	630	208
3	2, 9, 11, 12	423	24	151	182	619	58
4	5, 9, 11, 12	420			.		
5	8, 9, 11, 12				.		
	.				.		
	.				.		
	.				.		
	.				.		

† Corresponds to Landsat 1 sensor.

The particular study referred to in Table 7-1 was done prior the construction of the Landsat 1 multispectral scanner in order to evaluate the potential usefulness of the spectral bands selected. The particular subset of bands which corresponds to the Landsat 1 bands is so marked. This subset was determined to be second best in this case; however, the best set of spectral bands involved a spectral band (band 1) which is in the blue portion of the spectrum. For a spacecraft sensor, data from the blue portion of the spectrum were anticipated to be less useful than they would be in the aircraft data set because of the greater Rayleigh scattering present for a sensor at spacecraft altitudes.

We can now use the results of this study to demonstrate the relationship of the information classes to the other four parameters in the list. To see this, one has only to change the classes and note the dramatic change in channels selected. Because of the great complexity of 12-dimensional space, simulation techniques will be used here to illustrate this effect. A simple way to simulate a change in the interclass relationships is to impose a maximum on the entries in the class-pair separability table. Suppose, for example, all the entires in the class-pair table which are greater than 200 are changed to the value 200. This approximately simulates moving classes together which initially were greatly separated from their neighbors. An example is the wheat class shown in Fig. 7-16.

If, after doing this, one allows the feature-selection algorithm to redetermine the correct order of feature subsets, again based on average separability, the result is as shown in Table 7-2. The change in the overall results is dramatic. The Landsat-related subset of bands, for example, has now dropped from second to fifty-fifth place in the ranking. We thus illustrate the fact that the particular set of classes chosen for the analysis impacts the achievable classification accuracy and therefore that the classes and their class interrelationship is a valid member of our parameter list. By this illustration we also point to an inter-relationship of this parameter with at least one and, therefore, *all* of the previous

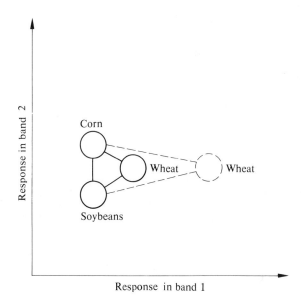

Figure 7-16 The effect of a maximum constraint imposed on the pairwise interclass distance measure, simulating moving the wheat class closer to corn and soybeans.

four. It is now obvious that the class interrelationships and the class separability in *n*-dimensional space are also directly related, for example, to the signal-to-noise ratio, in that if classes were selected which were in close proximity to one another in *n*-dimensional space, they would be more difficult to classify accurately when the signal-to-noise ratio was poorer. This observation is also supported by the results of the second experiment described above. Recall from Fig. 7-9 that the accuracy of identification of the soybean class, a more marginally identifiable

Table 7-2 Feature selection output with maximum pairwise distance constraint

			Class separability measure (maximum = 200)				
				Individual class separability			
Rank	Spectral bands	Average separability	SC	SW	SA	WR	WY...
1	1, 6, 10, 12	155	34	200	196	200	180
2	1, 6, 10, 11	154	34	200	192	200	183
3	1, 6, 9, 12	153	26	200	193	200	200
4	1, 6, 9, 11	153	27	200	190	200	200
	.	.		.			
	.			.			
	.			.			
55†	6, 9, 11, 12	145	26	177	200	200	200
	.			.			
	.			.			
	.			.			

† Corresponds to Landsat 1 sensor.

class, was more adversely affected than other classes by the decrease of the signal-to-noise ratio.

A sixth parameter: temporal sampling So far in this section we have given a set of five *system* parameters which are fundamental to the extraction of information about the scene. They provide a framework against which to judge whether a given sensor system design is adequate for a given application. From the fundamentals previously discussed, it may be seen that multitemporal observations can provide additional information. If one allows the use of a sensor system over a given scene at more than one time, a sixth parameter, data-collection time, can be added to the list.

In Chap. 1 we noted that knowledge of how the spectral response of the earth surface varies with time is useful in at least three different fashions. We will review and amplify upon these here. First, such knowledge can enable one to select the optimal time for data gathering or to assess the probable degree of success when one must accept data gathered at a less than optimal time. In this case we are simply pointing to the fact that temporal scene variation is an important factor in class interrelationships (the fifth parameter). Second, in many applications the information needed actually is contained in a time history of the change which has taken place in the scene. Change detection methods are used to capture and quantify this information.

The third use of temporal knowledge about the scene takes us still more directly to the use of temporal variations to increase the amount of information obtainable from the scene. We will proceed from the following observation. In an agricultural scene, we should somehow be able to increase the number of classes identifiable and the accuracy with which they can be identified, since the way a crop changes through the year bears information about what the crop is. In the United States cornbelt, for example, a field which is slightly green in color in the fall and develops to a deep green with the coming of spring has a high probability of being wheat.

One can use temporal knowledge in a quantitative fashion as follows. Suppose spectral measurements of a scene have been taken in two different spectral bands at two different times. Suppose also that there is a modest separability of the classes of interest in each of the two-dimensional spaces defined by the data gathered at each of these two times. By combining the two sets of data taken at two times·on a pixel-by-pixel basis, one can construct a four-dimensional data set, i.e., one in which the two bands at time 1 are dimensions 1 and 2 and the two bands at time 2 are dimensions 3 and 4. Thus the two original two-dimensional spaces are now subspaces of a new four-dimensional space. It is quite possible that, as a result of this combination, the modest separability that existed at each of the two times, respectively, can be made to be at least partially additive so that somewhat greater separability will exist in four-space than was present in either of the two-spaces. Greater classification accuracy could then be expected.

Having pointed out this possibility, we must add that it is also possible for less separability to result, for while the distance between the centers of the two distributions might indeed be greater in the four-space than in either of the two-spaces, so might the statistical spread within each of the two classes be greater. Said another way, having combined the two data sets in the hopes that the "signal" would be additive, it is also possible that the "noise" would be additive as well. We noted above that more spatial resolution and more spectral bands may actually decrease the classification accuracy, and the same is true for more temporal observations. At the very least it usually takes a larger number of training samples to adequately estimate higher-dimensional statistics. Thus multitemporal observations may require a greater amount of reference data in order to extract the increased information which the data contains.

There is an additional point about temporal observations which is important to keep in mind, relating to the previous comments about scene complexity. Usually a given informational class can, at a given time, exist in a variety of different conditions. For example, a given crop in an agricultural region may exist in several different development states due to different planting dates, different soil types, moisture conditions, etc. It is this variability which results in a given informational class having several spectral subclasses. Thus the number of permutations and combinations of spectral subclasses which can be present in a given informational class at *two* combined observation times is very great.

For example, suppose a scene contains five informational classes $\{\omega_n,$ $n = 1, \ldots, 5\}$ and it is observed at two times $\{t_m, m = 1, 2\}$. Suppose that each of the five classes is made up of three spectral subclasses $\{\omega_{ni}, i = 1, 2, 3\}$ and that this is true at both observation times. Then in the combined bitemporal data set, the number of spectrally distinct classes in $\{\omega_{nim}\}$ could be as large as $n \times i^m$ or 5×3^2 $= 45$. That is, a pixel in any given subclass at time t_1 could become a member of any one of three subclasses at time t_2. To achieve the possible improvement in performance which such a bitemporal classifier might provide could thus require considerably more training samples because up to 45 sets of statistics might need to be estimated.

Thus while multitemporal classification provides potential for improving performance, this improvement is not without a price. The amount of added information multitemporal data represents is sometimes very great, but the processing complexity needed to extract it cannot be ignored. In essence, multiple time observations can significantly increase the intrinsic dimensionality of the data.

Quantitative Ancillary Data

Our premise with regard to the sensor system is that it must make measurements in such a fashion as to adequately characterize the scene and that, based on the fundamentals of remote sensing, this implies proper characterization of the spectral, spatial, and temporal variations of the scene. Sometimes there is available quantitative data from sources other than the sensor system itself, which can prove useful in extracting information about the scene. For example, perhaps a topographic map is available, from an earlier survey of the region. The possible information-bearing attributes of such ancillary data are broad and varied. In mountainous terrains, for example, certain species are known to exist only in certain elevation ranges. In croplands some crops and soil types are found only within narrow ranges of land slopes. The registration of topographic information onto a multispectral data set can therefore result in a significant enhancement in the amount of information extractable from the data set.

Examples of such ancillary data types are topography (elevation, slope, and aspect), soil type, temperature, precipitation, aeromagnetic data, well logs, and various types of demographic data such as census tract boundaries and political subdivision boundaries. As is the case with multitemporal data, the use of ancillary data in this fashion requires the colocation or registration of the new data with the sensor-gathered data. From a fundamental standpoint, adding ancillary data in this way is again an attempt to increase the intrinsic dimensionality of the data and thus the information that can be extracted.

Summarizing, then, in this section we have treated the second portion of a remote sensing information system, i.e., the sensor. We have pointed to its job as being one of gathering data which adequately characterizes the scene from an information-bearing standpoint. In terms of the fundamentals of remote sensing, this implies an adequate characterization of the spectral, spatial, and temporal variations present in the scene to insure that the information which was present at the sensor aperture (the input) is still present in some form in the data it produces (the output). Thus we visualize the sensor as a sort of transducer, not altering the amount of information that is present but merely changing its form. Our study of the sensor in this context focused on the development and discussion of a fundamental list of parameters which relate directly to its design. The list contains five members with a sixth added if multitemporal use of the sensor is considered. A seventh element, the use of quantitative, geographically distributed ancillary data, was also discussed.

PROBLEMS

7-5 One of the sensor system specifications that is often of interest to the user is spatial resolution. Intuition dictates that the finer the spatial resolution the better the sensor system. This is not always the case. Consider the problem of examining your arm with a remote sensing instrument. What do you believe is the best spatial resolution to use if:

(*a*) The objective of the measurement is to detect the presence or absence of your arm?

1 meter_____   10 centimeters_____   1 centimeter_____
1 millimeter_____   0.1 millimeter_____   1 micrometer_____

(*b*) The objective of the measurement is to count the number of mosquito bites on your arm?

1 meter_____   10 centimeters_____   1 centimeter_____
1 millimeter_____   0.1 millimeter_____   1 micrometer_____

(*c*) The objective of the measurement is to enable your doctor to render an opinion as to whether or not additional tests should be made to determine whether you have skin cancer?

1 meter_____   10 centimeters_____   1 centimeter_____
1 millimeter_____   0.1 millimeter_____   1 micrometer_____

7-6 Imagine yourself in the position of preparing a briefing to convince your "management" (job supervisor, vice-president of your company, your professor) that currently available multispectral scanning systems are inadequate or inappropriate for the application your organization is interested in and that what is needed is a new sensor system.

(*a*) Write an introductory statement which describes the purpose and function of the sensor in a remote sensing system and how it relates to the scene of interest, the processing system, and the needs of your organization.

(*b*) Next prepare a list of important sensor parameters. Choose one or two design specifications which are not met by existing sensor system requirements for the application you have in mind. Then prepare a convincing argument illustrating the fact that the design parameters are not independent and that design trade-offs are necessary.

(*c*) Expand the discussion to include parameters which are not inherent in the electro-mechanical aspects of the sensor but do come into play when considering the planning of data-collection missions and the merging of ancillary data.

7-7 Consider approximating the spectral reflectance characteristics of green vegetation by a set of rectangular basis functions as shown in Fig. 7-14. Draw the approximation you would obtain if you used eight basis functions, each one-half the width of the Landsat 1 and 2 spectral bands. Specify a set of rectangular basis functions (give width and location of each) to give you a "truly good" representation of the curve.

Study objectives

Upon completion of Sec. 7-4 you should be able to:

1. Describe the primary function of the processing system.
2. Discuss the factors which influence the choice of
   (*a*) Algorithm   (*c*) Amount of human participation
   (*b*) Processor implementation   (*d*) Type of user interface

   in the design of a processing system.

7-4 THE PROCESSING SYSTEM

So far in this chapter we have divided the entire remote sensing information system into three parts and have discussed the first two. The scene was presented as a very complex physical observable one and the job of the sensor system was defined as gathering data to adequately characterize the spectral, spatial, and temporal variability factors of the scene which are important to the information desired.

The third portion of the system, the *processing system*, must be designed to successfully extract the desired information from the data and "deliver" it to the user. Since the processing system exists between the sensor and the user, its specifications will depend heavily upon the data and on the user's needs. They will also depend upon the system operator (the data analyst), because, as discussed earlier, this is the only part of the system over which the analyst may have real control.

For our purposes here, the discussion of the processing system will be divided into four parts: processing algorithms, processor implementation, human participation in processing, and the user interface.

Processing Algorithms

Usually the first step in designing a processing system is the selection or specification of the algorithms to be used. Note especially that we make a distinction between the *algorithms* and their *implementation*. We are speaking, for example, of the equations which govern the analysis operations as contrasted with the equipment or software through which they are implemented. In algorithm selection, consideration must be given to the abilities of the algorithms in terms of information extraction, the efficiency with which implementations of them can be used, and their complexity relative to use by the data analyst. In the design of any specific processing system, there will be a trade-off which must be reconciled with regard to these factors. A very powerful and sophisticated algorithm from an information-extraction standpoint may be difficult to operate efficiently or very complex to use from the data analyst's standpoint, or both.

There is also a relationship among the complexities of the algorithms, the data, and the information desired from the data. If the data available from the sensor are very simple, involving few spectral bands, few shades of gray, etc., a very advanced and complex analysis algorithm would be wasteful. In like manner, if only very simple information is required by the user, there again would be little need for choosing a complex or sophisticated analysis procedure.

Let us explore the matter of algorithm complexity further by listing some of the characteristics which different algorithms may possess. First of all, in terms of analysis algorithms one may approach the identifiability of classes either on an absolute or a relative basis. Perhaps the most straightforward approach to spectral identifiability is to pose the question in an absolute sense. The viewpoint is that the class of current interest is the target while everything else is background. For example, Fig. 7-17 shows the spectral reflectance for green vegetation. We might specify that anything which has this spectral response is vegetation; everything else is not. A very simple algorithm for this case is the straightforward

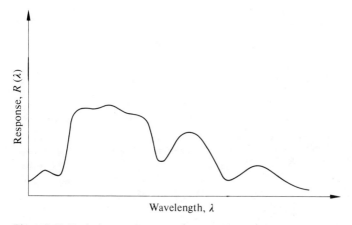

Figure 7-17 Typical spectral response for green vegetation.

instruction to a photointerpreter to locate everything in a color photograph which looks green. Note that in this case no concern is shown for other classes.

As compared to this, there is an important scientific (and pragmatic) principle that it is easier to make *relative* measurements than absolute ones, and it is easier to make relative decisions than absolute ones. For example, a voltage can more accurately be measured by reference to a standard voltage; decision making is better done when choosing among alternatives. As a further example, it is much easier at a paint store to pick out a color which is similar to your living room if you have a sample of that color with you, because then you can more easily choose between the various alternatives by direct comparison. Most of the analysis algorithms discussed in Chap. 3 make use of this important principle.

Beyond these considerations is the mathematical model one has in mind for the signal and the noise, i.e., the variations in response which convey the desired information and those which do not. We shall consider a simple hierarchy which illustrates the varying complexity and sophistication one may use in devising or selecting algorithms.

Deterministic or first-order statistical Again thinking in terms of the information being conveyed through the spectral distribution of energy, perhaps the simplest model is that in which each class is represented by a single, deterministic curve of response versus wavelength, such as we saw in Fig. 7-17. This curve may have been derived either in a deterministic manner or by use of first-order statistics (the average of a large number of spectra). Algorithms suggested by models of this complexity may be ad hoc in nature; e.g., in discriminating between soil and vegetation, a simple band-ratioing using the wavelengths marked λ_1 and λ_2 in Fig. 7-18 might be well suited.

First-order statistical plus noise Even though the curves of Fig. 7-18 may have been arrived at through statistical averaging, the model described above may be

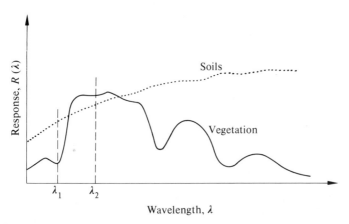

Wavelength, λ

Figure 7-18 Typical spectral response for green vegetation and soils.

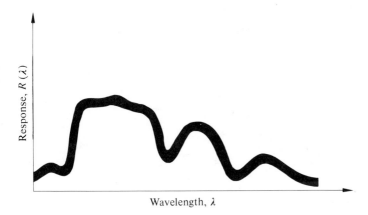

Figure 7-19 Spectral response for green vegetation showing random statistical variability about the mean response.

thought of as deterministic because it does not in any direct way account for the statistical variability of the data. A more complete characterization of actual spectral data would include the statistical variability about the mean spectral curves (see Fig. 7-19). In this case we are still thinking of the signal as the mean spectral response; however, it is recognized that a random component is present which is viewed as non-information-bearing, i.e., random noise. This was the viewpoint used with Fig. 1-14 when it was recognized that, in spite of the statistical spread present, successful classification of these data would still be possible using measurements in the region of 1.7 μm, because in this region there is no overlap between the two classes. Thus, a sample algorithm for this situation might be level slicing, perhaps applied sequentially in several bands if there were more than two classes and points of nonoverlap occurred at different wavelengths for different class pairs. The linear discriminant analysis of Fig. 1-10 is philosophically compatible with this model also; although the statistical spread of data from the classes is recognized, the decision boundaries are derived from only the mean values, i.e., the average spectral response.

Second-order statistical model A still more complete model of the manner in which information is conveyed by spectral variations would recognize that, though the average spectral response as a function of wavelength contains a great deal of information which can be used to discriminate between classes, the manner in which given classes vary about the average may be information bearing as well. In this case we are using a statistical model not only for the non-information-bearing variations (noise) but for the information-bearing ones (signal) as well. The typical response for green vegetation would now be thought of as shown in Fig. 7-20, rather than as in Fig. 7-19, and the fine structure of Fig. 7-20 is important because it, too, is information bearing. This model suggests algorithms which base discrimination not only upon the "average separation" between data

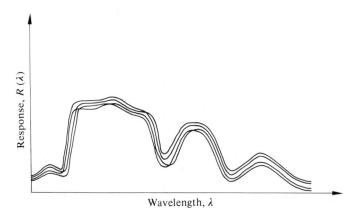

Figure 7-20 Spectral response for green vegetation showing second-order variations.

sets in n-dimensional space but also upon the shape of these distributions. It is this philosophy which justifies a maximum-likelihood scheme in which both first- and second-order multivariate statistics are used. The decision boundaries in this case usually turn out to be segments of second-order curves.

Since the characterization of the information borne in the spectral measurements becomes more complete as one proceeds through the hierarchy just described, it is reasonable to expect in general that classifier performance will improve. To obtain this improvement, the decision boundaries must be carefully located. This may require any or all of the following: greater accuracy in estimating the training statistics, more representative training samples, a greater number of training samples, higher quality data, etc. The accurate estimation of second-order statistics is more difficult to achieve than the accurate estimation of first-order statistics.

And, finally, recall (Sec. 1-3) that one may utilize the features in either a multivariate or a multiple univariate form. The discussion in connection with Fig. 1-15 illustrates the analysis of data in a multiple univariate form, while that for Fig. 1-17 shows the same data in a multivariate sense.

It is, of course, difficult to deal exhaustively with a subject as broad as the matter of algorithms for data analysis. We have attempted here to provide only some generalizations which undoubtedly will not apply to every situation. However, since they are based on well-established fundamentals, these generalizations do provide a useful point of depature in considering algorithms for analysis on a spectral basis.

Adding spatial information But what if the most complex data and most complex algorithm prove insufficient to derive the desired information? The illustrations we have used above have been based solely on spectral variations. The fundamentals of remote sensing suggest that information is contained in the spatial and temporal variations of the energy as well. The next step in the process might then be to employ an algorithm using spatial variations as an adjunct to spectral ones in order to extract information. Though we have not previously discussed such algorithms in this text, they are certainly feasible. They rely on the fact that the statistical dependencies between the spectral responses of neighboring pixels are information bearing in some way.

Think of the scene as being made up of a number of objects, an *object* being a contiguous region in the scene over which the desired class name does not change; e.g., an agricultural field might be considered an object. If the classes of the scene are distinguishable on spectral grounds alone, then the spatial resolution with which the scene must be scanned does not need to be significantly finer than the size of the smallest object in order to achieve identification of the class of that object. On the other hand, should finer spatial resolution be available so that there are a number of pixels per object, algorithms which use spatial and spectral information jointly may make possible the successful identification of objects (classes) which are not identifiable based on spectral data alone.[8] The instantaneous field of view must be fine enough so that the number of pixels per object is sufficient to successfully convey the distinguishing spatial variation characteristic of the class of the object.

The use of spatial information as just described makes use of the local spatial characteristics, i.e., the spatial variations within objects; however, when a human views a scene it is clear that he or she utilizes more than just the within-object variations to identify classes in the scene. It is quite possible to devise machine-processing algorithms which also make use of these more global characteristics. Measures of the size and shape of objects themselves and of the distribution of sizes and shapes in regions are all information-bearing attributes of the data. Although a detailed discussion of this type of processor is beyond the scope of this text, one can readily visualize the tremendous range of algorithm complexity which is possible and which will be utilized in future data-analysis systems.

Hierarchical Classification

As a further illustration of how added algorithm complexity can be helpful, let us review briefly the use of more complex decision logic. Recall the information-tree concept discussed earlier. Figure 7-2 shows an information tree based on classes of informational value. It is possible to design a classifier which reaches a final decision on a picture element by passing through several intermediate decisions; the information tree can serve as a map of the decision-making process. That is, a pixel is first classified according to the class list of the first layer of the tree. The pixel is then further classified through successive layers in accordance with the branches of the tree.[9] There are a number of possible advantages of this more complex decision logic. First, certain types of illogical errors can be eliminated. A pixel in the midst of an agricultural field would not be misclassified into a forest class, for example.

A second possible advantage arises from the fact that at each node of the tree a smaller number of alternative classes exists. As a result of this, the feature set used at that node can be tailored specifically to the smaller number of alternatives. This can result not only in increased accuracy but also in computational efficiency, since a smaller feature set can often be used.

A third possibility for such layered decision logic is that the classification algorithm itself can be changed at each branch point (node) if desired. For example, suppose one is processing a data set which consists of a combination of multispectral data for which a multivariate Gaussian assumption is valid and topographic data for which it is not. Classifications can be carried out using the multispectral data alone at some nodes and simple level slicing on the topographic data at others. Varying the classification algorithm or the feature set might be useful also in the case of multitemporal data. For example, the layers of the decision tree might represent the individual data-collection times, allowing for classification at later times to be conditioned upon the classification at previous data-collection times.

The purpose of this discussion has been simply to make it clear that there are many choices available to the system designer or operator in selecting processing algorithms. And surely the number of choices will increase as research continues. Thus it is important not only to be familiar with a certain specific set of algorithms but also to understand the more general principles upon which they are based.

Processor Implementation

Once the algorithms to be used in processing the data have been selected, the next step is to consider the manner in which they will be implemented. This choice usually rests heavily on economic factors, which include the processing capacity needed to handle the maximum volume of data which must be processed per unit time (per day), and the total period of time for which the implementation (whatever its form) will be used or over which it will be amortized. From a very generalized viewpoint, the following list of implementation types samples the spectrum of possibilities:

1. Services purchased from a vendor
2. General-purpose software on a general-purpose computer
3. Special-purpose software on a general-purpose computer
4. Special-purpose hardware

When only a limited amount of data is to be processed or if data are to be processed only for a very limited period of time, purchase of services from a vendor might be considered. This will minimize the initial investment required which, in the case of small amounts of data to be analyzed, may be the dominant cost of implementation.

Considered next might be the use of a general-purpose computer. The flexibility and convenience of such machines mean that software implementations of the needed algorithms can be achieved rather quickly and inexpensively. If the data volume is expected to be fairly large, it may be advantageous to spend resources for preparing the software to optimize machine efficiency. For example, instead of implementing a Fortran-written pattern classifier, one may wish to use a table-look-up procedure written in a basic machine language and, in the process, to take special precautions that the software is written to use each of the computer system components (main memory, disc memory, channels, etc.) in the most efficient way. This is an example of special-purpose software, and, while the cost of achieving the implementation would be greater, significantly lower processing costs per unit classification might be achievable.

And finally in the list, if a very large volume of data must be dealt with, even greater initial expenditure might be appropriate in order to achieve the highest possible processing efficiency. Generally this would be achieved by procuring special-purpose hardware implementations of the desired algorithms.

We can see from the above that as we proceed down the list of implementation types, one trade-off involved is the initial investment required as compared to the per-unit processing cost after beginning operation. Further, it is often true that, in order to achieve high efficiency in processing costs, we must sacrifice flexibility. The more carefully we can define the set of problems to be dealt with, the more safely we can decide to give up flexibility in favor of efficiency.

A third factor to keep in mind is the dynamism of the technology. So long as the potential for further development of remote sensing capability remains high in terms of new sensor systems and data-processing algorithms, the question of obsolescence must be carefully considered. The great, world-wide needs for better information about resources, the relatively primitive development of current satellites with regard to complete characterization of the scene, and the potential for further development in data-processing systems (not only in the area of algorithms as outlined above but also for processing hardware technology) suggest that the remote sensing field will continue to be a rapidly developing technology for some time into the future. Thus, in addition to data volume, initial investment in processing systems, per-unit processing costs, and flexibility, the possibility of obsolescence of the data-processing capability must be carefully weighed when deciding on a means of implementation.

Human Participation in Processing

Earlier we drew attention to the fact that it is in the processing (rather than scene or sensor) portion of the system that there is the greatest opportunity for human participation. It is appropriate that we explore the nature of this participation at least briefly.

In attempting to deal with this question in a generalized way, i.e., outside the context of any specific problem or application, it is helpful for us to categorize the various degrees of human participation in data processing. Processing systems have been designed which range from being fully manual, to machine-aided manual, to man-aided machine, and finally to fully machine, or automatic, systems. Examples illustrating these categories follow.

The field of remote sensing began, naturally enough, with fully manual data-analysis techniques. As a most extreme example one might envision a photointerpreter interpreting a photographically produced image and hand-drawing the map product. However, one can quickly and inexpensively increase the system throughput and effectiveness by moving first to more sophisticated ways of producing the imagery for human interpretation and then providing machine-implemented aids to the analyst. As a simple example, computer-generated calibration signals could be applied to produce imagery of more consistent radiometric quality. Geometric manipulation of the imagery by machine could improve the cartographic quality of the data.

Continuing more deeply into the machine-aided manual category, it might be helpful to apply various image-enhancement techniques to make the desired information more easily perceivable to the human interpreter.

An example of a man-aided machine-processing step might occur in the registration of two multitemporal images of a scene. The human participation might involve locating a series of common points in the two data sets through use of an interactive display device; this would be followed by the application of a machine-implemented, two-dimensional image-correlation algorithm in order to achieve precise image alignment between the manually located reference points. The pattern-recognition approach discussed in this text is an example of man-aided-machine processing. An iterative combination of man and machine steps is used to define a list of classes, locate training data, and compute the class statistics after which the pixels of a scene are classified by a supervised pattern-recognition algorithm.

Fully automatic processing generally occurs only in the preprocessing phase where the processing algorithms are not data dependent and are simple enough so that human talents are not really needed. The use of fully automatic data analysis is frequently difficult to justify, even as a goal, since human reasoning powers and ability to use contextual information and the like are difficult to achieve in a machine-implemented algorithm. The value of ancillary information of a subjective nature is yet to be quantified, but its positive impact on processing efficiency and the quality of the results cannot really be doubted.

Thus we see that the matter of human participation in processing extends well beyond the choice of interactive hardware. It really is a question of teaming man with machine, and learning which tasks man can do better and which the machine. The decisions depend greatly upon the precise design objectives of the information system designer and whether it is appropriate to weigh more heavily the accuracy of the information produced, the throughput rate, the repeatability of results, the initial system cost, the per-unit operation (or product) cost, the range of problems which can be dealt with, the range of data types which can be dealt with, or any of the many other factors which we mentioned in the discussion of system evaluation in Chap. 4.

The User Interface

And finally there is the matter of how the rest of the system should be interfaced with the consumer of the information. Again, the range of possibilities is great and the possibility for a system failure at this interface is perhaps even greater. The format, quality, timeliness, and cost of these output products must be well mated to the user's needs. This is often a challenging question because not infrequently the user is unable to state precisely what the needs are. Not all potential users can be expected to be knowledgeable enough about remote sensing information systems that they can correctly write quantitative specifications for the products needed. On the other hand, at the outset the producer of the products most likely will not understand the user's problem well enough to be able to advise properly.

One technique which has been used to alleviate this problem is involving the user directly in deriving the needed results. This increases the opportunity for dealing iteratively with the questions: "What does the user need?" and, "What can the system produce?"

From the vantage point of the broad systems overview of Fig. 7-1, we must conclude (1) that the system is very centralized at the sensor but (2) that it must become very decentralized at some point "down the data stream" toward the user, because there must be enough users and uses to justify the creation and operation of the system. It is clear that each user cannot have his or her own satellite-borne sensor, and thus that a given sensor must serve many users. The point in the system at which this decentralization takes place has a considerable impact on the interface of the system with the user. Economies of scale which may result from processing large quantities of data tend to suggest that the decentralization should take place as far downstream as possible.

PROBLEMS

7-8 The average user of information derived from remote sensing data has more choice in the design and selection of the processing system than any other system component. Discuss the evolution of a data-processing system for the management of a half billion hectares (1.2 billion acres) of forest and rangeland as the program progresses through the stages of:

(a) A research operation
(b) A small-scale feasibility study (single forest district)
(c) A sizeable demonstration project (100 million hectares; 247 million acres)
(d) Full-scale operation

Include in your discussion the choice of algorithms, processor implementation, amount of human participation in the analysis process, and the amount and type of human interface at each stage of the evolution.

7-9 What type of processor implementation do you feel would be most appropriate for a user who requires frequent updating of his information base but whose primary interest is in a relatively small geographic area? An example would be a private land owner, say a farmer with 1000 hectares (2500 acres). Also comment on the degree of algorithm flexibility and type of user interface you would recommend. Support your recommendations by pointing out the advantages of the system you propose.

7-10 A county zoning board is investigating the use of remote sensing to provide it with annually updated land-use maps. Consider the role of the data-processing system for this type of application. What characteristics would you look for if you had the responsibility of recommending the procurement of equipment or services for this application?

REFERENCES

1. Knuth, D. E.: "The Art of Computer Programming," vol. 1, "Fundamental Algorithms," Addison-Wesley Publishing Co., Reading, Mass., 1975.
2. National Aeronautics and Space Administration: *Proc. of the Third Earth Resources Technology Satellite Symposium*, 3 vols., Special Publications SP-351, SP 356, SP-357, NASA Goddard Space Flight Center, Greenbelt, Md., 1973.
3. MacDonald, R. B., M. E. Bauer, R. D. Allen, J. W. Clifton, J. D. Erickson, and D. A. Landgrebe: Results of the 1971 Corn Blight Watch Experiment, in *Proc. of the Eighth International Symposium on Remote Sensing of the Environment*, vol. 1, Environmental Research Institute of Michigan, Ann Arbor, Mich., pp. 157–189, 1972.
4. Hughes, G. F.: On the Mean Accuracy of Statistical Pattern Recognizers, *IEEE Trans. on Information Theory*, vol. IT-14, no. 1, pp. 55–63, 1968.
5. Fu, K. S., D. A. Landgrebe, and T. L. Phillips: Information Processing of Remotely Sensed Agricultural Data, *Proc. of the IEEE*, vol. 57, no. 4, pp. 639–653, 1969.
6. Ready, P. J., P. A. Wintz, S. J. Whitsitt, and D. A. Landgrebe: Effects of Compression and Random Noise on Multispectral Data, in *Proc. of the Seventh International Symposium on Remote Sensing of the Environment*, Environmental Research Institute of Michigan, Ann Arbor, Mich., pp. 1321–1343, 1971.
7. Courant, R., and D. Hilbert: "Methods of Mathematical Physics," vol. I, Interscience Publishers, New York, 1953.
8. Kettig, R. L., and D. A. Landgrebe: Classification of Multispectral Image Data by Extraction and Classification of Homogeneous Objects, *IEEE Trans. Geoscience Electronics*, vol. GE-14, pp. 19–26, January, 1976.
9. Swain, P. H., and H. Hauska: The Decision Tree Classifier: Design and Potential, *IEEE Trans. Geoscience Electronics*, vol. GE-15, pp. 142–147, July, 1977.

GLOSSARY

This glossary contains terms commonly encountered in quantitative remote sensing, and the definitions given here are in some cases slanted toward the subject matter rather than being completely general. A term set in boldface type within a definition can also be found as a separate entry in the glossary.

Absolute zero   A temperature of 0 K or $-273°C$.

Absorptance   A measure of the ability of a surface to absorb incident energy, often at specific wavelengths.

Absorption   A process of attenuation through which radiant energy is absorbed and converted into other forms of energy as it passes through the atmosphere or other media (also see **energy-balance equation**).

Absorption band   A range of wavelengths over which radiant energy is absorbed by a specific material which may be present on the earth's surface or in the atmosphere. For example, a "water-absorption band" is located in the vicinity of 2.6 μm; solar energy near that wavelength is absorbed by the water present both in the atmosphere and in the earth surface materials it strikes.

Albedo   The fraction of the total energy incident on a reflecting surface which is reflected back in all directions.

Algorithm   A series of well-defined steps used in carrying out a specific process; e.g., the classification algorithm.

Alphanumeric   Using or composed of a character set which is made up of both letters and digits, e.g., a computer line printer and its output products.

Analog-to-digital conversion (*A/D conversion*)   The process of sampling continuous analog signals in order to convert them into a stream of digital values. Multispectral scanner data, if collected in analog form, undergoes this conversion prior to digital analysis (see Sec. 2-7).

Analyst   See **data analyst.**

Ancillary data   In remote sensing, secondary data pertaining to the area or classes of interest, such as topographic, demographic, or climatological data. Ancillary data may be digitized and used in the analysis process in conjunction with the primary remote sensing data.

Aperture stop   The physical element of an optical system which partially determines the light-gathering ability or "speed" of the lens but does not affect the size of the field imaged (see Sec. 2-3).

Atmospheric attenuation   The reduction of radiation intensity due to absorption and/or scattering of energy by the atmosphere; usually wavelength dependent; may affect both solar radiation traveling to the earth and reflected/emitted radiation traveling to the sensor from the earth's surface.

Atmospheric effects   See **atmospheric attenuation.**

Atmospheric windows   Those wavelength ranges where radiation can pass through the atmosphere with relatively little attenuation; in the optical portion of the spectrum, approximately 0.3 to 2.5, 3.0 to 4.0, 4.2 to 5.0, and 7.0 to 15.0 μm.

Blackbody   An ideal body which, if it existed, would be a perfect absorber and a perfect radiator, absorbing all incident radiation, reflecting none, and emitting radiation at all wavelengths. In remote sensing, the exitance curves of blackbodies at various temperatures can be used to model naturally occurring phenomena like solar radiation and terrestrial emittance (see Sec. 2-1).

Calibration data   Measurements pertaining to the spectral and/or geometric characteristics of the sensor and/or of the radiation source; calibration data are obtained through the use of a fixed energy source such as a calibration lamp, a temperature plate, or a geometric test pattern. Calibration data are one kind of **ephemeral data.**

Cathode ray tube (CRT)   A vacuum tube capable of producing a black-and-white or color image by beaming electrons onto a sensitized screen. As a component of a data-processing system, the CRT can be used to provide rapid, pictorial access to numerical data (see Fig. 4-9).

Classification map   The map-like representation that shows the class assignment of each data vector, often taking the form of a CRT image or computer line-printer output, e.g., Fig. 6-20.

Clustering   The analysis of a set of measurement vectors to detect their inherent tendency to form clusters in multidimensional measurement space (see Sec. 3-10).

Color infrared film   Photographic film which is sensitive to energy in the visible and near-infrared wavelengths, generally from 0.4 to 0.9 μm; usually used with a minus-blue (yellow) filter which results in an effective film sensitivity of 0.5 to 0.9 μm. Color infrared film is especially useful for detecting changes in the condition of the vegetative canopy which are often manifested in the near-infrared region of the spectrum. Note that color infrared film is *not* sensitive in the thermal infrared region and therefore cannot be used as a heat-sensitive detector.

Data-acquisition system   The collection of devices and media which measures physical variables and records them prior to input to the data-processing system.

Data analyst   In a production-oriented system, the operator of a data-analysis system whose job it is to monitor the flow of products and control the operation of the system; in a research-oriented system, a researcher who interacts closely with the data and uses the data-analysis system as a tool of his research.

Data bank   A well-defined collection of data, usually of the same general type, which can be accessed by a computer.

Data compression   Any technique that condenses the available data so as to make data storage or transmission more efficient with minimal loss of information (see Sec. 4-4).

Data dimensionality   The number of variables (e.g., channels) present in the data set. The term "intrinsic dimensionality" refers to the smallest number of variables which could be used to represent the data set accurately.

Data flow   The path and rate of transfer of the data through the remote sensing system, from the collection of the data through its analysis and use.

Data set   A specific collection of related data elements used for a particular task; may include data from many sources and in many formats.

Decision region   A region in the measurement space corresponding to a specific class; defined by means of **discriminant functions** and used to classify data vectors of unknown class association.

Decision rule (or *classification rule*)   The criterion used to establish discriminant functions for classification; e.g., nearest-neighbor rule, minimum-distance-to-means rule, maximum-likelihood rule.

Decision tree   A logical branching arrangement of decision processes (involving classes, features, and decision rules) which allows class-membership decisions to be made in sequential steps and among relatively fewer alternatives at each node of the tree.

Density   In an optical image a point-by-point measure of the degree of blackness; often used in a more general sense to refer to relative intensities in remote sensing data.

Density slicing (or *level slicing*)   A general class of electronic or digital techniques used to assign image points or data vectors to particular classes based on the density or level of the response in a single image or channel; classification by thresholds.

Discriminant function   One of a set of mathematical functions which in remote sensing are commonly derived from training samples and a decision rule and are used to divide the measurement space into **decision regions** (see Sec. 3-6).

Display   An output device that produces a visible representation of a data set for quick visual access; usually the primary hardware component is a **cathode ray tube** (see Fig. 4-9).

Edge enhancement   See **enhancement.**

Electromagnetic spectrum   The array of all electromagnetic radiation that moves with the velocity of light, characterized by wavelength or frequency (see Fig. 1-6). The optical wavelengths (0.3 to 15.0 μm) are the ones most used in remote sensing. Energy at these wavelengths can be reflected and refracted with solid materials like mirrors and lenses.

Electrostatic printer   A hardcopy output device which uses electronic charges to attract carbon particles onto a plate which is subsequently used for printing.

Emissive infrared   See **infrared.**

Emittance (*emissivity*)   The ratio of the radiation given off by a surface to the radiation given off by a **blackbody** at the same temperature; a blackbody has an emissivity of 1, other objects between 0 and 1.

Energy-balance equation   The affirmation that all incident energy at a particular wavelength is either absorbed by the object it strikes, reflected from it, or transmitted through it: $I_\lambda = A_\lambda + R_\lambda + T_\lambda$. For remote sensing purposes, it is convenient to think of the equation in terms of reflectance: $R_\lambda = I_\lambda - (A_\lambda + T_\lambda)$; that is, reflected energy at a particular wavelength equals the incident energy at that wavelength minus the sum of the energy that is absorbed by the object and transmitted through it. The energy-balance concept is basic to understanding why various objects reflect energy differently (see Sec. 5-2).

Enhancement   Data filtering and other processes which improve the visual quality of the pictorially presented data or which visually accentuate a characteristic of the data; e.g., edge enhancement, noise reduction.

Ephemeral data   Data which help to characterize the conditions under which the remote sensing data were collected; may be used to calibrate the sensor data prior to analysis; include such information as the positioning and spectral stability of sensors, sun angle, platform attitude, etc.

Far infrared   See **infrared.**

Feature   In pattern recognition, one of the measurements of a pattern or a mathematical transformation of such measurements; in remote sensing, often the reflectance measurement in one channel of the sensor. The number of features associated with a pattern defines its dimensionality.

Feature extraction   In pattern recognition, any process of performing transformations on the set of measurements of a pattern in order to accentuate the characterizing features of the pattern. Often used to reduce the dimensionality of the data; e.g., feature selection (see Sec. 3-9).

Feature selection   See **feature extraction.**

Field stop   The physical element of an optical system which directly controls the size of the field of view and, ultimately, the portion of the scene that can be imaged by the system (see Sec. 2-3).

Filter   Any mechanism which modifies optical, electrical, or digital signals in accordance with specified criteria. Often a filter is a means of extracting a particular subset of data from a larger set containing irrelevant data. An optical filter passes only desired optical wavelengths of energy. A digital filter is an arithmetic procedure that operates on a digitized data stream in much the same way as an electric filter operates on a continuous electrical signal; its purpose is generally to eliminate irrelevant data or noise.

Gaussian assumption   The assumption that the probability function for any class of interest can be approximated by a Gaussian (or normal) probability density function; through the use of this assumption, a class of multivariate patterns can be characterized by two sets of statistics: the vector of feature means and the covariance matrix (see Sec. 3-5).

Geometric transformations   Adjustments made in the image data to change its geometric character, usually to improve its geometric consistency or cartographic utility (see Sec. 4-4).

Graybody   An energy absorber/radiator which has a **spectral radiant exitance** curve of the same functional form as an ideal blackbody radiation curve, but reduced by a constant factor for every wavelength. Equivalently, a graybody has an emissivity which is less than 1.0 and the same at all wavelengths.

Ground resolution element   See **resolution cell.**

Ground truth   See **reference data.**

Hardware   The physical components of a data-processing system, including the central processor, data carriers, and remote terminals. Compare with **software.**

Histogram   The graphical display of a set of data which shows the frequency of occurrence (along the vertical axis) of individual measurements or values (along the horizontal axis); a frequency distribution.

IFOV   See **instantaneous field of view.**

Image enhancement   See **enhancement.**

Infrared   Pertaining to energy in the 0.7 to 100-μm wavelength region of the electromagnetic spectrum. For remote sensing, the infrared wavelengths are often subdivided into *near infrared* (0.7 to 1.3 μm), *middle infrared* (1.3 to 3.0 μm), and *far infrared* (7.0 to 15.0 μm). Far infrared is sometimes referred to as *thermal* or *emissive* infrared (see Sec. 2-1).

Instantaneous field of view   When expressed in degrees or radians, the smallest plane angle over which an instrument (e.g., a scanner) is sensitive to radiation; when expressed in linear or area units such as meters or hectares, it is an altitude-dependent measure of the ground resolution of the scanner, in which case it is also called "instantaneous viewing area." Compare to **total field of view.**

Instantaneous viewing area   See **instantaneous field of view.**

Intrinsic dimensionality   See **data dimensionality**.

Irradiance   The measure of radiant power per unit area that is incident on an object or surface (see Sec. 2-1).

Kirchhoff radiation law   The law stating that at a given absolute temperature and wavelength, the ratio of **spectral radiant exitance** to spectral absorptance for any body is equal to the spectral radiant exitance of an ideal **blackbody** at the same absolute temperature and wavelength; and that because an ideal blackbody is both an ideal absorber and emitter, it follows that absorptance equals emittance.

Lambertian surface An ideal, perfectly diffusing surface which reflects energy equally in all directions.

LARSYS A data-analysis system for remote sensing research developed by the Laboratory for Applications of Remote Sensing, Purdue University, West Lafayette, Indiana. LARSYS includes a software system, a hardware system, a library of multispectral earth resources data, comprehensive documentation, and educational materials to train users.

Level slicing See **density slicing.**

Light pen A pointer-like device, used in conjunction with a display screen, which provides a signal to identify for later reference particular elements displayed on the screen (see Fig. 4-9).

Line printer A computer system output device which produces a line-by-line printed record of the data it receives, usually in discrete symbols such as letters and numbers.

Matrix A rectangular table containing x rows and y columns used to display an array of $x \times y$ numbers. Each term in the array can be referred to by position; for example, $A_{3,5}$ is the term in row 3, column 5 of matrix A.

Measurement complexity In digital multispectral data, an index of how detailed the measurements of spectral energy are, based on the number of spectral channels used and the number of brightness levels recorded in each channel.

Middle infrared See **infrared.**

Mie scattering The scattering of electromagnetic energy by particles in the atmosphere which are comparable in size to the wavelength of the scattered energy.

Multispectral scanner A line-scanning sensor which uses an oscillating or rotating mirror, a wavelength-selective dispersive mechanism, and an array of detectors to measure simultaneously the energy available in several wavelength bands, often in several spectral regions. The movement of the platform usually provides for the along-track progression of the scanner.

Multivariate analysis A data-analysis approach which makes use of multi-dimensional interrelationships and correlations within the data for effective discrimination (see Sec. 1-3).

Near infrared See **infrared.**

Noise Random or regular interfering effects in the data which degrade its information-bearing quality.

Nonselective scattering The scattering of electromagnetic energy by particles in the atmosphere much larger than the wavelengths of the energy, causing all wavelengths to be scattered equally.

Normal assumption See **Gaussian assumption.**

Optical mechanical scanner See **multispectral scanner.**

Optical stop See **aperture stop** and **field stop.**

Optical wavelengths See **electromagnetic spectrum.**

Pass band   A wavelength range over which a sensor is capable of receiving and measuring electromagnetic energy. In a photographic system, sensitivity may be controlled through the use of optical filters and the sensitivity characteristics of the film; in a scanner, through the design of the sensor's dispersive system and detectors.

Pattern recognition   The automated process through which unidentified patterns can be classified into a limited number of discrete classes through comparison with other class-defining patterns or characteristics.

Pixel (derived from "picture element")   A data element having both spatial and spectral aspects. The spatial variable defines the apparent size of the resolution cell (i.e., the area on the ground represented by the data values), and the spectral variable defines the intensity of the spectral response for that cell in a particular channel.

Planck's radiation law   A fundamental law of radiation which correlates the temperature of a blackbody to the intensity of the radiation it emits as a function of wavelength (see Sec. 2-1, Fig. 2-4).

Preprocessing   In a remote sensing system, the processing of the data received from the sensor to a form acceptable by the data bank and subsequent processing functions; may also include geometric and radiometric calibration, enhancements, and other transformations.

Primary data   The data collected by the remote sensors (also see **ancillary data, ephemeral data, reference data**).

Probability function   A function indicating the relative frequency with which any measurement vector may be expected to occur. In a remote sensing application, generally associated with each of the classes to be recognized.

Processing system   That part of a remote sensing system following the scene and the sensor system which includes data handling, data analysis, and results presentation.

Radiance   The spatial distribution of radiant power density.

Radiant energy   The electromagnetic radiation transmitted by waves through space or other media.

Radiant exitance   The measure of radiant energy per unit area that leaves the object or surface of interest (see Sec. 2-1).

Radiant flux   The time rate of the flow of radiant energy; radiant power.

Radiometric transformations   Adjustments made in the data to convert the raw multispectral data to a radiometrically consistent set of measurements. Radiometric transformations may be used to compensate for sensor system irregularities or environmental variations (see Sec. 4-4).

Rayleigh-Jeans radiation law   A classically derived law of radiation intensity which, for energy radiated in long wavelengths by bodies of high temperature, approximates Planck's experimentally derived law (see Sec. 2-1).

Rayleigh scattering   The wavelength-dependent scattering of electromagnetic radiation by particles in the atmosphere much smaller than the wavelengths scattered.

Real-time processing   A mode of data processing in which all analysis operations are performed at the same time and at the same rate as the data are collected.

Reference data   Data about the physical state of the earth obtained from sources other than the primary remote sensing data source and used in support of the remote sensing data analysis. May typically include maps and aerial photographs, topographic information, temperature measurements and other field measurements, and data on atmospheric conditions. Also include some types of ancillary and ephemeral data. In most cases reference data are more useful if collected concurrently with the primary data-collection mission. Also referred to as *ground truth, ground data, ground-based measurements.*

Reflectance   A measure of the ability of a surface to reflect energy; specifically the ratio of the reflected energy to the incident energy. Reflectance is affected not only by the nature of the surface itself, but also by the angle of incidence and the viewing angle.

Reflective infrared   The portion of the electromagnetic spectrum from approximately 0.72 to 3.0 μm; often is subdivided into *near infrared* and *middle infrared.*

Reflective wavelengths   Those wavelengths in the optical portion of the electromagnetic spectrum in which the energy available for remote sensing results primarily from reflection of the sun's radiation. Range of wavelengths is 0.3 to 3.0 μm.

Registration   The process of geometrically aligning two or more sets of image data such that resolution cells for a single ground area can be digitally or visually superimposed. Data being registered may be of the same type, from very different kinds of sensors, or collected at different times.

Resolution (or *resolving power*)   A measure of the ability of an optical system to distinguish between signals that are spatially near or spectrally similar. *Spatial resolution* is a measure of the smallest angular or linear separation between two objects (usually expressed in radians or meters) with a smaller resolution parameter denoting greater resolving power. *Spectral resolution* is a measure of both the discreteness of the bandwidths and the sensitivity of the sensor to distinguish between gray levels; e.g., in Landsat 1 and 2, three of the sensors were sensitive over 0.1-μm wavelength ranges and could distinguish among 128 spectral intensity levels; the fourth sensor was sensitive over a 0.3-μm range and could distinguish among 64 intensity levels. (The term *thermal resolution* may also be used to describe the sensitivity of sensors measuring emitted radiation.) The spatial resolution of a remote sensing system is also a function of the spectral contrast between objects in the scene and their background, of the shape of the objects, and of the signal-to-noise ratio of the system.

Resolution cell   The smallest area in a scene considered as a unit of data. For Landsat 1 and 2, the resolution cell approximates a rectangular ground area of 0.44 hectares or 1.1 acres (see **pixel, instantaneous field of view**).

Sampling rate   The temporal, spatial, or spectral rate at which measurements of physical quantities are taken. *Temporally*, sampling variables may describe how often data are collected or the rate at which an analog signal is sampled for conversion to digital format; the *spatial* sampling rate describes the number, ground size, and position of areas where spectral measurements are made; the *spectral* sampling rate refers to the location and width of the sensor's spectral channels with respect to the electromagnetic spectrum.

Scattering   The reflection and refraction of electromagnetic energy by particles in the atmosphere; frequently wavelength dependent (see **Rayleigh scattering, Mie scattering, nonselective scattering**).

Scene   In a passive remote sensing system, everything that occurs spatially or temporally before the sensor, including the earth's surface, the energy source, and the atmosphere which the energy passes through as it travels from its source to the earth and from the earth to the sensor.

Sensor   Any device that is sensitive to levels or changes in physical quantities (such as light intensity or temperature) and converts these phenomena into a form suitable for input to an information-gathering system. An active sensor system (such as radar) produces the energy needed to detect these phenomena; a passive sensor system (multispectral scanner, aerial photographic camera) depends on already existing energy sources, e.g., the sun.

Signal   The effect (e.g., pulse of electromagnetic energy) conveyed over a communication path or system. Signals are received by the sensor from the scene and converted to another form for transmission to the processing system.

Signal-to-noise ratio   The ratio of the level of the information-bearing signal power to the level of the noise power. Abbreviated as S/N (see Sec. 2-6).

Signature   See **spectral signature.**

Signature extension   The use of training statistics obtained from one geographic area to classify data from similar areas some distance away; includes consideration of changes in atmosphere and other geographic and temporal conditions which can cause differences in signal level for single classes of interest (see **spectral signature**).

Software   The computer programs which drive the hardware components of a data-processing system; includes system monitoring programs, programming language processors, data-handling utilities, and data-analysis programs.

Spatial information   Information conveyed by the spatial variations of spectral response (or other physical variables) present in the scene.

Spectra   Data that result from spectral scanning; measurements of the variations in spectral response over a range of wavelengths for a single, constant viewing area.

Spectral band   A well-defined, continuous range of wavelengths in the electromagnetic spectrum; e.g., the 0.5 to 0.6-μm spectral band. Also called *wavelength band.*

Spectral information   Information conveyed by the spectral response of individual resolution cells in the scene.

Spectral irradiance   See **irradiance.**

Spectral map   A **classification map** in which the classes are based on relative spectral properties rather than the ground cover type represented. Often results from **unsupervised classification**.

Spectral radiance   The radiance of an object or surface described with respect to the distribution of the power across the spectrum.

Spectral radiant exitance   The wavelength-related measure of radiant power per unit area that is emitted by the object or surface of interest.

Spectral regions   Conveniently designated ranges of wavelengths subdividing the electromagnetic spectrum; e.g., the visible region, x-ray region, infrared region, middle-infrared region.

Spectral response   The response of a material as a function of wavelength to incident electromagnetic energy, particularly in terms of the measurable energy reflected from and emitted by the material.

Spectral signature   The spectral characterization of an object or class of objects on the earth's surface. Often used in a way which naively oversimplifies the complexity of the spectral representation problem in a natural scene (see Sec. 5-5).

Spectrometer   An optical instrument used to measure the apparent electromagnetic radiation emanating from a target in one or more fixed wavelength bands or sequentially through a range of wavelengths.

Spectroscopy   The branch of physics concerned with the production, transmission, measurement, and interpretation of electromagnetic spectra.

Specular reflection   The reflectance of electromagnetic energy without scattering or diffusion, as from a surface that is smooth in relation to the wavelengths of incident energy. Also called *mirror reflection*.

Stefan-Boltzmann radiation law   A radiation law that states that the total energy radiated from a blackbody is proportional to the fourth power of the absolute temperature of the body.

Stochastic model   A model that uses laws of probability to define or describe random variables with known distributions. Stochastic models are especially appropriate for use with data that have high variability or uncertainty associated with them.

Supervised classification   A computer-implemented process through which each measurement vector is assigned to a class according to a specified decision rule, where the possible classes have been defined based on representative **training samples** of known identity (also see **unsupervised classification**).

Swath width   See **total field of view.**

Synoptic view   The ability to see or otherwise measure widely dispersed areas at the same time and under the same conditions; e.g., the overall view of a large portion of the earth's surface which can be obtained from satellite altitudes.

System designer   The person responsible for determining in detail the components and operation of a data-processing system by specifying software and hardware in order to achieve the most effective system design for a specific purpose.

System user (or simply *the user*)   The consumer of the information produced by a data-processing system.

Target   The portion of the earth's surface which produces by reflection or emission the radiation measured by the remote sensing system.

Thermal infrared   See **infrared.**

Total field of view (or *swath width*)   The overall plane angle or linear ground distance covered by a multispectral scanner in the across-track direction (transverse to the direction of travel of the sensor platform).

Training samples   The data samples of known identity used to determine decision boundaries in the measurement or feature space prior to classification of the overall set of data vectors from a scene (see Sec. 3-3).

Unsupervised classification   A computer-implemented process through which each measurement vector is assigned to a class according to a specified decision rule, where, in contrast with **supervised classification**, the possible classes have been defined based on inherent data characteristics rather than on **training samples** (see **clustering**).

Visible wavelengths   The radiation range in which the human eye is sensitive, approximately 0.4 to 0.7 μm.

Wien radiation law   A classically derived law of radiation intensity which, for energy radiated in the shorter wavelengths by low-temperature blackbodies, approximates Planck's experimentally derived law (see Sec. 2-1).

Window function   A rectangular bandpass function typically used to specify the spectral bands of a multispectral scanner, defining the band width, band location, and dynamic sensitivity (see Sec. 7-3).

INDEX